Edge-of-Things in Personalized Healthcare Support Systems

Cognitive Data Science in Sustainable
Computing

Edge-of-Things in Personalized Healthcare Support Systems

Edited by

RAJESWARI SRIDHAR
Department of Computer Science and Engineering,
National Institute of Technology, Tiruchirappalli, India

G.R. GANGADHARAN
Department of Computer Applications,
National Institute of Technology, Tiruchirappalli, India

MICHAEL SHENG
Department of Computing, Macquarie University,
Sydney, NSW, Australia

RAJAN SHANKARAN
Department of Computing, Macquarie University,
Sydney, NSW, Australia

Series Editor

ARUN KUMAR SANGAIAH

ACADEMIC PRESS
An imprint of Elsevier

ELSEVIER

Academic Press is an imprint of Elsevier
125 London Wall, London EC2Y 5AS, United Kingdom
525 B Street, Suite 1650, San Diego, CA 92101, United States
50 Hampshire Street, 5th Floor, Cambridge, MA 02139, United States
The Boulevard, Langford Lane, Kidlington, Oxford OX5 1GB, United Kingdom

MATLAB® is a trademark of The MathWorks, Inc. and is used with permission.
The MathWorks does not warrant the accuracy of the text or exercises in this book. This
book's use or discussion of MATLAB® software or related products does not constitute
endorsement or sponsorship by The MathWorks of a particular pedagogical approach or
particular use of the MATLAB® software.

Notices

Knowledge and best practice in this field are constantly changing. As new research and
experience broaden our understanding, changes in research methods, professional practices,
or medical treatment may become necessary.

Practitioners and researchers must always rely on their own experience and knowledge in
evaluating and using any information, methods, compounds, or experiments described
herein. In using such information or methods they should be mindful of their own safety
and the safety of others, including parties for whom they have a professional responsibility.

To the fullest extent of the law, neither the Publisher nor the authors, contributors, or
editors, assume any liability for any injury and/or damage to persons or property as a matter
of products liability, negligence or otherwise, or from any use or operation of any methods,
products, instructions, or ideas contained in the material herein.

ISBN: 978-0-323-90585-5

For Information on all Academic Press publications
visit our website at https://www.elsevier.com/books-and-journals

Publisher: Mara Conner
Editorial Project Manager: Zsereena Mampusti
Production Project Manager: Maria Bernard
Cover Designer: Vicky Pearson Esser

Typeset by MPS Limited, Chennai, India

Contents

5. A recommendation system for the prediction of drug—target associations 115

Simone Contini and Simona Ester Rombo

6. Towards building an efficient deep neural network based on YOLO detector for fetal head localization from ultrasound images 137

M. Ramla, S. Sangeetha and S. Nickolas

7. FunNet: a deep learning network for the detection of age-related macular degeneration   157

Anju Thomas, P.M. Harikrishnan and Varun P. Gopi

8. An improved method for automated detection of microaneurysm in retinal fundus images   173

Avinash A., Biju P., Prapu Premanath, Anju Thomas and Varun P. Gopi

List of contributors

Avinash A.
Department of ECE, College of Engineering Thalassery, Thalassery, India

Rohit Kumar Bondugula
School of Computer and Information Sciences, University of Hyderabad, Hyderabad, India

Ling Chen
Institute of Hospital and Health Care Administration, National Yang Ming Chiao Tung University, Taipei, Taiwan

Simone Contini
Department of Mathematics and Computer Science, University of Palermo, Palermo, Italy

Uma Gandhi
Department of Instrumentation and Control Engineering, National Institute of Technology Tiruchirappalli, Tiruchirappalli, India

Varun P. Gopi
Department of Electronic and Communication Engineering, National Institute of Technology Tiruchirappalli, Tiruchirappalli, India

Christos Goumopoulos
Department of Information and Communication Systems Engineering, University of the Aegean, Samos, Greece

Ajay Kr. Gupta
Department of Computer Science and Engineering, M.M.M. University of Technology, Gorakhpur, India

P.M. Harikrishnan
Department of Electronic and Communication Engineering, National Institute of Technology Tiruchirappalli, Tiruchirappalli, India

Lei Li
Department of Information Technology, Kennesaw State University, Marietta, GA, United States

Umapathy Mangalanathan
Department of Instrumentation and Control Engineering, National Institute of Technology Tiruchirappalli, Tiruchirappalli, India

Ahmet Anıl Müngen
Department of Software Engineering, OSTIM Technical University, Ankara, Turkey

Mohammad Nasajpour
Department of Information Technology, Kennesaw State University, Marietta, GA, United States

S. Nickolas
Department of Computer Applications, National Institute of Technology Tiruchirappalli, Tiruchirappalli, India

Biju P.
Department of ECE, College of Engineering Thalassery, Thalassery, India

Palanisamy P.
Department of Electronic and Communication Engineering, National Institute of Technology Tiruchirappalli, Tiruchirappalli, India

Reza M. Parizi
Department of Software Engineering and Game Development, Kennesaw State University, Marietta, GA, United States

Seyedamin Pouriyeh
Department of Information Technology, Kennesaw State University, Marietta, GA, United States

S. Prabavathy
Department of Information Technology, G. Narayanamma Institute of Technology and Science (For Women), Hyderabad, India

D. Pradeep
Department of Computer Science and Engineering, M.Kumarasamy College of Engineering, Karur, India

Prapu Premanath
Department of ECE, College of Engineering Vadakara, Vatakara, India

Nashrah Rahman
Faculty of Engineering and Technology, Jamia Millia Islamia, New Delhi, India

M. Ramla
Department of Computer Applications, National Institute of Technology Tiruchirappalli, Tiruchirappalli, India

P. Ramya
Department of Information Technology, Christian College of Engineering and Technology, Oddanchatram, India

Nivetha B. Ramya Sri Bilakanti
Department of Electronic and Communication Engineering, National Institute of Technology Tiruchirappalli, Tiruchirappalli, India

I. Ravi Prakash Reddy
Department of Information Technology, G. Narayanamma Institute of Technology and Science (For Women), Hyderabad, India

Simona Ester Rombo
Department of Mathematics and Computer Science, University of Palermo, Palermo, Italy

S. Sangeetha
Department of Computer Applications, National Institute of Technology Tiruchirappalli, Tiruchirappalli, India

Fabio Alberto Schreiber
Department of Electronics, Information and Bioingeneering, Politecnico of Milano, Milan, Italy

Udai Shanker
Department of Computer Science and Engineering, M.M.M. University of Technology, Gorakhpur, India

Kaushik Bhargav Sivangi
School of Computer and Information Sciences, University of Hyderabad, Hyderabad, India

M.B. Smithamol
Department of Computer Science and Engineering, LBS College of Engineering, Kasaragod, India

N. Sruthi
Department of Computer Science and Engineering, LBS College of Engineering, Kasaragod, India

Nikolaos G. Stergiopoulos
Department of Information and Communication Systems Engineering, University of the Aegean, Samos, Greece

C. Sundar
Department of Computer Science and Engineering, Christian College of Engineering and Technology, Oddanchatram, India

Gau-Jun Tang
Institute of Hospital and Health Care Administration, National Yang Ming Chiao Tung University, Taipei, Taiwan

Geerthy Thambiraj
Centre for Healthcare Entrepreneurship, Indian Institute of Technology, Hyderabad, India

Anju Thomas
Department of Electronic and Communication Engineering, National Institute of Technology Tiruchirappalli, Tiruchirappalli, India

Hsuan-Ming Tsao
Division of Cardiology, National Yang Ming Chiao Tung University Hospital, Yi-Lan, Taiwan

Vincent S. Tseng
Department of Computer Science, National Yang Ming Chiao Tung University, Hsinchu, Taiwan

Siba K. Udgata
School of Computer and Information Sciences, University of Hyderabad, Hyderabad, India

Maria Elena Valcher
Department of Information Engineering, University of Padova, Padova, Italy

S. Venkatesh Babu
Department of Computer Science and Engineering, Christian College of Engineering and Technology, Oddanchatram, India

Swatthi Vijay Sanker
Department of Electronic and Communication Engineering, National Institute of
Technology Tiruchirappalli, Tiruchirappalli, India

Heba M. Wagih
Information Systems Department, The British University in Egypt, Cairo, Egypt

Liang Zhao
Department of Information Technology, Kennesaw State University, Marietta, GA,
United States

CHAPTER 1

Exploring the dichotomy on opportunities and challenges of smart technologies in healthcare systems

S. Prabavathy and I. Ravi Prakash Reddy
Department of Information Technology, G. Narayanamma Institute of Technology and Science (For Women), Hyderabad, India

1.1 Introduction

In the information era, healthcare systems use the advancement of digital revolution. The need for digitization in healthcare system are leveraging medical data, improving patient-care service and reducing the cost of medical service (Velthoven et al., 2019). The key revolutionizing technologies in healthcare are: Internet of Things (IoT), Artificial Intelligence, and Cloud computing. These technologies not only provide digitization but also made multidimensional change in the healthcare systems. It changes the medical model, that is, from disease-centered care to patient-centered and changes in treatment concept, that is, the focus from disease treatment to focus towards the preventive healthcare (Jayaratne et al., 2019). In addition, there was also change in information collection and management including clinical data to regional medical data and general management to personalized management (Hassanalieragh et al., 2015). These multilevel changes improved the efficiency of healthcare system and supports people to lead a healthier life.

The vision of IoT is to deploy intelligent sensor and using Internet Protocol for communication to provide pervasive and ubiquitous environment and experience (Prabavathy et al., 2018a,b). IoT has a potential impact in healthcare system. In smart healthcare, it improves the convenience of physicians and patients by using real-time monitoring with efficient information processing. It enables an effective treatment for the patients using data collected through remote monitoring of IoT devices used by patients (Ungurean & Brezulianu, 2017). A variety of body sensor

Edge-of-Things in Personalized Healthcare Support Systems
DOI: https://doi.org/10.1016/B978-0-323-90585-5.00001-1

1

devices have been used for monitoring patients, providing physical activity consciousness and maintaining individual fitness.

The real-time data from the IoT medical devices need to be stored, managed, and transported with security and privacy. This need is satisfied by cloud computing which provides computing services and analytics over the internet (Pattnaik et al., 2017). Integrating IoT and cloud computing in healthcare allows accessing of shared medical data in common infrastructure transparently by providing on-demand services through the network. Healthcare system needs availability of critical information and processing of real-time data with low latency. Edge computing addresses this need by storing and processing critical data near to the devices where it is collected. Edge computing is highly valuable in healthcare where data is needed immediately and delay for processing critical data on cloud is not efficient. The large amount of data collected from the IoT medical sensors, which are stored on the cloud, need to be analyzed for providing insights that is equivalent to human level of accuracy. This is possible by integrating artificial intelligence (AI) along with IoT and cloud computing in healthcare systems. The goal of AI is to mimic the cognitive functions of human beings (Martinez-Miranda & Aldea, 2005). AI is receiving greater interest in healthcare system because of its capability in analyzing large volume of medical data. The massive progress in AI techniques such as deep learning, robotics, computer vision, and natural language processing has brought a paradigm shift in the medical field (Bartoletti, 2019). Fig. 1.1 describes the integration of smart technologies in healthcare sector.

This chapter provides recent breakthrough in integrating IoT, AI, and cloud computing in healthcare systems along with applications and

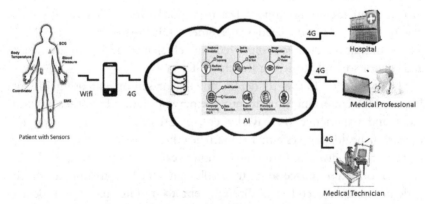

Figure 1.1 Smart technologies integrated healthcare.

challenges for further progress in healthcare systems. The main objectives of this chapter are

- motivation for integration of IoT, AI Cloud Computing, and Edge computing with healthcare systems
- challenges in integration of IoT, AI Cloud Computing, and Edge computing in healthcare applications
- future research directions in healthcare systems.

The remainder of this chapter is organized as follows—Section 1.2 presents the opportunities and risk in integrating IoT in healthcare systems. In section 1.3 the applications and challenges in integrating Cloud computing in healthcare system are elaborated. Section 1.4 provides the opportunities and challenges in integrating artificial intelligence in healthcare system. Section 1.5 presents the challenges and opportunities in integrating edge computing in healthcare system. Section 1.6 presents the forthcoming research technologies and future trends in healthcare system. Finally, we conclude the chapter in Section 1.7.

1.2 Internet of things in healthcare system

IoT is an emerging field of research, and its use in healthcare area is still in its early stage (Shah et al., 2019). In this section, IoT technology is studied and its applicability for healthcare system is analyzed. Several researches towards building healthcare IoT systems are reviewed to analyze the opportunities and risks in involving IoT in healthcare systems.

1.2.1 Internet of things

IoT is a virtual network that interacts with real world, using wireless sensor network and Internet as its core technology (Prabavathy et al., 2018a, 2018b). The goal of IoT is to use intelligent sensors and actuators for enabling pervasive environment with ubiquitous experience. It allows automation in a large range of industries with collection and processing voluminous data. Fig. 1.2 shows the layered architecture of IoT. Recently, IoT technology is penetrating into commercial application such as smart home (Stojkoska & Trivodaliev, 2017), agriculture (Ruan et al., 2019), smart parking (Khanna & Anand, 2016), smart grid (Al-Turjman & Abujubbeh, 2019), etc. Healthcare system is different from the aforementioned fields because the outcome of IoT-based healthcare system should be plausible and reliable. IoT technology in other commercial areas have proven that real-time remote monitoring with detection, collection, and

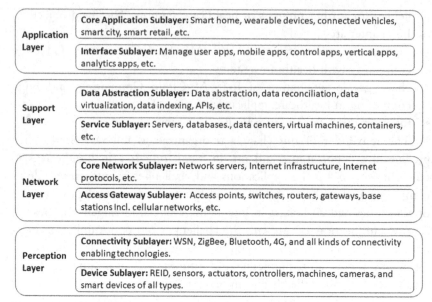

Figure 1.2 Internet of things layered architecture.

reporting of data are achievable (Butler et al., 2019; Saravanan et al., 2018; Yu et al., 2005). This capability of IoT can be expanded and used in healthcare systems for monitoring patient health and reporting it to healthcare related people such as caretakers, doctors, or emergency services.

1.2.2 Opportunities of internet of things in healthcare

IoT provides a promising solution to digitize healthcare systems, and many researches are being performed in IoT healthcare system (Farahani et al., 2020; Negash et al., 2018; Strielkina et al., 2017). Healthcare is a vast ecosystem hence it facilitates numerous IoT applications in patient monitoring (Archip et al., 2016), smart pills (Goffredo et al., 2015), robots in disease treatment (Cianchetti & Menciassi, 2017), wearable sensors (Rodgers et al., 2014) etc. The data collected by the IoT devices in healthcare system provide detailed contextual analysis and clear insights about the disease and patients' health. IoT not only helps people lead healthier life but also supports doctors to improve treatment and researches.

Remote patient monitoring is the most important application for healthcare system because it enables real-time personalized healthcare. It also helps them reduce cost and time (Kumar, 2017). The patients can be monitored using IoT wearable medical devices such as heart rate monitoring cuffs, sleep analyzer, blood pressure measuring bands, fitness bands, glucometer, etc., which reduces the burden of medical professionals (Khan et al., 2016). It is very much suitable for tracking the health conditions of elderly patients to detect the risk in advance.

The data from the IoT medical devices has many applications. Fig. 1.3 gives the benefits of IoT in healthcare. The doctors can use patient data from smart medical devices for faster treatment. The doctor can alter the dosage of medicines accurately using those data (Laranjo et al., 2012). It also helps in analyzing risk and benefit of particular drug. The data is available anywhere and anytime in the hospital IT infrastructure helps the physicians to make reliable decisions and services. These data also helps government to predict the viral spread, track patient journey for medical research and resource allocation.

The success of remote health monitoring have initiated more researches in integrating IoT in various healthcare applications. There are several works where IoT healthcare systems have been built for specific purposes, including rehabilitation (Celesti et al., 2020), diabetes management (Ara & Ara, 2017), and assisted ambient living for elderly persons (Abdelgawad et al., 2016). Though these systems are designed for many different purposes, they are each strongly related through their use of similar enabling technologies.

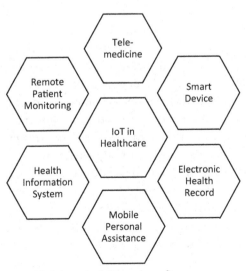

Figure 1.3 Healthcare with internet of things benefits.

1.2.3 Risks of integrating internet of things in healthcare

IoT is incredible to revolutionize the healthcare systems with its technology. But there are some challenges in integrating it within healthcare systems. Two of the most important risks in integrating IoT are data security and privacy issue. IoT medical devices monitor and detect patients' real-time data, which are vulnerable to multiple data security and privacy threats (Chacko & Hayajneh, 2018). As the IoT medical devices continuously detect and share personal data of patient, it is essential to ensure security and privacy. The impact of attack on these devices may lead to disaster or loss of life (Devi & Muthuselvi, 2016). Technologies for IoT healthcare are in the developing stage and are subjected to increased technical implications. Hence these new technologies of IoT healthcare have new challenges in ensuring security and privacy. The cybercriminals can compromise the patients' IoT medical device and use that personal health data for fraudulent health insurance claims and sometimes for buying and selling medical drugs illegally using fake IDs.

The next major challenge is integration of multiple IoT devices and protocols in healthcare systems. Most of the IoT devices are in developing stage and manufacturers do not follow standardization in communication protocols and device development (Knickerbocker et al., 2018). Hence there can be interoperability issues when multiple devices are involved in healthcare application. The various disciplines captured by healthcare IoT devices are regulated by a diverse group of regulatory agencies (Huycke & All, 2000). This creates complexity in healthcare system wherein medical standards are subjected to strict regulations. The lack of interoperability among the heterogeneous device platforms and standards that exist for authentication is an important vulnerability that leads to data privacy threats, compliance regulation issues, and backward compatibility with legacy systems. The IoT medical devices must be compatible with many transmission formats and protocols for authentication and encryption.

IoT integrated healthcare system involves data management challenges. The real-time monitoring data from body sensors will be voluminous and it is in various data formats. The healthcare system should be developed using data-driven learning techniques to handle these voluminous and variety of data formats (Reda et al., 2018). In healthcare system, data are collected from various sources, therefore proper authentication is needed to assure that healthcare data are submitted from registered clinics, hospitals, and medical institutions. Overloading of data can affect the accuracy in decision-making during the treatment. IoT-based healthcare are not always easy to use by medical professionals and physicians involved in it. The vast number of features involved in it could sometimes make healthcare system complicated which in turn discourage healthcare workers in learning.

1.3 Cloud computing in healthcare system

Cloud computing is not a new term in healthcare systems. In the last few years, the impact of cloud computing technology is increasing at faster rate in healthcare industry. In this section, cloud computing technology and its influence on healthcare has been analyzed. Recent researches in adopting cloud technology in healthcare systems are reviewed to understand the opportunities and challenges in taking up cloud based healthcare systems.

1.3.1 Cloud computing

Cloud computing has modified the basic building blocks of computing and revolutionized the ability of computing in wide variety of applications. Cloud computing facilitates a shared pool of computing resources containing networks, servers, storage applications, and service (Velte et al., 2009). These computing resources can be configurable and used with minimal effort. Based on demand of the customers the computing resources are allocated and released from the shared resources. This guarantees efficient and optimized resource allocation. The users need not invest on IT infrastructure and they can pay based on their utilization, hence it is cost effective too. Cloud computing facilitates customers with multiple technologies such as virtualization (Lombardi & Di Pietro, 2011) and multitenancy (AlJahdali et al., 2014). Using the web applications, the shared computing resources are managed and customers use the shared physical resources through virtual environment.

Based on a deployment model, the cloud model is classified as public, private, hybrid, and community cloud (Savu, 2011). Public cloud holds vast amount of resources and has high scalability which can be accessed by multiple organizations. Private cloud is designated to a single organization with high security. Hybrid cloud is a combination of private and public cloud and allows seamless interaction between the platforms. Community cloud is a multitenant platform provided to multiple organizations for sharing the same applications. Fig. 1.4 describes cloud deployment models. Based on the service provided, cloud models are classified as Infrastructure-as-a-Service (IaaS), Platform-as-a-Service (PaaS), and Software-as-a-Service (SaaS) (Weinhardt et al., 2009). Fig. 1.5 shows the various Cloud Service Models. Further these models can be specific to Storage, Database, Information, Process, Application, Integration, Security, Management, and Testing-as-a-service

1.3.2 Opportunities of Cloud computing in healthcare

The generation, consumption, storage, and sharing of data in healthcare system has increased tremendously. The digitization of data in healthcare industry has made massive shift in the data management processes (Fabian et al., 2015). The adoption of cloud computing technology has revolutionized the data management in healthcare system. Fig. 1.6 gives the important features of cloud computing. The extensive cloud adoption in

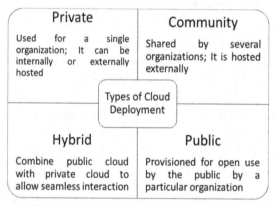

Figure 1.4 Cloud deployment models.

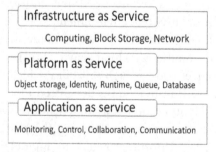

Figure 1.5 Cloud service models.

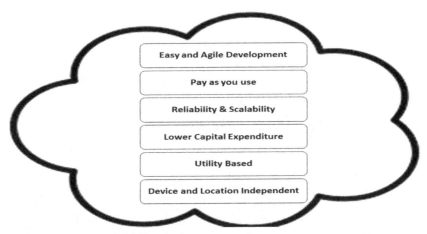

Figure 1.6 Features of cloud computing.

healthcare system have gained benefits not only in data storing but also in other areas such as workflow optimizations, cost benefits, and personalized care.

The cloud adoption in healthcare system provides significant cost benefits for patients as well as hospital management. High capital investments are avoided and operating cost alone is considered because the required computing resources are obtained on demand and released in cloud adopted healthcare system (Sultan, 2014). Since the computing and data management resources are maintained by cloud service providers, the related cost for IT staff resources in healthcare management is reduced.

Cloud services offers interoperability for data integration at every point of data management. The cloud adoption in healthcare provides access to patient data from multiple sources and share among the healthcare stakeholders to provide efficient treatment. The insights gained from multiple sources of patient's data facilitates efficient planning and healthcare service. The geographic limitations are avoided and provides seamless data transfer among various stakeholders such as pharmaceuticals, labs, and insurance companies, owing to interoperability of cloud service (Lupşe et al., 2012). Fig. 1.7 shows the cloud integrated healthcare system. Thus cloud service facilitates the data transformation from volume to value in healthcare.

Scalability of cloud services supports the healthcare system abundantly by providing on-demand resources scaling (Ahn et al., 2013). In recent

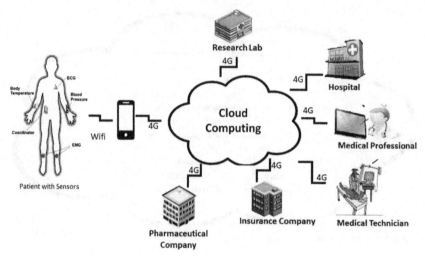

Figure 1.7 Cloud integrated healthcare system.

times large number of medical sensors are used in healthcare industry. These sensors are sampled at high frequency which leads to vast amount of data. Cloud services supports heterogeneous data of varying size by providing high storage and computing capabilities. Cloud adoption in healthcare system provides remote accessibility which aids numerous healthcare functionalities such as high personal care schemes, telemedicine, and posthospitalization monitoring schemes. Telehealth application involving cloud computing enables convenient sharing of data for patients during the treatment as well as posttreatment (Wang et al., 2017). Cloud services offer wide variety of capabilities for healthcare management staff to enable improved ways of working and provide new services to patients. Cloud adoption in healthcare supports faster upgrading of services with zero service interruption and minimal cost.

Cloud services can integrate with the emerging IoT devices which enables rapid development in the innovation of new pervasive and ubiquitous services in healthcare environment. Cloud technology is based on internet which implements standard protocols hence connects to healthcare applications without much technical complications. The medical mistakes can be avoided by using cloud services with cognitive capabilities such as intelligent business process management suites and case management frameworks. The latest analytical techniques in cloud service enable improved knowledge on disease diagnosis, treatment techniques, and patient care using the vast amount of data in the cloud (Iyengar et al., 2018).

1.3.3 Risks of integrating cloud computing in healthcare

Cloud adoption in healthcare is no doubt exciting, but it has various potential problems that need to be addressed. Security is a major concern in cloud based healthcare system because it stores and processes sensitive data held by healthcare systems (Khattak et al., 2015). The most common security problems are data breach and data loss, account hijacking, denial of service, and insecure interfaces. The patient data should be accessed with effective authorization else the attackers will compromise the system to access the patient data. There are possibilities for ransomware attack in which the attacker deliberately corrupts or deletes the patient data which may lead to data loss (Spence et al., 2018). Sometimes attacker may also perform identity theft to access the account of any stakeholders involved in the healthcare system. Adversaries seek to hijack the account of healthcare system stakeholders by stealing their security credentials and then eavesdropping on their activities and transactions (Flynn et al., 2020).

As the number of connected devices are growing in healthcare system, they are more prone to distributed denial of service attacks (DDoS) (Latif et al., 2014). During a DDoS attack, the attacker will flood healthcare IT systems by using a myriad of connections to overwhelm the system. The attacker use bots to generate numerous attacks from the connected devices, and it is hard to block. In healthcare, DDoS attackers can shut the cloud access for the stored data and services, which could prevent healthcare professionals from accessing critical patient information.

The security of cloud adopted healthcare mainly depends on the security of interfaces and application program interfaces (APIs) through which the stakeholders of healthcare system interact and manage the healthcare cloud infrastructure (AbuKhousa et al., 2012). The configuration of cloud infrastructure must be shipped with security else these configurations for API can be altered and the entire healthcare cloud infrastructure can be compromised by attackers.

The cloud providers offer services by combining them from subcontractors. Multiple subcontractors may be involved in cloud adopted healthcare system because of its complexity. Sometimes, service provided by these subcontractors will not be under the control of healthcare professionals hence there exist lack of influence and they alter the service during course of contract. This may lead to inefficiency in providing the health service.

The healthcare professionals do not have complete information about processing methods of cloud service providers (Al-Issa et al., 2019). This will affect the healthcare service during the maintenance phase. Sometimes there exist inadequacy of tools for managing and accessing cloud adopted healthcare data due to the dynamic and voluminous growth of healthcare data. Patients' personal data privacy is very hard to maintain in cloud adopted healthcare system because the cloud service provider has the physical control over the data.

1.4 Edge computing in healthcare system

The critical information storing and processing is one of the major requirement of digital healthcare system which can achieved efficiently by edge computing. In this section, edge computing technology is reviewed. Several researches towards involving edge computing in healthcare systems are reviewed to analyze the opportunities and challenges in integrating edge computing in healthcare systems (Chen et al., 2018; Ray et al., 2019).

1.4.1 Edge computing

In edge computing data is processed closer to location where it was generated such as devices or sensors. It reduces the amount of data transfer between the devices to cloud. The network load is reduced by localizing data storage and processing (Shi et al., 2016). It also reduces the data processing delay since it is performed near to location where it was generated instead of transferring to cloud back and forth. The localized data storage in edge computing allows the data to be accessed even when network is offline. The data are stored closer, hence it reduces data transmission cost.

Edge computing model allows even complex image processing algorithms to run locally on edge system having increased processing power. The real-time applications involving quick processing and response such as autonomous vehicles, and augmented reality, can be implemented using edge computing. The edge devices include simple sensor, smart phone, security camera, laptop, and gateways. Fig. 1.8 provides the basic architecture of edge computing.

1.4.2 Opportunities of edge computing in healthcare system

Edge computing facilitates healthcare sector to identify and analyses data as it is produced from IoT devices for quick understanding and action.

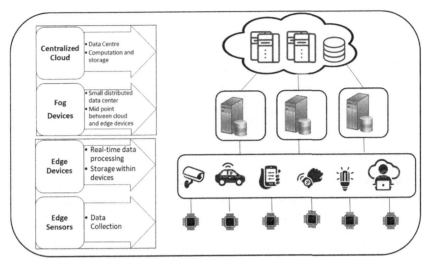

Figure 1.8 Edge computing architecture.

This moves the healthcare industry towards improved outcome for healthcare stakeholders. Edge computing allows to store and process critical information task on the edge devices which helps the healthcare industry to perform fast real-time medical data analysis and quickly respond to emergencies (Oueida et al., 2018). The noncritical medical data can be stored and processed in cloud environment.

Cloud computing provides services to store and compute data, still most of the healthcare stakeholders prefer to handle their data in on-premises data centers. Storing data in on-premises data centers provides more control over the data in terms of security and compliance. This also reduces the risk of downtime when compared cloud based systems (Shi et al., 2016). It allows autonomous control and security policies over the medical data. These security policies ensure that only authorized access of data and services are provided. Data is stored on-premises, hence the potential security breaches can be identified and prevented earlier, and thus it enables on-site security in high-risk environment.

Healthcare industry involves media-rich contents such as X-ray, and scan videos. There could be increase in delay, when these data are stored and accessed using conventional network. This issue is solved edge computing by storing and retrieving the data locally in on-premises data centers (Cha et al., 2018). Thus it provides fast access to critical media-rich data which supports on-time diagnosis and treatment. Fig. 1.9 shows the integration of Edge computing in healthcare sector.

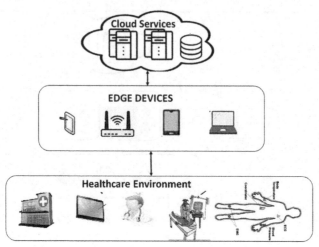

Figure 1.9 Edge computing in healthcare.

Interconnectivity is a major problem in healthcare infrastructure. Healthcare systems was suffering with recordkeeping and incompatible medical equipment and systems. Edge computing along with IoT devices solved this problem by providing communications quickly across the devices and rapid access and processing of data (Hassan et al., 2018). In addition to it, healthcare industry faces other challenges such as limited availability of skilled staff, data management and administration problems, and complex computing system and network which becomes life threatening for patients. Edge computing handles these issues by providing support to staff, bringing structure to the unstructured data collected from the IoT devices for efficient data management.

1.4.3 Challenges in integrating edge computing in healthcare system

Edge computing will revolutionize the healthcare systems with its technology, but still there are some challenges in integrating it with healthcare systems. The major challenge in integrating edge computing with healthcare system is interoperability (Sigwele et al., 2018). Healthcare infrastructure involves heterogeneous edge devices and systems from different vendors and using different communication protocols. Interoperability between these edge devices are required for implementing edge computing in healthcare system.

Edge computing architecture involves computing among heterogeneous devices with different computational power and storage (Cicirelli et al., 2017). Therefore load distribution among these heterogeneous devices with efficient resource allocation and usage is one of major challenge to process media-rich data of healthcare system.

Edge computing involves distributed process which enlarges the scope of the attack surface (Xiao et al., 2019). The healthcare edge devices are made smarter which will increases the vulnerability of the device. Not all edge devices are filled with resources hence implementing complex security algorithms are not feasible with those devices. The distributed dynamic architecture of the edge devices in healthcare environment increases the dimension of the attack surface.

Edge computing is in the development stage with nonstandard protocols and technologies. It has lot of implications in implementing in critical infrastructure like healthcare systems.

1.5 Artificial intelligence in healthcare system

AI is developed to mimic human cognitive capabilities (Müller & Bostrom, 2016). It is changing the medical field drastically using the digitized medical data and various machine learning algorithms. It assist the medical professionals in disease diagnosis and treatment efficiently. The voluminous healthcare data and vast growth in big data analytics using AI algorithms has made paradigm shift in healthcare system (Jiang et al., 2017). It is necessary to study the potential capabilities of AI to transform healthcare system and the risks involved in applying AI for healthcare system.

1.5.1 Artificial intelligence

The goal of AI is to emulate the cognitive capabilities of human intelligence such as ability to perceive, reasoning, planning, decision-making, learning and understanding (Bogue, 2014). The recent advancements in AI is leading it as potential contributor of fourth industrial revolution. Fig. 1.10 shows the important components of AI. Initially AI had some limitations due to limited computation capability, data availability, and storage but now the cloud and IoT has powered AI with high computing resources and voluminous data. The developments in AI have generated algorithms which can learn and predict from voluminous complex raw data. Presently, decision-making using AI involves huge volume of data

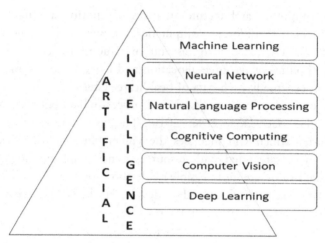

Figure 1.10 Components of artificial intelligence.

and high computation so the errors are reduced and decision are made faster with higher accuracy. The AI algorithms can efficiently detect the unforeseen relationship and complex nonlinear interactions in the voluminous data which cannot be done with standard statistics (Hofmann et al., 2017). AI algorithms have been implemented successfully in many application areas such as agriculture, business, healthcare, and science.

1.5.2 Opportunities of artificial intelligence in healthcare

AI in healthcare uses huge amount of data for analysis and interpretation to assist medical professional for making accurate decisions faster (Bennett & Hauser, 2013). Various pattern recognition AI algorithms help the doctors to handle complicated health conditions more efficiently. AI enables early detection of diseases using predictive algorithms. The health conditions of patients are diagnosed and decisions are made faster because time is a life–altering parameter for patients.

AI supports information management in healthcare systems. It helps healthcare stakeholder for efficiently managing the information. The AI enabled telemedicine helps both doctor and patients to save time and money with better treatment (Lysaght et al., 2019). Thus it takes the strain off of healthcare professionals in information management and improves the comfort of patients. AI also helps in improving the quality of Electronic Health Record (EHR) without human error (Koppel & Lehmann, 2015). Using deep learning combined with speech recognition technology, doctor interactions with patients is recorded along with clinical diagnosis and treatments provided are documented more accurately.

Presently AI has the ability to analyses voluminous patient data to detect the various treatment options using the natural language processing algorithms to provide personalized medication. Drug discovery is one of the greatest opportunity of healthcare where AI can be used efficiently to reduce long and expensive process of drug discovery (Smalley, 2017). Various AI algorithms are used in discovering new drugs with reduced risk in developing and testing. AI technology helps researchers to identify suitable patients to involve in the experiments and assists in monitoring medical response of patient more accurately.

The hospital operations such as managing emergency rooms, handling in-patient wards and scheduling doctor appointments are automated using AI algorithms. AI enabled Chatbots and virtual assistants are provided through telehealth (Sharmin et al., 2006). It also has preliminary diagnosis also provided through machine learning algorithms which helps patients to save time and money. The high risk patients are efficiently treated using AI algorithms by analyzing large dataset from various sources using personalized drugs dosage.

In addition to the above, the other applications of AI include, surgical robots provides surgeons support in surgical procedures (Palep, 2009), the virtual nursing assistants help patients and care providers in communication 24/7. AI also helps the health insurance companies to detect the false claims through the connected healthcare system. Fig. 1.11 gives the list of applications in integrating AI within healthcare sector.

1.5.3 Risks in integrating artificial intelligence in healthcare

AI in healthcare has more development in technologies but still it has some limitations. The main challenge in integrating AI with healthcare system is lack of standard training datasets (Iliashenko et al., 2019).

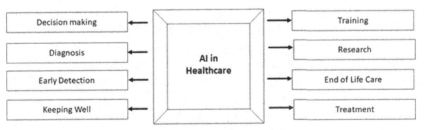

Figure 1.11 Applications of artificial intelligence in healthcare.

Many AI algorithms need training dataset to train itself for detection or prediction mechanism. The accuracy of result in AI algorithms depend on training data used to train the said algorithms. The training dataset may contain bias which may lead to incorrect decision-making. The genetic and behavioral data are required for some diseases which are hard to collect. If some information is underrepresented in training data then AI algorithms will not provide accurate results. AI algorithm-based decisions may be accurate but it may not be always optimal. Collecting complete patient data for decision-making is a daunting task.

The basic AI algorithms for healthcare system involves tracking and recording of daily activities of patients. It will be useful to generate efficient AI algorithms but it may affect patient privacy. Some of the automated healthcare applications installed in patient's mobile phone may collect patient's data without consent. AI technology is not completely free from errors. Sometime a small flaw may lead to wrong diagnosis and life risk. Patients' EHR are stored and maintained by hospitals. There are possibility for data breaches and these data can be used by hackers for false claim from insurance company (Thornton et al., 2015). The AI technology is growing every day, to meet these technology requirements the hardware and software should be updated frequently. This leads to increased maintenance cost as healthcare system involves complex and costly machines.

1.6 Future trends and research challenges in healthcare system

There are numerous advanced devices and algorithms are developed in incorporating smart technologies in healthcare sector. Apple Watch developed with advanced health features such as electrocardiogram (EKG) and menstrual health-tracking feature. AVA is a smart ovulation monitoring bracelet which gives better insights into the health of women. Smart Stop detects the craving for cigarettes by using sensors that keep on monitoring the changes happening in the body and provide medicine to reduce their addiction. Enlitic, a deep-learning algorithm uses patient's health record to optimize physician decision-making and to identify diseases. A cloud based platform for healthcare stakeholders have been developed by Flatiron which uses patients medical history, experiences, and scan reports to enhance oncological care real-time. Atomnet analyses the protein structure which caused the disease using deep-learning algorithms and designs drugs based on its analysis and insights. Zebra Medical Vision is a diagnostic program that collects and analyzes data to provide information about location and size of tumor, cardiovascular disease level risk, etc.

Various research groups all over world are working on integrating these smart technologies in healthcare sector. University of Pittsburgh Medical Center, United States, in research partnership with Microsoft Healthcare Next is working on healthcare research projects applying cloud and AI techniques. This research group focuses on integrating robots, voice recognition, and cognitive services into healthcare applications. University of Oxford, United Kingdom, has Computational Health Informatics Lab focusing on integrating AI in healthcare, called Clinical AI. The major research by this group is applying deep learning and Bayesian inference related methods on healthcare dataset to get new insights from healthcare data. Indian Institute of Science, Bangalore, India in collaboration with WIPRO GE Healthcare is working on healthcare diagnostics with deep learning, AI to facilitate sophisticated diagnostics and medical imaging reconstruction techniques for better medical imaging. Duke University School of Medicine, Stanford Medicine, and Google are jointly working on healthcare research called Project Baseline. The aim of this project is to find baseline for good health and important risk factors involved for various kinds of diseases.

The future trends in involving smart technologies for healthcare are Blockchain, wearable devices, big data, and advanced mobile applications. Blockchain technology is a permanent digital record in the form of open immutable ledger (Nofer et al., 2017). It has started to revolutionize healthcare system by providing secured sharing of patient data to increase the quality of diagnostics and treatment with transparency (Mettler, 2016). It improves the security of information between the healthcare stakeholders and also provides a better technique to manage the supply chain workflows.

Wearable IoT devices have great impact on healthcare system but with more technology advancement involving combination of predictive analytics and improved hardware, it is redefining disease diagnostics and patient-care techniques (Jiang et al., 2020). It provides constant flow of patient data for quicker and accurate diagnostics. Big data from the healthcare devices has revolutionized healthcare data analytics (Cui et al., 2020). It paves a way for the patients to reduce the cost of treatment significantly (Shafqat et al., 2020). It also provides information about preventable diseases to improve the life of people with more care. Advanced medical mobile applications have completely changed patients' way of receiving treatment and doctors' way of diagnosing and providing medical care (Qudah & Luetsch, 2019). It enables patients to search medical services in their mobile devices and provides easy-to-use environment for both patients and doctors, making their life easier.

The future research challenges in involving these smart technologies in healthcare sector includes virtual prescriptions, virtual reality, and augmented reality to teach medical students about healthcare courses, ambulance services embedded with sensor along the road junctions to reduce the traffic delay, emergency vehicular networks and developing algorithms to receive high-quality images in shorter-time to minimize exposure during the scan in radiology department.

1.7 Conclusion

The rapid advancement in information and communication technologies redefines the healthcare system. The integration of IoT, Cloud, and AI into healthcare system provides huge benefit to healthcare stakeholders by providing high resources and improved services. In this chapter, some of the major opportunities of integrating IoT, Cloud, and AI into healthcare sector were discussed. Though integration of these technologies have improved the healthcare sectors significantly but still we derived there are challenges existing in using these technologies for healthcare. Furthermore, the future technology trends such as blockchain, big data, and advanced mobile application for healthcare system are addressed.

The future technologies for healthcare has to be analyzed to understand the opportunities and challenges in applying them to healthcare system.

References

Abdelgawad, A., Yelamarthi, K., & Khattab, A. (2016). *IoT-based health monitoring system for active and assisted living. International Conference on Smart Objects and Technologies for Social Good* (pp. 11−20). Cham: Springer.

AbuKhousa, E., Mohamed, N., & Al-Jaroodi, J. (2012). e-Health cloud: Opportunities and challenges. *Future internet, 4*(3), 621−645.

Ahn, Y. W., Cheng, A. M., Baek, J., Jo, M., & Chen, H. H. (2013). An auto-scaling mechanism for virtual resources to support mobile, pervasive, real-time healthcare applications in cloud computing. *IEEE Network, 27*(5), 62−68.

Al-Issa, Y., Ottom, M. A., & Tamrawi, A. (2019). eHealth cloud security challenges: A survey. *Journal of Healthcare Engineering,* 2019.

AlJahdali, H., Albatli, A., Garraghan, P., Townend, P., Lau, L., & Xu, J. (2014). *Multi-tenancy in cloud computing. 2014 IEEE 8th International Symposium on Service Oriented System Engineering* (pp. 344–351). IEEE.

Al-Turjman, F., & Abujubbeh, M. (2019). IoT-enabled smart grid via SM: An overview. *Future Generation Computer Systems, 96*, 579–590.

Ara, A., & Ara, A. (2017). *Case study: Integrating IoT, streaming analytics and machine learning to improve intelligent diabetes management system. 2017 International conference on energy, communication, data analytics and soft computing (ICECDS)* (pp. 3179–3182). IEEE.

Archip, A., Botezatu, N., Şerban, E., Herghelegiu, P.C. and Zală, A., 2016, May. An IoT based system for remote patient monitoring. In 2016 17th International Carpathian Control Conference (ICCC) (pp. 1–6). IEEE.

Bartoletti. (2019). *AI in healthcare: Ethical and privacy challenges. Conference on Artificial Intelligence in Medicine in Europe* (pp. 7–10). Cham: Springer.

Bennett, C. C., & Hauser, K. (2013). Artificial intelligence framework for simulating clinical decision-making: A Markov decision process approach. *Artificial Intelligence in Medicine, 57*(1), 9–19.

Bogue. (2014). The role of artificial intelligence in robotics. *Industrial Robot: An International Journal*.

Butler, M., Angelopoulos, M., & Mahy, D. (2019). *Efficient IoT-enabled landslide monitoring. In 2019 IEEE 5th World Forum on Internet of Things (WF-IoT)* (pp. 171–176). IEEE.

Celesti, A., Lay-Ekuakille, A., Wan, J., Fazio, M., Celesti, F., Romano, A., Bramanti, P., & Villari, M. (2020). Information management in IoT cloud-based tele-rehabilitation as a service for smart cities: Comparison of NoSQL approaches. *Measurement, 151*, 107218.

Cha, S. J., Jeon, S. H., Jeong, Y. J., Kim, J. M., Jung, S., & Pack, S. (2018). *Boosting edge computing performance through heterogeneous many core systems. 2018 International Conference on Information and Communication Technology Convergence (ICTC)* (pp. 922–924). IEEE.

Chacko, A., & Hayajneh, T. (2018). Security and privacy issues with IoT in healthcare. *EAI Endorsed Transactions on Pervasive Health and Technology, 4*(14).

Chen, M., Li, W., Hao, Y., Qian, Y., & Humar, I. (2018). Edge cognitive computing based smart healthcare system. *Future Generation Computer Systems, 86*, 403–411.

Cianchetti, M., & Menciassi, A. (2017). *Soft robots in surgery. Soft robotics: Trends, applications and challenges* (pp. 75–85). Cham: Springer.

Cicirelli, F., Guerrieri, A., Spezzano, G., Vinci, A., Briante, O., Iera, A., & Ruggeri, G. (2017). Edge computing and social internet of things for large-scale smart environments development. *IEEE Internet of Things The Journal, 5*(4), 2557–2571.

Cui, Y., Kara, S., & Chan, K. C. (2020). Manufacturing big data ecosystem: A systematic literature review. *Robotics and computer- integrated manufacturing, 62*, 101861.

Devi, K. N., & Muthuselvi, R. (2016). *Parallel processing of IoT health care ap plications. 2016 10th International Conference on Intelligent Systems and Control (ISCO)* (pp. 1–6). IEEE.

Fabian, B., Ermakova, T., & Junghanns, P. (2015). Collaborative and secure sharing of healthcare data in multi-clouds. *Information Systems, 48*, 132–150.

Farahani, B., Firouzi, F., & Chakrabarty, K. (2020). *Healthcare IoT. In Intelligent Internet of Things* (pp. 515–545). Cham: Springer.

Flynn, T., Grispos, G., Glisson, W., & Mahoney, W. (2020). Knock! knock! who is there? investigating data leakage from a medical internet of things hijacking attack. *Proceedings of the 53rd Hawaii International Conference on System Sciences*.

Goffredo, R., Accoto, D., & Guglielmelli, E. (2015). Swallowable smart pills for local drug delivery: present status and future perspectives. *Expert Review of Medical Devices, 12*(5), 585–599.

Hassan, N., Gillani, S., Ahmed, E., Yaqoob, I., & Imran, M. (2018). The role of edge computing in internet of things. *IEEE Communications Magazine, 56*(11), 110–115.

Hassanalieragh, M., Page, A., Soyata, T., Sharma, G., Aktas, M., Mateos, G., Kantarci, B. and Andreescu, S. (2015, June). Health monitoring and management using Internet-of-Things (IoT) sensing with cloud-based processing: Opportunities and challenges. In 2015 IEEE International Conference on Services Computing (pp. 285–292). IEEE.

Hofmann, M., Neukart, F. and Bäck, T. (2017). Artificial intelligence and data science in the automotive industry. arXiv preprint arXiv:1709.01989.

Huycke, L., & All, A. C. (2000). Quality in health care and ethical principles. *Journal of Advanced Nursing, 32*(3), 562–571.

Iliashenko, O., Bikkulova, Z. and Dubgorn, A. (2019). Opportunities and challenges of artificial intelligence in healthcare. In E3S Web of Conferences (Vol. 110, p. 02028). EDP Sciences.

Iyengar, A., Kundu, A., Sharma, U., & Zhang, P. (2018). *A trusted healthcare data analytics cloud platform. 2018 IEEE 38th International Conference on Distributed Computing Systems (ICDCS)* (pp. 1238–1249). IEEE.

Jayaratne, M., Nallaperuma, D., De Silva, D., Alahakoon, D., Devitt, B., Webster, K. E., & Chilamkurti, N. (2019). A data integration platform for patient-centered e-healthcare and clinical decision support. *Future Generation Computer Systems, 92*, 996–1008.

Jiang, F., Jiang, Y., Zhi, H., Dong, Y., Li, H., Ma, S., Wang, Y., Dong, Q., Shen, H., & Wang, Y. (2017). Artificial intelligence in healthcare: past, present and future. *Stroke and vascular neurology, 2*(4), 230–243.

Jiang, N., Mück, J. E., & Yetisen, A. K. (2020). The regulation of wearable medical devices. *Trends in Biotechnology, 38*(2), 129–133.

Khan, Y., Ostfeld, A. E., Lochner, C. M., Pierre, A., & Arias, A. C. (2016). Monitoring of vital signs with flexible and wearable medical devices. *Advanced Materials, 28*(22), 4373–4395.

Khanna, A. and Anand, R. (2016, January). IoT based smart parking system. In 2016 International Conference on Internet of Things and Applications (IOTA) (pp. 266–270). IEEE.

Khattak, H. A. K., Abbass, H., Naeem, A., Saleem, K., & Iqbal, W. (2015). *Security concerns of cloud-based healthcare systems: A perspective of moving from single-cloud to a multi-cloud infrastructure. 2015 17th International Conference on E-health Networking, Application & Services (HealthCom)* (pp. 61–67). IEEE, October.

Knickerbocker, J., Budd, R., Dang, B., Chen, Q., Colgan, E., Hung, L. W., Kumar, S., Lee, K. W., Lu, M., Nah, J. W., & Narayanan, R. (2018). *Heterogeneous integration technology demonstrations for future healthcare, IoT, and AI computing solutions. 2018 IEEE 68th electronic components and technology conference (ECTC)* (pp. 1519–1528). IEEE.

Koppel, R., & Lehmann, C. U. (2015). Implications of an emerging EHR monoculture for hospitals and healthcare syst ems. *Journal of the American Medical Informatics Association, 22*(2), 465–471.

Kumar. (2017). *IoT architecture and system design for healthcare systems. 2017 International Conference on Smart Technologies for Smart Nation (SmartTechCon)* (pp. 1118–1123). IEEE.

Laranjo, I., Macedo, J., & Santos, A. (2012). Internet of things for medication control: Service implementation and testing. Procedia. *Technology (Elmsford, N.Y.), 5*, 777–786.

Latif, R., Abbas, H., & Assar, S. (2014). Distributed denial of service (DDoS) attack in cloud-assisted wireless body area networks:a systematic literature review. *Journal of medical systems, 38*(11), 128.

Lombardi, F., & Di Pietro, R. (2011). Secure virtualization for cloud computing. *Journal of network and computer applications, 34*(4), 1113–1122.

Lupşe, O. S., Vida, M. M., & Tivadar, L. (2012). Cloud computing and interoperability in healthcare information systems, April *The first international conference on intelligent systems and applications*, 81–85.

Lysaght, T., Lim, H. Y., Xafis, V., & Ngiam, K. Y. (2019). AI-assisted decision-making in healthcare. *Asian Bioethics Review, 11*(3), 299−314.

Martinez-M iranda, J., & Aldea, A. (2005). Emotions in human and artificial intelligence. *Computers in Human Behavior, 21*(2), 323−341.

Mettler. (2016). *Blockchain technology in healthcare: The revolution starts here. 2016 IEEE 18th international conference on e-health networking, applications and services (Healthcom)* (pp. 1−3). IEEE, September.

Müller, V. C., & Bostrom, N. (2016). *Future progress in artificial intelligence: A survey of expert opinion. In Fundamental issues of artificial intelligence* (pp. 555−572). Cham: Springer.

Negash, B., Gia, T. N., Anzanpour, A., Azimi, I., Jiang, M., Westerlund, T., Rahmani, A. M., Liljeberg, P., & Tenhunen, H. (2018). *Leveraging fog computing for healthcare IoT. In Fog computing in the internet of things* (pp. 145−169). Cham: Springer.

Nofer, M., Gomber, P., Hinz, O., & Schiereck, D. (2017). Blockchain. *Business & Information Systems Engineering, 59*(3), 183−187.

Oueida, S., Kotb, Y., Aloqaily, M., Jararweh, Y., & Baker, T. (2018). An edge computing based smart healthcare framework for resource management. *Sensors, 18*(12), 4307.

Palep. (2009). Robotic assisted minimally invasive surgery. *Journal of minimal access surgery, 5*(1), 1.

Pattnaik, P. K., Rautaray, S. S., Das, H., & Nayak, J. (2017). Progress in computing, analytics and networking. *Proceedings of ICCAN*, 710.

Prabavathy, S., Sundarakantham, K., & Shalinie, S. M. (2018a). Design of cognitive fog computing for intrusion detection in Int ernet of Things. *Journal of Communications and Networks, 20*(3), 291−298.

Prabavathy, S., Sundarakantham, K., & Shalinie, S. M. (2018b). Design of cognitive fog computing for autonomic security system in critical infrastructure. *J. UCS, 24*(5), 577−602.

Qudah, B., & Luetsch, K. (2019). The influence of mobile health applications on patient-healthcare provider relationships: a systematic, narrative review. *Patient Education and Counseling, 102*(6), 1080−1089.

Ray, P. P., Dash, D., & De, D. (2019). Edge computing for Internet of Things: A survey, e-healthcare case study and future direction. *Journal of Network and Computer Applications, 140*, 1−22.

Reda, R., Piccinini, F., & Carbonaro, A. (2018). Towards consistent data representation in the IoT healthcare landscape. *Proceedings of the 2018 International Conference on Digital Health*, 5−10.

Rodgers, M. M., Pai, V. M., & Conroy, R. S. (2014). Recent advances in wearable sensors for health monitoring. *IEEE Sensors Journal, 15*(6), 3119−3126.

Ruan, J., Jiang, H., Zhu, C., Hu, X., Shi, Y., Liu, T., Rao, W., & Chan, F. T. S. (2019). Agriculture IoT: emerging trends, cooperation networks, and outlook. *IEEE Wireless Communications, 26*(6), 56−63.

Saravanan, K., Anusuya, E., & Kumar, R. (2018). Real-time water quality monitoring using Internet of Things in SCADA. *Environmental Monitoring and Assessment, 190*(9), 556.

Savu. (2011). *Cloud computing: Deployment models, delivery models, risks and research challenges. 2011 International Conference on Computer and Management (CAMAN)* (pp. 1−4). IEEE.

Shafqat, S., Kishwer, S., Rasool, R. U., Qadir, J., Amjad, T., & Ahmad, H. F. (2020). Big data analytics enhanced healthcare sy stems: A review. *The Journal of Supercomputing, 76* (3), 1754−1799.

Shah, S. T. U., Yar, H., Khan, I., Ikram, M., & Khan, H. (2019). *Internet of things-based healthcare: Recent advances and challenges. Applications of Intelligent Technologies in Healthcare* (pp. 153−162). Cham: Springer.

Sharmin, M., Ahmed, S., Ahamed, S. I., Haque, M. M., & Khan, A. J. (2006). *Healthcare aide: Towards a virtual assistant for doctors using pervasive middleware. Fourth Annual IEEE*

International Conference on Pervasive Computing and Communications Workshops (PERCOMW'06) (pp. 6–pp). IEEE.

Shi, W., Cao, J., Zhang, Q., Li, Y., & Xu, L. (2016). Edge computing: Vision and challenges. *IEEE internet of things journal, 3*(5), 637–646.

Sigwele, T., Hu, Y. F., Ali, M., Hou, J., Susanto, M., & Fitriawan, H. (2018). *An intelligent edge computing based semantic gateway for healthcare systems interoperability and collaboration. 2018 IEEE 6th International Conference on Future Internet of Things and Cloud (FiCloud)* (pp. 370–376). IEEE.

Smalley. (2017). AI-powered drug discovery captures pharma interest.

Spence, N., Niharika Bhardwaj, M. B. B. S., & Paul, D. P., III (2018). Ransomware in healthcare facilities: A harbinger of the future? *Perspectives in Health Information Management,* 1–22.

Stojkoska, B. L. R., & Trivodaliev, K. V. (2017). A review of internet of things for smart home: Challenges and solutions. *Journal of Cleaner Production, 140,* 1454–1464.

Strielkina, A., Uzun, D. and Kharchenko, V. (2017, September). Modelling of healthcare IoT using the queueing theory.

Sultan, N. (2014). Making use of cloud computing for healthcare provision: Opportunities and challenges. *International Journal of Information Management, 34*(2), 177–184.

Thornton, D., Brinkhuis, M., Amrit, C., & Aly, R. (2015). Categorizing and describing the types of fraud in healthcare. *Procedia Computer Science, 64,* 713–720.

Ungurean, I., & Brezulianu, A. (2017). An internet of things framework for remote monitoring of the healthcare parameters. *Advances in Electrical and Computer Engineering, 17* (2), 11–16.

Velte, T., Velte, A., & Elsenpeter, R. (2009). *Cloud computing, a practical approach.* McGraw-Hill, Inc.

Velthoven, M. H., Cordon, C., & Challagalla, G. (2019). Digitization of healthcare organizations: The digital health landscape and information theory. *International Journal of Medical Informatics, 124,* 49–57.

Wang, J., Qiu, M., & Guo, B. (2017). Enabling real-time information service on telehealth system over cloud-based big data platform. *Journal of Systems Architecture, 72,* 69–79.

Weinhardt, C., Anandasivam, A., Blau, B., Borissov, N., Meinl, T., Michalk, W., & Stößer, J. (2009). Cloud computing—A classification, business models, and research directions. *Business & Information Systems Engineering, 1*(5), 391–399.

Xiao, Y., Jia, Y., Liu, C., Cheng, X., Yu, J., & Lv, W. (2019). Edge computing security: State of the art and challenges. *Proceedings of the IEEE, 107*(8), 1608–1631.

Yu, L., Wang, N. and Meng, X. (2005, September). Real-time forest fire detection with wireless sensor networks. In Proceedings. 2005 International Conference on Wireless Communications, Networking and Mobile Computing, 2005. (Vol. 2, pp. 1214–1217). IEEE.

CHAPTER 2

The architecture of smartness in healthcare

S. Venkatesh Babu[1], P. Ramya[2], C. Sundar[1] and D. Pradeep[3]
[1]Department of Computer Science and Engineering, Christian College of Engineering and Technology, Oddanchatram, India
[2]Department of Information Technology, Christian College of Engineering and Technology, Oddanchatram, India
[3]Department of Computer Science and Engineering, M.Kumarasamy College of Engineering, Karur, India

2.1 Introduction

The traditional healthcare industry transforms into e–Health due to the development of the internet and its related applications as stated by Qi et al. (2017). The entry of inexpensive devices like sensors, wearable, and mobile devices fasten the development of internet of things (IoT)-based healthcare industry, and these are internet enabled. According to Sangeetha et al. (2018), in healthcare, IoT is already available in the form of smart equipment that can be worn by patients to collect information like blood pressure, heartbeat, and other metabolic-based data. The IoT uses information and communication technology as its backbone along with pervasive computing, ubiquitous communication, and intelligence which facilitate faster growth of the IoT industry as given in Kühner and Daniel (2007). The IoT can operate by closely integrating with its environment and can establish communication to its near and far peers. In most scenarios the devices are equipped with intelligence and such objects are called smart objects as coined by Information Society Technologies Advisory Group ISTAG (2009).

The concept of integrating various resources through the internet under the common infrastructure is the widely used definition for IoT as given by Shamila et al. (2019). This common infrastructure provides better solutions for business and scientific researches. The IoT will be part of the next industrial revolution and future healthcare according to Ibarra-Esquer et al. (2017), Dang et al. (2019). The IoT device is mostly resource scare devices that is they depend mostly on the cloud for various computing activities, the growing number of IoT in the current scenario will

Edge-of-Things in Personalized Healthcare Support Systems
DOI: https://doi.org/10.1016/B978-0-323-90585-5.00002-3

directly impact the performance of cloud so to scale this situation Fog computing is employed. The Fog is an architecture which acts as a mini cloud and it can be deployed in IoT denser areas that reduces the communication cost and improves the response time considerably. Edge computing still improves the performance of the IoT ecosystem by taking the computing capability to closer to IoT, that is, most of the processing is done at the location where the data is generated.

2.2 Healthcare

2.2.1 Internet of things-based healthcare

The IoT model mostly works with any device that it connects as stated by Neill (2013), IoT can connect people, any network, and any service at any time. Then more specifically the connection of devices is mainly done through the internet as given in Meola (2019). The connectivity is simple and they are similar to connecting a smartphone to smart TV as given in Chiuchisan et al. (2014). Then more specifically the connection of devices is mainly done through the internet as given in Meola (2019), they are similar to connecting a smartphone to smart TV as given in Chiuchisan et al. (2014).

The popularity of IoT is because it changes the web-based virtual cyberspace to a network of physical commodity devices and the characteristics such as,

- ability to create a new network with customized infrastructure
- it can provide and utilize new services in a heterogeneous environment
- machine-to-machine communication is one of the popular characteristics of IoT as given in Gigli and Koo (2011).

In the healthcare ecosystem, the data produced by the devices are collectively sent to the data collecting centers which are maintained by a healthcare organization, and these data are utilized by the medical practitioners to meet their goal. Then to make the ecosystem more pervasive Electronic Medical Records (EMRs) are integrated with the cloud as given in Bates et al. (2003). The EMRs mostly consist of data related to health and personal information about the patient. The integration of IoT with EMRs will improve the quality of care, reduce the cost for accessing the data, and provide the necessary care as given in Kulkarni and Sathe (2014). According to Pang (2013), healthcare is one of the gorgeous application areas of IoT and it gives rise to remote healthcare, fitness activities and elderly care, chronic diseases. The treatment and medication which are provided at home with the help of various medical devices, sensors, and imaging and diagnostic devices will form a core part of IoT-based healthcare.

IoT-based healthcare provides many benefits that reduce cost, increase quality of life and enhance user experience. In the view of healthcare providers, the IoT provides effortless communication between individual patient and the clinics or healthcare organizations which reduces cost and failure rate. The wireless technologies enable the healthcare network to be alive always so the information generated in this network is always up-to-date. This network can be utilized to provide expert support to chronic diseases, real-time monitoring, early diagnosis, and medical emergencies. The medical industry is well equipped with various biological sensors, these sensors are collectively used to form a Wireless Sensor Network (WSN) as given in Ko et al. (2010). The WSN is the initial stage of IoT-based healthcare and the adaptation of an IP-based network for sensor networks facilitates the faster growth of IoT.

2.2.1.1 Layered internet of things architecture

The physical objects which are located near and far allowed to connect to gather, analyze, and monitor information is a typical IoT network. The layered architecture as shown in Fig. 2.1 is adopted for this type of network to facilitate seamless management and according to many researchers the IoT network consists of three layers such as,

- perception layer
- network layer
- application layer

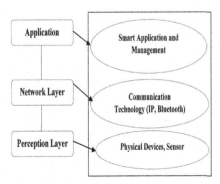

Figure 2.1 Internet of things layer architecture.

2.2.1.1.1 Perception layer

This layer act as the information collector from the things which are deployed in the environment, the things are identified by their unique identifiers as given in Silva et al. (2017). The collected information is mostly passed to the higher layer for processing through the network layer.

2.2.1.1.2 Network layer

This layer act as the brain for any typical IoT architecture and the main function of this layer is to collect information from one layer and propagate it to other layers. Logically this layer is the combination of internet and communication technologies. This layer incorporates many connecting technologies like 4G, Bluetooth, WLAN, and gateway nodes for interoperability. The IPv6 low power Wireless Personal Area Networks 6LoWPAN is a technique that applies internet protocol to low power devices which enables their communication capability.

2.2.1.1.3 Application layer

The application layer act as an interface between end-user and IoT, the user can utilize the data and manage the device through the set of application that runs in this layer. This IoT layer can be tailored according to the requirement and it is possible to run a high-level smart application in this layer.

As given in Suo et al. (2012), the bottom perception layer is called as *object layer* and it's responsible to collect information from all available heterogeneous IoT devices like a mobile phone is used to collect information from various sensors deployed in the patient body or simply from Wireless Body Area Network. The application layer is also classified into three-layer such as *Service management layer*, *Application layer*, and *Business layer*. In this, the service management layer takes responsibility to process the information, decision-making, and taking care of service requests, and the business layer is used to run business logic. The more interesting development in the IoT architecture according to Loiselle and Ahmed (2017), is the addition of a *support layer* which is placed in between the network and application layer. In the support layer, the technologies which facilitate the effective processing of information like cloud, intelligent, Fog, and Edge computing are employed. Then to make the application layer simple, data accumulation layer is used to store the vast amount of data produced by IoT.

2.2.2 Healthcare ecosystem

The traditional medical industry followed the practice of making the patient visit the hospital. This kind of practice is not suitable in some situations like in the battlefield, emergency, and the healthcare of the elderly. In this situation the medical service will not be available immediately. In the olden days radio communication was employed to assist the medical-related service at the battlefields and sailing ships, this paved the way for telehealth. The growth of telehealth technology leads to the concepts of connected health as given in Sharma et al. (2017). Connected healthcare employs smartphone and mobile applications along with other communication technologies. It leads to the development of smart health in which wearable medical devices such as blood pressure monitors, smartwatches, smart contact lenses, and other IoT devices that enhance the availability of medical service on time as given in Johri et al. (2014).

The devices in smart health are equipped to measure biological signals from the human body and these data are sent to a remote destination. The collected data are actively stored in the database which is maintained in the cloud and these data are used by the physicians for real-time monitoring and analysis and in research view, these data are used to study the nature of health, diseases, and other biological conductions around the various parts of the world. The amount of data contributed by smart health devices and IoT is huge and they are referred to as big data. Big data are mostly helping to provide meaningful information when processed as stated in Johri et al. (2014).

2.2.2.1 Internet of things-based healthcare network

The network is the collection of nodes that can be configured based on requirements like the IoT-based healthcare network has to be configured accordingly. As the network consists of *topology*, *architecture*, and *platform*. The topology is used to arrange the various medical equipment according to the requirements. The network is populated with heterogeneous devices such as laptops, smartphones, and medical terminals, various sensors devices like blood pressure monitoring devices, electrocardiograms (ECG) which are configured as hybrid grids as given in Viswanathan et al. (2012).

Fig. 2.2 shows a typical case in which a patient in emergency condition is being taken to a hospital in an ambulance equipped with IoTs, the vital signature are captured and sent to caretakers to act accordingly.

Figure 2.2 Internet of things smart environment.

The network topology will incorporate devices based on the scenario like the healthcare IoT collected ultrasound images or videos then the network should include high-speed networking devices like 4G, or satellite network to stream the data. The cloud can be a part of this topology which is to provide storage facilities and processing capabilities as given in Doukas and Maglogiannis (2012). The architecture is mostly designed like layers and the layer according to Zhang et al. (2012), are physical, adaptation, network, transport, and application layer.

- The link and physical layer deal with the physical components and their connectivity using IEEE 802.15.4 standard.
- The adaptation layer provides support for data transmission using IPv6 and 6LoWPAN.
- The sensor nodes use User Datagram Protocol.
- The application layer runs the required application and this application communicates using HTTP, SSL, and other application-level protocol.

The IoT healthcare platform specifies the various entity that required to execute the network and application seamlessly, like providing various software packages, Application Program Interface, networking tools to build and maintain the networks. The network mostly consists of heterogeneous nodes so whatever thing that gets inside the network must interoperate with all other devices so as stated in Rasid (2014), some set of standardization is required.

The IoT-based healthcare solutions are provided by using the following three configurations, sensor-based healthcare solutions, smartphone, and microcontroller-based solutions. In the early stage of healthcare, environment sensors are used to monitor human biological vital signals. The body temperature sensors (Max 30205), pulse rate sensors (BME 680) are the commonly used sensors as given in Tsakalakis and Bourbakis (2014). The authors, Ahouandjinou et al. (2016), developed a blueprint that includes humidity sensor, biochemical sensing sensor like glucometer, body movement detecting sensor, CO_2 level sensing device, and respiration monitoring sensor, these sensors can be baked according to the requirement as a wearable device and they form a Wearable Body Area Network. Radio-frequency identification is used to develop a smart real-time Intensive Care Unit system as shown in Ahouandjinou et al. (2016). Grossi (2018) points out that the modern smartphone consists of various useful resources that is they contain 15 or more sensors. These sensors can be used to monitor biological events and the collected data are used to make the decision. The modern smartphone comes with more storage space and high-speed internet connectivity, which makes it an ideal healthcare solution. According to Kumar et al. (2017), smartphone is used to measure the ECG and these data are used to raise the alarm if any variation is seen in ECG. The microcontroller is more power full device that can process raw data faster, as shown in Wang et al. (2004), the Arduino-based health monitoring system is used to measure the body temperature and pulse rate.

2.2.2.2 Healthcare services based on internet of things

The IoT-based healthcare services are limited and they vary based on the requirements, each service will provide some set of healthcare solutions. The internet of mobile health is the simplest service that is provided with the help of mobile, medical sensors, and communication technologies. The smartwatch monitors the temperature and heartbeat rate and they are communicated to the paired mobile as shown in Pang, Chen, et al. (2013). In Istepanian et al. (2004), some of the challenges are discussed to some extent. The drugs which are prescribed will sometimes lead to some adverse reaction due to the patient conduction, to tackle this problem in Istepanian (2011), proposed an iMedBox which scans the NFC enabled medicine when they are brought from the medical shop and it will be compared with the patient medical record and based on the results the medicines are allowed to be taken by the patient as given in Yang (2014).

An Ambient Assisted Living (AAL) is a kind of environment that integrates the various sensors and services they are used to provide medical support to elderly and special need people. The AAL becoming mandatory in the modern world where the life of human beings increases as the result elder population goes up. This situation leads to the requirement of cost-effective solutions to provide healthcare solutions to those elder populations. The ambition of the AAL is to provide wellbeing and safety to elderly people and based on this various applications are developed such as safety and security, social contacts, telehealth, sleep pattern monitoring, and tracking the daily life activities for better improvement of health. The data produced in this industry is becoming a more valuable asset for the scientific community and in the recent time domain, the IoT is moving towards AI to make the environment smarter.

2.3 Technology based smartness

Medical science has begun to embed with digital technology as a result of collected and stored information increasing in huge amount. These data are processed in some statistical way to dig out more useful information that can be used to bring out various enhancements in medical science. The IoT device is not resource-rich, so they depend on some platform for the processing and storage of information. The Quality of Service (QoS) is one of the main requirement of digital-based medical service such as real-time patient monitoring system should not experience any delay or jitter. The application that requires QoS demands dynamic allocation of resources during its operation in the cloud which has been a cheap and scalable platform that can be used flexibly to support the demand made by the application.

2.3.1 Integration of artificial intelligent

The way in which information is processed and utilized is changed from normal traditional algorithm to AI-based algorithms. The Internet of Medical Things (IoMT) is beginning to transform into smart medical sensors which are operated intelligently, for example, the smart heartbeat wearable smart device can automatically make a warning about the irregularity that happens in the heartbeat rate, and it can alert the user based on the users' walking pattern which leads to heartbeat irregularity. The IoMT which is attached to a blind person to facilitate to guide him is one kind of AI-based device, it used Wireless Body Area Network which is the collection of sensors attached with the person. The motion sensor will monitor the movement of the persons and the direction of the location of are pinpointed with the help of GPS devices and high-resolution camera which act as eyes will capture images in real-time and process it in association with other sensors and the results are input to the person as audio information and in some cases vibrations are used. If the AI wants the person to turn left side means the vibrator in the left hand vibrates and right one for the right direction, if the both vibrator vibrates then it indicates the presence of the object in the path or the person may taken the wrong direction.

In Jara et al. (2010), real-time IoT-based ECG remote monitoring is proposed. In which an intelligent algorithm is deployed in the smartphone which gets input from the wearable devices and makes real-time evaluation and produces an efficient suggestion about the healthiness of the heart based on his physical activities, the AI algorithm can acquire knowledge from the historical data available in the cloud. The IoMT can be deployed in two different ways such as static monitoring and dynamic monitoring. In static monitoring, the data, the patient, and the associated devices will not move such as home, hospital, and the dynamic monitoring is done at any place, that is, the patient and associated devices are in motion. The static and dynamic model of operation requires a different type of computing and communication requirements and to manage this different types of processing architecture are employed as stated in Satija et al. (2017). The different levels are Edge, Fog, and cloud computing.

- The cloud process processing and data storage support as remote support.
- The fog is the small framework of the cloud and it operates near the IoMT deployment area which aims to reduce communication cost and load on the network.
- The Edge in which the processing capability is provided to the device or placed at the gateway server, smartphone, advanced sinks, etc.

The three architecture can be used collaboratively, for example, in a static monitoring environment the data from the sensor devices are collected by fog nodes, these data can be sent to the cloud for processing and they can remain in the fog database. In dynamic monitoring, environment fog is replaced by edge devices that would directly interact with cloud services. In IoMT the data collected by the monitoring device are allowed to reach the end-user through the cloud. The cloud is a more congested environment where the response time will not be guaranteed since much medical-related application requires timely delivery of information, so to tackle this case the mini cloud is developed which are termed as fog, that acts like a real cloud. When a temperature monitoring system is integrated with the fog, the fog will gather information continually and this information is aggregated. These aggregated values will give a meaning full messages and it is sent to the cloud, this will reduce the load on the cloud and the application does not experience any delay in getting responses. The cloud and fog are used to balance the processing capability and to increase the response time, but some critical healthcare application requires immediate responses such kind of application are pushed to the edge.

2.3.1.1 Edge computing in healthcare

The era of edge begins with the entry of active and wearable sensors as given in Al-Fuqaha et al. (2015), and it also presents an application in which the twelve kinds of human behavior are gathered by the devices and sent to fog where an Long Short-Term Memory (LSTM) based algorithm is used to analyze the data and can be used to make any prediction related to health. The machine learning (ML) techniques can be applied with the physiological data and they can be used to detect the anomalies in the physiological parameter as given in Al-Fuqaha et al. (2015). The performance of this ML-based application can be improved with the integration of edge. As given in Poniszewska-Maranda et al. (2019), the Hierarchical Temporal Memory algorithm which is distributed across the edge nodes is used for analysis.

In Sood and Mahajan (2017), a fog-based system is described in which the system analyzes the environment and health symptoms which are used to predict the presence of the virus and makes use of it to alert the people in that particular location. The users of smartphones can find the risk level of diabetic patients using a decision tree classifier as given in Devarajan et al. (2019). Thus the AI and ML-based approaches are useful to detect anomalies, predictive risk monitoring, decision, and treatment support. The limited capability of an edge can be overcome by properly dividing the functionality among edge, fog, and cloud.

2.3.2 Semantic objects

Semantics is the concept that deals with connecting entities based on their relationships defined by some concepts. In a different concept, the entities will combine differently, for example, an entity like blood pressure will have a relationship with the heart, since blood pressure and heartbeat are related to each other. The AI has the capability of learning and doing things in a better manner, but the AI has a shortfall in making things better when there is no appropriate data that is the quality of data is very low. In such a case the semantic and AI can be combined to get better results when applied to classification and recommendation systems. The semantic knowledge will act as the brain for AI so that it will provide quality outcomes.

The knowledge processing based on semantic will enrich the measured data and gives out more fine-grain information. The method described in Wang et al. (2004), Web Ontology Language (OWL) encoded context ontology is the first knowledge-based processing. Table 2.1 shows some of the systems built based on ontology.

Table 2.1 Ontology-based system.

Source	Findings
Avancha et al. (2004)	The process of determining the expected behavior of sensor networks based on conditions using ontology.
Matheus et al. (2006)	The various sensor data are gathered and processed to provide service based on the context with the support of ontology.
Witt et al. (2008) and Schadow and McDonald (2009)	Sensor Model Language (SensorML) and Unified Code for Units of Measure (UCUM) are the ideal company for ontology-based IoT.
Hu et al. (2007)	It is the first proof of concept, which describes the use of web ontology, semantic web, and SensorML in Ontology-based Wireless Sensor Network.
Herzog et al. (2008)	Device-Agent Based Middleware Approach for Mixed Mode Environment is shown, in which heterogeneity-based measurements are interpreted based on the device ontology.
Bowers et al. (2008)	The measurement data are merged semantically and discovered using Web Ontology Language, Description Logic (OWL-DL).
Stevenson et al. (2009)	Antonym-Sensor describes the core concepts of location, time, people, and sensing and integration of these entities based on semantic.
Calder et al. (2010)	Coastal Environmental Sensing Network (CESN) uses Marine based sensor data and rule-based reasoning to the hideout the anomalies.

Most AI calculations function admirably either with text or with organized information, yet those two sorts of information are seldom joined in most cases. Semantic information models can overcome this issue. Connections and relations among business and information objects of all configurations like XML, social information, personal health record, and unstructured content can be made accessible for additional examination. This permits us to connect information even across heterogeneous information sources to give information objects as preparing informational indexes which are made out of data from organized information and text simultaneously.

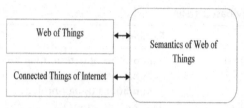

Figure 2.3 Semantics of things.

Semantic AI is the cutting edge Artificial Intelligence. AI can assist with expanding information diagrams that is the corpus and consequently, information diagrams can assist with improving ML calculations. This coordinated methodology at last prompts frameworks that work such as self-streamlining machines after an underlying arrangement is made and it is transparent to the basic information models as shown in Fig. 2.3.

The concepts in semantic can be devices, sensors and the communication channel, workstation, data storage servers. The semantic will decide which concepts are related together and the data are collected from only such concepts and they are given to AI for processing. The semantics will enhance the processing of the data concepts using rule-based and logical interpretation, like when the human pulse rate falls below a certain level then other parameters which are related to heartbeat are measured so that correct response action can be taken immediately. The semantic-based AI model will pinpoint the required service which will enhance the treatment given. The adoption of semantic comes with certain difficulties which are given below:

- The generation of semantic information at the sensor level is not practically possible.
- To specify which concepts are required to be grouped to perform a particular service in a heterogeneous environment is difficult.
- The integration of different domains is done by mapping the semantic concepts and there is a question when this integration is needed and why.
- The integration of semantic with AI will consume time, which are not suitable for time sensitive applications.
- The development of sharable model for the semantic is highly complex work.

The following are the semantic technology that are used to interconnect the IoT platform to the operation domains.

- Hypercat (Lea, 2013), provides format and API for interacting, gathering, and searching IoT entities. The Hypercat provides APIs that are build using universal adapter language and it can access any resource using the URL.

- OpenIoT is the technology that enables the connection of sensor devices to the software to facilitate context and semantic discovery. The OpenIoT uses the Hypercar API for interoperability and it can be used as Sensing-as-a Service as given in Kim and Lee (2014).
- In Fortino et al. (2018), Generic Ontology of IoT platform is developed which solves the problem of semantically integration of devices, middleware, and data to ensure efficient interoperability.

2.3.2.1 Healthcare with semantic

The ambient living environment is mostly used to detect the health conduction of any individuals that is to check they are healthy or not, but they are not able to predict the development of disease or disorder. This is addressed in Devarajan et al. (2019), where the Activities of Daily Living (ADL) of over 60 years are used to study the nature of occurrence of diseases like neurodegenerative diseases and other genetic disorders. This ADL data analysis can be integrated with the IoT to automatically detect disease or disorder alert system as given in Katz (2019), Wu (2019). The ADL, when integrated with IoT-based healthcare, will enhance the ambient healthcare ecosystem as given in Zgheib (2019), in which a continuously reducing activity level of an individual will indicate fatigue. The author showcased the integration of semantic interoperability with the IoT system with the help of ontologies. As given in Sareen (2017), given a Disease Ontology which has the description symptoms and disease in an IoT environment.

The development of any system should be scalable and for all time the chronic diseases are more frequent among elderly people and children living in close proximity as given in Schriml (2020). The environment is very vigorous that it spread the infections among those living in such closed proximity due to the sharing of air, food, water, and healthcare facility from the common source which is seen today in the case of COVID-19. As in Ziakas (2011), the epidemic outbreaks of infectious disease will increase the mortality rate so an early warning system based on the IoT and ADL will provide a better result by analyzing the present disease spreading pattern with the help of IoT and the support of ADL data. The activities of human are studied at various levels, that is, they can be done by simply analyzing the person's physical conduction and analyzing the various biological activities which take place in a human body as given in Abdallah (2018). The ADL will use to detect the habit of eating, grooming and drinking, and taking medicine.

The next level of activity involves active detection by involving analyzing capability of IoT-based semantic system. The high level of understanding the human behavior will narrow down the time to classify the normal and abnormal human behavior. To support the integration of intelligence to IoT, Perriot (2014) proposed an algorithm to classify the physical and abnormal activities in a chronic obstructive pulmonary disease patient. The smart environment is assisting to identify the development of neurodegenerative disorders such as dementia and Alzheimer's by examining the activities with the support of ADL as given in Sareen (2017). The semantic and AI plays an important role in healthcare analytics and it increases the knowledge of medical practitioner's knowledge and with the help of such a system decision are taken with the help of semantic-based AI expert system at the time of any medical emergency.

Smart Appliances REFerence for Health (SA-REF4Health) is developed to handle ECG which is embedded on smart wearable devices. The ECG signals are serialized to ensure they are compatible with healthcare software as shown in Moreira et al. (2018). In Paganelli and Giuli (2011), the ontology-based context management system is used to support a home-based care system. The system uses rule-based reasoning to figure out the risk based on the conduction, alarms, and social conduction and alert the nearby medical facility or caretaker. As stated in Enshaeifar et al. (2018), Technology Integrated Health Management is developed to support home-based dementia care with the help of machine learning-based information extraction and aggregation on environmental and physiological data.

2.3.2.2 Internet of things big data and healthcare: a discussion

The concept of Precision Medicine Initiative was launched in the USA in the year 2015. It aims to improve health and disease therapy through altering the treatment and performing preventative action based on the specification of individual patient needs. Every patient will have unique health issues so the doctor will have different types of treatment to best fit the individual. In the cancer ecosystem, the genetic profile plays an important role, that is cancer and many other diseases and disorders are mainly dependent on the genetic profile. Genetic information is huge and comes under the concept of big data and they are collected and maintained in any big data storage architecture. The IoT can be integrated into a big data environment and it can be used to perform various analysis activities and act as an expert system.

2.4 Security

The internet acts as the backbone for IoT and other real-time monitoring devices and since the internet is an open environment, security-related problems will occur and the IoT devices are mainly small battery-operated devices in which it is more complex to execute any security-related modules, so the problem of providing security is a major research challenge. The fast detection of potential security threats remains a challenge because of the number and complexity of emerging software and hardware vulnerabilities. This issue is getting worse as an increasing number of devices are being connected to the Internet. Today, default authentication remains prevalent, and insecure web-based interface access further increases the attack surface. Additionally, we have also seen a surge in the proliferation of wearable devices (including different types of embedded sensors and implanted medical devices) in recent years. The lack of security standards for the devices along with the availability of powerful search engines such as which is possible to locate the internet-connected devices make these wearable devices vulnerable to all kinds of attacks. The following are the common type of attacks that are possible in the IoT ecosystem:

- An attack that challenges the information distribution, which includes interruption, modification, and replay.
- The attack compromises the host system by attacking any one or more of the following system components like hardware, software, and user.
- The attack that happens through the networking side such as altering the standard protocol, and network stack attack.

Dimitrov and Dimite (2016) detail the application of big data and medical IoT healthcare industry-based security glitches. The sensors, manufactured by different manufacturers, are the heart of IoT which collect data and these data are transmitted to the servers for further processing. These heterogeneous types of sensors use a different type of communication techniques to send information which implements common security service. In Kang et al. (2019), the problem of security enhancement is narrowed down to a specific location, that is, the digital hospital is divided into three categories such as patient, worker, and medical environment. In this category the securing of devices that perform a vital biological signal measurement, transmission, and processing are more important than other areas. Agrawal (2014) demonstrates privacy mechanisms such as encryption, secure routing, and secure authentication in the healthcare industry. The use of traditional encryption techniques depends on the efficiency of key management, and it's important to make it a hotspot for the attacker. To overcome this vulnerability homomorphic encryption is used.

Three areas need to be secured for a threat-free environment, like network, data, operating system (OS), or the system software, and the security to these areas are provided by the means of security services. The author highlighted the possibilities of gaining access to the IoT devices remotely through networking interfaces and protocols and the possibilities of eradicating are to use Virtual Private Networks which encrypts the data traffic that passes through it. The data integrity can be ensured by using encryption but at a high cost since the IoT, the environment is a heterogeneous one with different types of devices with a different configuration. The OS is a prime target for the attackers, the applicability of security to OS is the learning and enhancing process, which is an early version of OS will come with some basic security, and it will be patched with solutions when new threats are learned from smart attackers. The Denial of Service attack is the more serious and commonly taking place in IoT environment, this can be overcome by limiting the use of various nodes, network ports, and serviceable interfaces.

2.5 Conclusion

The IoT is becoming a part of everyday life in all most every field of science and technology, in which medical science is one of the most important areas where the IoT enters the field from simple monitoring devices. The demand for quality of life is increasing due to the degradation of the environment and increasing of population along with pollution makes the necessity of technology to be embedded with the medical field. The IoT is an cost-effective one so it is widely adopted in the medical field. The healthcare based wearable entities and small monitoring devices are widely popular in e-health care sectors. These devices are connected which enables timely response to take place in case of an emergency and the integration of these IoT with the intelligence system makes the environment more sophisticated. The intelligence helps to identify the diseases or the disorder by acting as the expert system and assist the healthcare provider. Improvement in the security area will help the IoT-based healthcare industry grow faster.

References

Abdallah, Z. S. (2018). Activity recognition with evolving data streams: a review. *ACM Comput Surv (CSUR)*, *51*(4), 71.

Agrawal, V. (2014). *Security and privacy issues in wireless sensor networks for healthcare.* International Internet of Things Summit (pp. 223−228). Springer.

Ahouandjinou, A.S., Assogba, K. and Motamed, C. (2016) Smart and pervasive ICU based-IoT for improving intensive health care. *International Conference on Bioengineering for Smart Technologies (BioSMART)*, https://doi.org/10.1109/BIOSMART. 2016.7835599, 1−4.

Al-Fuqaha, A., Guizani, M., Mohammadi, M., Aledhari, M., & Ayyash, M. (2015). Internet of things: A survey on enabling technologies, protocols, and applications. *IEEE Communications Surveys & Tutorials*, *17*(4), 2347−2376.

Avancha, S., Joshi, A. and Patel, C. (2004) Ontology-driven adaptive sensor networks. In Proceedings of the Te1st Annual IEEE International Conference on Mobile and Ubiquitous Systems: Networking and Services (MOBIQUITOUS), Boston, Mass, USA, 194−202.

Bates, D. W., Ebell, M., Gotlieb, E., Zapp, J., & Mullins, H. C. (2003). A proposal for electronic medical records in US primary care. *Journal of the American Medical Informatics Association*, *10*(1), 1−10.

Bowers, S., Madin, J.S. and Schildhauer, M.P. (2008) A conceptual modeling framework for expressing observational data semantics. In Proceedings of the International Conference on Conceptual Modeling, Springer, Barcelona, Spain, 5231, 41−54.

Calder, M., Morris, R. A., & Peri, F. (2010). Machine reasoning about anomalous sensor data. *Ecological Informatics*, *5*(1), 9−18.

Chiuchisan, I., Costin, H.N. and Geman, O. (2014) Adopting the internet of things technologies in health care systems. *IEEE International Conference and Exposition on Electrical and Power Engineering (EPE)*, 532−535.

Dang, L. M., Piran, M. J., Han, D., Min, K., & Moon, H. (2019). A survey on internet of things and cloud computing for healthcare. *Electronics*, *8*, 768.

Devarajan, M., Subramaniyaswamy, V., Vijayakumar, V., & Ravi, L. (2019). Fog-assisted personalized healthcare-support system for remote patients with diabetes. *J. Ambient Intell. Humaniz. Comput*, *10*, 3747−3760.

Dimitrov, V., & Dimite, V. (2016). Medical internet of things and big data in healthcare. *Healthcare informatics research*, *22*(3), 156−163.

Doukas, C. and Maglogiannis, I. (2012) Bringing IoT and cloud computing towards pervasive healthcare. in *Proceedings from International Conference on Innovative Mobile and Internet Services in Ubiquitous Computing (IMIS)*, 922−926.

Enshaeifar, S., Barnaghi, P., & Skillman, S. (2018). The internet of things for dementia care. *IEEE Internet Computing*, *22*(1), 8−17.

Fortino, G., Savaglio, C., & Palau, C. E. (2018). *Towards multi-layer interoperability of heterogeneous IoT platforms. Integration, Interconnection, and Interoperability of IoT Systems* (pp. 199−232). Springer.

Gigli, M., & Koo, S. (2011). Internet of Things: Services and Applications Categorization. *J. Adv. Internet Things*, *1*, 27−31.

Grossi, M. (2018) A sensor-centric survey on the development of smartphone measurement and sensing systems Measurement, https://doi.org/10.1016/j.measurement.2018. 12.014, 135, 572−592.

Herzog, A., Jacobi, D. and Buchmann, A. (2008) A3ME - an agent based middleware approach for mixed mode environments. In *Proceedings of the 2nd International Conference on Mobile Ubiquitous Computing, Systems, Services and Technologies*, IEEE, Valencia, Spain, 1−5.

Hu, Y., Wu, Z. and Guo, M. (2007) Ontology driven adaptive data processing in wireless sensor networks. In *Proceedings of the 2nd International ICST Conference on Scalable Information Systems*, Suzhou, China, 46.

Ibarra-Esquer, J.E., González-Navarro, F.F., Flores-Rios, B.L., Burtseva, L. and Astorga-Vargas, M.A. (2017) Tracking the Evolution of the Internet of Things Concept Across Different Application Domains Sensors. **17**, 1379.

Information Society Technologies Advisory Group (ISTAG). (2009) Revising Europe's ICT Strategy: Report from the Information Society Technologies Advisory Group (ISTAG).

Istepanian, R.S.H. (2011) The potential of Internet of Things (IoT) for assisted living applications. In *Proc. IET Seminar Assist. Living*, 1–40.

Istepanian, R. S. H., Jovanov, E., & Zhang, Y. T. (2004). Guest editorial introduction to the special section on m-health: Beyond seamless mobility and global wireless healthcare connectivity. *IEEE Transactions on Information Technology in Biomedicine: a Publication of the IEEE Engineering in Medicine and Biology Society, 8*(4), 405–414.

Jara, A.F., Belchi., Alcolea, A.F., Santa, J., Zamora-Izquierdo, M.A. and Gomez-Skarmeta, A.F. (2010) A pharmaceutical intelligent information system to detect allergies and adverse drugs reactions based on Internet of Things. *In Proc. IEEE Int. Conf. Pervasive Comput. Commun. Workshops (PERCOM Workshops)*, 809–812.

Johri, P., Singh, T., Das, S. and Anan, S. (2014) Vitality of big data analytics in healthcare department. *IEEE International Conference on Infocom Technologies and Unmanned Systems*, Dubai.

Kang, S., Baek, H., Jung, E., Hwang, H., & Yoo, S. (2019). Survey on the demand for adoption of internet of things (IoT)-based services in hospitals: Investigation of nurses' perception in a tertiary university hospital. *Applied Nursing Research, 47*(2), 18.

Katz, S. (2019). Progress in development of the index of adl. *The Gerontologist, 10*, 20–30.

Kim, J. and Lee, J.W. (2014) OpenIoT: an open service framework for the internet of things. In *Proceedings of the IEEE World Forum on Internet of Things*, Seoul, South Korea, 89–93.

Ko, J., Lu, C., Srivastava, M. B., Stankovic, J. A., Terzis, A., & Welsh, M. (2010). Wireless sensor networks for healthcare. *Proc. IEEE, 98*(11), 1947–1960.

Kulkarni, A., & Sathe, S. (2014). Healthcare applications of the internet of things: A review. *International Journal of Computer Science and Information Technologies, 5*(5), 6229–6232.

Kumar, S.P., Samson, V.R.R., Sai, U.B., Rao, P.M. and Eswar, K.K. (2017) Smart health monitoring system of patient through IoT. *International Conference on SMAC (IoT in Social, Mobile, Analytics and Cloud)(SMAC)*, https://doi.org/10.1109/SMAC.2017.8058240, 551–556.

Kühner and Daniel (2007) Internet der Dinge Telekommunikationsinfrastruktur. Edited by Seminar band: Mobile und Verteilte Systeme. *Ubiquitous Computing Teil IV. Seminar band: Mobile und Verteilte Systeme - Ubiquitous Computing Teil IV*, Universität Karlsruhe - Fakultät für Informatik.

Lea, R. (2013). Hypercat: an iot interoperability specifcation. *IoT ecosystem demonstrator interoperability working group*, 1–9.

Loiselle, C. G., & Ahmed, S. (2017). Is connected health contributing to a healthier population. *Journal of Medical Internet Research, 19*(11), 386.

Matheus, C. J., Tribble, D., Kokar, M. M., Ceruti, M. G., & McGirr, S. C. (2006). *Towards a formal pedigree ontology for level-one sensor fusion.* Mechanicsburg, PA: Versatile Information Systems Inc.

Meola, A. (2019) What is the Internet of Things (IoT)? Meaning & definition. *Business Insider*, Available from: https://www.businessinsider.com/internet-ofthings-definition.

Moreira, J., Ferreira Pires, L., Sinderen, M.V. and Daniele, L. (2018) SAREF4health: IoT standard-based ontology-driven health care systems. In *Proceedings of the Formal Ontology in Information Systems (FOIS)*, Cape Town, South Africa, 239–252.

Neill, D. B. (2013). Using artificial intelligence to improve hospital inpatient care. *IEEE Intelligent Systems, 28*(2), 92—95.

Paganelli, F., & Giuli, D. (2011). An ontology-based system for context-aware and configurable services to support home-based continuous care. *IEEE Transactions on Information Technology in Biomedicine, 15*(2), 324—333.

Pang, Z. (2013) Technologies and architectures of the Internet-of-Things (IoT) for health and well-being. *M.S. thesis, Department of Electronic and Computer Systems, KTH—Royal Institute of Technology,* Stockholm, Sweden.

Pang, Z., Chen, Q., Tian, J., Zheng, L. and Dubrova, E. (2013) Ecosystem analysis in the design of open platform-based in-home healthcare terminals towards the Internet-of-Things, in *Proceedings of International Conference on Advanced Communication Technology. (ICACT),* 529—534.

Perriot, B. (2014). Characterization of physical activity in copd patients: validation of a robust algorithm for actigraphic measurements in living situations. *IEEE Journal of Biomedical and Health Informatics, 18*(4), 1225—1231.

Poniszewska-Maranda, A., Kaczmarek, D., Kryvinska, N. and Xhafa, F. (2019) Studying usability of AI in the IoT systems/paradigm through embedding NN techniques into mobile smart service system, **101**(11), 1661—1685.

Qi, J., Yang, P., Min, G., Amft, O., Dong, F., & Xu, L. (2017). Advanced internet of things for personalized healthcare systems: A survey. *Pervasive and Mobile Computing, 41,* 132—149.

Rasid, M.F.A. (2014) Embedded gateway services for Internet of Things applications in ubiquitous healthcare, in *Proceedings of 2nd International Conference on Information and Communication Technology (ICoICT),* 145—148.

Sangeetha, R., Jegadeesan, R., Ramya, M. P., & Vennila, G. (2018). Health Monitoring System Using Internet of Things. *International Journal of Engineering Research and Advanced Technology (IJERAT), 4*(3).

Sareen, S. (2017). Secure internet of things-based cloud framework to control zika virus outbreak. *International Journal of Technology Assessment in Health Care, 33*(1), 11—18.

Satija, U., Ramkumar, B., & Manikandan, M. S. (2017). Real-time signal quality-aware ECG telemetry system for IoT-based health care monitoring. *IEEE Internet Things, 4*(3), 815—823.

Schadow, G. and McDonald, C.J. (2009) The Unifed Code for Units of Measure. *Regenstrief Institute and UCUM Organization,* Indianapolis, IN, USA.

Schriml, L.M. (2020) Symptom ontology, Available from: https://www.ebi.ac.uk/ols/ontologies/symp.

Shamila, M., Vinuthna, K. and A.K. Tyagi (2019) A Review on Several Critical Issues and Challenges in IoT based e-Healthcare System. *Proceedings of the International Conference on Intelligent Computing and Control Systems,* 1036—1043.

Sharma, S., Tripathi, M.M. and Mishra, V.M. (2017) Survey paper on sensors for body area network in health care. *The IEEE International Conference in Emerging Trends in Computing and Communication Technologies (ICETCCT).*

Silva, B. N., Khan, M., & Han, K. (2017). Internet of Things: A Comprehensive Review of Enabling Technologies, Architecture, and Challenges. *IETE Technical Review,* 1—16.

Sood, S. K., & Mahajan, I. (2017). A fog-based healthcare framework for chikungunya. *IEEE Internet Things, 5*(2), 794—801.

Stevenson, G., Knox, S., Dobson, S. and Nixon, P. (2009) Ontonym: a collection of upper ontologies for developing pervasive systems. In *Proceedings of the 1st Workshop on Context, Information and Ontologies (CIAO),* ACM, Heraklion, Greece.

Suo, H., Wan, J., Zou, C. and Liu, J. (2012) Security in the internet of things: a review. (2012) *International conference in Computer Science and Electronics Engineering (ICCSEE),* 648—651.

Tsakalakis, M. Bourbakis, N.G. (2014). Health care sensor-based systems for point of care monitoring and diagnostic applications: A brief survey. *36th Annual International Conference of the IEEE Engineering in Medicine and Biology Society*, https://doi.org/10.1109/EMBC.2014.6945061, 6266—6269.

Viswanathan, H., Lee. and Pompili, D. (2012) Mobile grid computing for data- and patient-centric ubiquitous healthcare. In *Proceedings of 1st IEEE Workshop on Enabling Technologies for Smartphone and Internet of Things*, 36—41.

Wang, X.H., Zhang, D.Q., Gu, T. and Pung, H.K. (2004) Ontology based context modeling and reasoning using OWL. In *Proceedings of the 2nd IEEE Annual Conference on Pervasive Computing and Communications, Workshops (PerCom '04)*, Orlando, FL, USA, 18—22.

Witt, K., Stanley, J., Smithbauer, D., Mandl, D. and Ly, V. (2008) Enabling sensor webs by utilizing swamo for autonomous operations. *In Proceedings of the 8th NASA Earth Science Technology Conference*, Largo, MD, USA, 263—270.

Wu, J. (2019). Sensor fusion for recognition of activities of daily living. *Sensors, 18*(11), 4029.

Yang, G. (2014). A health-IoT platform based on the integration of intelligent packaging, unobtrusive bio-sensor, and intelligent medicine box. *IEEE Transactions on Industrial Informatics, 10*(4), 2180—2191.

Zgheib, R. (2019). *Semantic middleware architectures for iot healthcare applications. Enhanced living environments* (pp. 263—294). Springer.

Zhang, G., Li, C., Zhang, Y., Xing, C. and Yang. (2012) SemanMedical: A kind of semantic medical monitoring system model based on the IoT sensors. *in Proceedings of IEEE International Conference on E-health Networking, Application & Services*, 238—243.

Ziakas, P.D. (2011) Prevalence and impact of clostridium difcile infection in elderly residents of long-term care facilities. 95(31), 4187.

CHAPTER 3

Personalized decision support for cardiology based on deep learning: an overview

Ling Chen[1], Vincent S. Tseng[2], Hsuan-Ming Tsao[3] and Gau-Jun Tang[1]

[1]Institute of Hospital and Health Care Administration, National Yang Ming Chiao Tung University, Taipei, Taiwan
[2]Department of Computer Science, National Yang Ming Chiao Tung University, Hsinchu, Taiwan
[3]Division of Cardiology, National Yang Ming Chiao Tung University Hospital, Yi-Lan, Taiwan

3.1 Introduction

Cardiovascular disease remains the leading global cause of death and is expected to account for more than 22.2 million deaths by 2030, according to the 2020 update of the American Heart Association. In the United States, the leading causes of death attributable to cardiovascular diseases in 2017 were coronary heart disease (42.6%), stroke (17.0%), high blood pressure (10.5%), heart failure (9.4%), and diseases of the arteries (2.9%) (Virani et al., 2020). To diagnose these heart diseases in time, a number of different tests are commonly used, including regular electrocardiogram (ECG), Holter monitoring (a portable and wearable ECG device for continuously recording heart rhythm), echocardiogram, stress tests, cardiac catherization (a short tube inserted into a vein or artery in a person's leg or arm to help the doctor to check for problems via X-ray images), cardiac computerized tomography (CT) scan, and cardiac magnetic resonance imaging (MRI). These tests generate cardiac data in two major formats, namely waveforms and images.

With the advances in technology, the increasingly large volume of cardiac data routinely collected every day is posing a challenge to the healthcare system. First, interpreting images takes time. According to a study based on more than 1.5 million cross-sectional imaging studies from 1999 to 2010, an average radiologist needs to interpret a CT or MRI image every 3—4 s per day to meet workload demands

45

(McDonald et al., 2015). Second, the quality of image interpretation depends heavily on the experience of the radiologists and complexity of the image. Third, it is almost impossible for human eyes to monitor waveform signals recorded by the devices due to the speed and volumes of the incoming data. Last but not the least, it is desirable to combine all the information available for cardiologists to make informed and personalized decisions for a particular patient, but the reality is that they do not have the luxury of spending so much time on a single patient.

Computerized clinical decision support systems (CDSSs) have long been developed to assist clinicians in their complex decision-making process since the 1980s. They provide functions ranging from diagnosis, disease management, alarm systems, and prescription to drug control, rendering suggestions, reminders, alerts, guidelines, report generation, patient data summarization, and many more aids (Sutton et al., 2020). Generally speaking, the progress of CDSSs followed the advances of artificial intelligence (AI) technology. Early, CDSSs were rule-based (or knowledge-based), defined by a set of IF-THEN rules. They can be perceived as the direct implementation of expert knowledge by programmers. Later, CDSSs relied on training machine learning models on top of a set of carefully designed hand-crafted features (Afshar et al., 2019). That was the time when feature extraction skills determined the winner. For example, finding SIFT features was a big success in computer vision (Lowe, 2004). The rise of deep learning provided a solution to the limitation of hand-crafted features. By layers of abstraction, deep learning is able to automatically extract features from the data and therefore outperforms traditional machine learning methods in complex learning tasks such as machine translation and image classification and has shown promising results in medical applications.

In this chapter we aim to give an overview of recent developments of deep learning in personalized clinical decision support for cardiology. Particularly, we will look at different modalities of cardiac data, their characteristics, major learning tasks, and the deep learning models developed in recent years. Since deep learning in cardiology is a fast growing research field, this article aims to serve as a complement to prior review articles, for example, Bizopoulos and colleagues' thorough review on deep learning in cardiology (Bizopoulos & Koutsouris, 2018), with up-to-date references.

The organization of the chapter is as follows. Section 3.2 introduces major modalities of data in cardiology, namely waveform and image. Section 3.3 gives a brief introduction to typical deep learning architecture and discusses important concepts such as transfer learning and model interpretability. Section 3.4 provides in depth discussion on the latest development of deep learning for cardiological tasks. Section 3.5 discusses the challenges and limitations, followed by a conclusion in Section 3.6.

3.2 Big Data in cardiology

When people talk about Big Data, it is often defined as a number of Vs. Although there have been many more Vs introduced over time, the core is the three Vs introduced by Laney et al. (2001), namely volume, velocity, and variety. Data in cardiology are considered "big" in the following senses:

- First, cardiac data are big in volume. A typical medical image stored in the Digital Imaging and Communications in Medicine (DICOM) format is 512×512 pixels, and a typical cardiac CT scan can generate tens to hundreds of image slices, which can be several hundred megabytes for just one patient's single scan.
- Second, cardiac data can be "big" in velocity, in the sense that incoming cardiac signals such as ECGs can be recorded in real time.
- Finally, cardiac data is "big" in variety. In additional to patient demographics and lab test results in categorical, numerical, or text formats, waveforms and images are two important data modalities for cardiac diagnosis and decision-making.

3.2.1 Cardiac waveforms

The waveform of a signal is the shape of the changes in amplitude of the signal over a period of time. The most commonly seen waveforms related to the functions of the heart are ECG and photoplethysmography (PPG).

3.2.1.1 Electrocardiography

The standard 12-lead ECGs are the most commonly used complementary examination for evaluating cardiac electrical activity across a wide range of clinical settings from primary care centers to intensive care units (ICUs). From ECG signals, the health conditions of the heart can be evaluated, such as cardiac chamber hypertrophy and enlargement, arrhythmias, acute coronary syndromes, and conduction disturbances. Fig. 3.1 shows an example of an ECG lead 1 waveform. Fig. 3.2 shows a complete

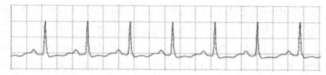

Figure 3.1 An example of electrocardiogram lead 1 drawn from the PhysioNet MIMIC III waveform data (Moody et al., 2020).

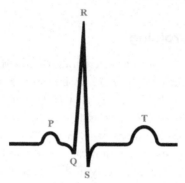

Figure 3.2 A typical PQRST heartbeat cycle.

heartbeat in ECG with its segments marked letters, including a P wave, a QRS complex, and a T wave. ECG data typically came as a multivariate time series of numerical values. However, in some cases it was in an image format of a waveform graph.

3.2.1.2 Photoplethysmography

PPG is a noninvasive and low-cost optical technique that measures the changes in blood volume in peripheral circulation at the surface of the skin. A PPG waveform is generated during a cardiac cycle and is commonly used for monitoring and detecting changes in arterial and venous circulation. Many physiological parameters, such as heart rate, oxygen saturation, and blood pressure, can be derived from PPG. These parameters might enable early screening of heart conditions, including arrhythmia. Commonly used PPG measurement devices can range from wrist-worn, smart phone, fingertip, ring-type, and facial to ear clip. PPG representations can be of the format of R-R intervals, time series, or image representation (Pereira et al., 2020).

3.2.1.3 Public datasets

The most well-known public waveform dataset is the PhysioNet MIMIC III Waveform dataset (Moody et al., 2020). It contains 67,830 records of about 30,000 patients at the ICUs. Each record of the waveform data contains digitized signals such as ECG, respiration, average blood pressure, and PPG, as well as time series of periodic measurements. They present a near-continuous recording of a patient's vital signs throughout an ICU stay. PhysioNet also provides the MIMIC III Waveform Matched Subset that linked the waveform data to patient information such as diagnoses, medications, lab tests, and demographics. This subset contains 22,247 numerical records and 22,317 waveform records of 10,282 ICU patients. Another commonly cited combined dataset with a much smaller scale is the "IEEE Signal Processing Cup 2015" dataset (Zhang et al., 2015). It contains 12 training sets of one-channel ECG signals, two-channel PPG signals, and three-axis acceleration signals. Another large-scale combined dataset is the "PPG−DaLiA" dataset (Reiss et al., 2019) for heart rate estimation. This dataset contains 8,300,000 multivariate time series collected from 15 subjects, including ECG, respiration, three-axis acceleration, blood volume pulse, electrodermal activity, and body temperature.

For ECG specific datasets the "PhysioNet Computing in Cardiology Challenge 2017" (Clifford et al., 2017) made available 8528 ECG recordings collected from mobile devices for AF prediction. The "2018 China Physiological Signal Challenge" dataset contains 9831 ECG 12-lead records of 7−60 min from 9458 patients. The challenge was to classify ECGs into normal or eight abnormal cardiac conditions such as atrial fibrillation (AF), left bundle brunch block, right bundle brunch block, first-degree atrioventricular block (I-AVB), etc. The "2019 China Physiological Signal Challenge," on the other hand, made available a total of 2000 single-lead ECG recordings from patients with cardiovascular disease for the detection of QRS complex in a cardiac cycle (see Fig. 3.2). The PTB-XL dataset (Wagner et al., 2020) is a new addition to the public dataset pool in 2020. It is a clinical 12-lead ECG dataset with multiple labels that comprises 21,837 10-s records from 18,885 patients. The diagnostic labels were further divided into subclasses.

3.2.2 Cardiac images

Cardiac images often come in the format of DICOM, which is the standard for communicating and managing medical imaging data and related information (DICOM, 2021). These DICOM files can be loaded for visualization on many commercial products as well as free software such as 3DSlicer (2021). They can also be easily read programmably via packages such as Pydicom for Python programming language (Pydicom WWW

Document, 2020). A wide range of information is accessible through DICOM tags, regarding the patient, study, equipment, image plane, image pixel, contrast, etc. The standard size of a DICOM image is 512×512 pixels and it is accessible via the "pixel_array" tag if read using Pydicom. The most commonly used cardiac image modalities are echocardiography (echo), cardiac CT, and cardiac MRI.

3.2.2.1 Echocardiography

While echocardiogram is often considered as a relatively subjective modality due to its reliance on the expertise of the cardiologist, it is one of the most widely employed tests for diagnosis in cardiology. It is routinely used to diagnose and follow-up patients with suspected or known heart diseases. The outputs of an echocardiography test contain a series of videos, images, and also Doppler measurements that are recorded from a variety of different viewing orientations. Commonly used standard views include apical long-axis (ALAX), apical two-chamber (A2C), parasternal long-axis (PLAX), apical four-chamber (A4C), parasternal short-axis (PSAX), vena cava inferior (SCVC), and subcostal four-chamber (SC4C). These can be obtained through conventional two-dimensional (2D) grayscale echocardiography. Fig. 3.3A shows an example of a four-chamber view echocardiogram. These 2D echocardiograms can be identified by the "Image Type" DICOM tag with the value "2D imaging."

3.2.2.2 Computerized tomography

Cardiac CT scans use radiation detectors, X-rays, and computers to produce slices of images through the body. Three major types of views are axial, coronal, and sagittal views. CT images are mostly in the axial plane, with a view looking down through the body. Some faster multidetector

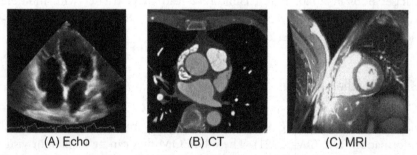

(A) Echo   (B) CT   (C) MRI

Figure 3.3 Examples of cardiac images in different modalities: (A) echo, (B) CT, and (C) MRI.

CT scanners can generate alternate views by capturing eight or more slices at once. An example of an axial view CT slice can be found in Fig. 3.3B. The "Rescale Intercept" and "Rescale Slope" DICOM tags can be used to normalize pixel values to Hounsfield units (HU). Values in HU space are useful for identifying and filtering different types of body tissues, such as air (−1000), fat (−120 to −90), soft tissue (+100 to +300), bone cancellous (+300 to +400), blood (+13 to +57), and water (0), although the use of contrast agents can modify HU values in some body parts. Generally speaking, in the literature, a single CT slice or image is called a 2D CT, a full volume CT with slices is referred as 3D CT, and 4D CT is often referred to as CT volumes recorded in a period of time, for example, within a cardiac cycle.

3.2.2.3 Magnetic resonance imaging

MRI is a commonly used medical imaging technique in radiology to visualize the anatomic structure and physiological processes of the body. It utilizes the interaction between atoms and a magnetic field to generate a picture of the interior of an object. Different from the previous modalities, cardiac MRI is an advanced cardiovascular imaging modality, generally considered as the reference standard method for evaluating cardiac structure and function, for example, volumes of the atria and ventricles, and the thickness of the myocardial wall (De Roos & Higgins, 2014). Fig. 3.3C shows an example of an MRI slice.

3.2.2.4 Public datasets

A dataset worth noting is the recently published public dataset for echocardiography called EchoNet. It contains 2.6 million echocardiogram four-chamber view images from 2850 patients (Ghorbani et al., 2020). Another recent dataset is the Cardiac Acquisitions for Multi-structure Ultrasound Segmentation (CAMUS) dataset (Leclerc et al., 2019) that contains two-chamber and four-chamber views of 500 patients.

As for cardiac MRI, the "Cardiac MRI" dataset that contains 7980 cardiac MR images collected from 33 subjects. The sequence of each subject consists of 20 frames along the long axis (Andreopoulos & Tsotsos, 2008). The challenges of Medical Image Computing and Computer Assisted Intervention (MICCAI, 2021) also made many cardiac MRI datasets available to the public, for example, the "Automatic Cardiac Diagnosis Challenge" (ACDC) dataset (Bernard et al., 2018). UK

Biobank is also on its way to acquiring 100,000 cardiac MRI records by 2023 (Raisi-Estabragh et al., 2020).

3.3 Deep learning

Deep learning is at the leading edge of AI and has played an increasingly important role in medical data analysis in recent years (Chassagnon et al., 2020; Litjens et al., 2019). As a subfield of machine learning, deep learning at its core is based on deep neural networks (DNNs)—neural networks with more than one hidden layer. The term "deep" comes from the network architecture of many layers between the input and output layer. The main advantage of deep learning is its ability to automatically extract high-quality features through layers of abstraction. By contrast, conventional machine learning heavily relies on the quality of hand-crafted features carefully designed by human experts (Chen & Jain, 2020; Goodfellow et al., 2016). DNNs with a series of convolutional layers are known as convolutional neural networks (CNNs). Convolutional layers have become the building block of almost all deep learning algorithms and have shown promising performance in numerous computer vision tasks from image classification, visual object detection, and semantic segmentation to visual question answering and many more (Lateef & Ruichek, 2019; Malinowski et al., 2017; Wang & Deng, 2018). Below, we will first briefly introduce typical DNN architecture and the concept of utilizing pretrained models, followed by a discussion on model interpretability, a very important aspect for the applicability of deep learning models in medicine.

3.3.1 Typical network architecture

Depending on the type of learning task and input data format, typical DNNs architecture includes conventional CNNs, autoencoders, recurrent neural networks (RNNs), and generative adversarial networks (GANs). There is a good introductory book called "Deep Learning" by Goodfellow et al. (2016).

3.3.1.1 Convolutional neural networks

A CNN is known for its convolutional layers. A typical CNN contains blocks of convolutional layers, often accompanied by pooling layers, and a number of fully connected/dense layers at the end. It is commonly used for classification tasks. A fully convolutional network (FCN) is a CNN with no fully connected layer. FCN models typically contain multiple blocks of convolutional layers and pooling layers, but do not have any dense layer to constrain the input image size. Because of these characteristics, they enjoy the advantage of being applicable to any input image size after the training, although when the training and testing image sizes differ too much, the performance could suffer. CNN and FCN models are commonly used for classification, detection, and regression types of tasks.

Another popular family of models is the region-based CNNs (R-CNNs) for object detection or object localization on images. These models are designed to classify objects within a region on an image. The key is to design an efficient region proposal mechanism that suggests which regions the objects might be at. Well-known examples of R-CNN-based models are fast R-CNN, faster R-CNN, and mask R-CNN (Zhao et al., 2019).

3.3.1.2 Autoencoders

An autoencoder contains two main components, an encoder and a decoder. The encoder encodes the input into a fixed length feature space, and the decoder decodes the feature space to a target space. The output of the encoder is often of a much smaller dimension than the input. For this reason, this type of network is commonly used for dimensionality reduction. Since the input dimension and output dimension are often the same, it is also commonly used for pixelwise prediction on images, where a prediction value is assigned to every pixel of the input image. For example, a popular variation is called U-Net (Ronneberger et al., 2015), which is basically an FCN with an encoder–decoder architecture plus some additional symmetric residual connections. It was introduced for the segmentation of electron microscopic images.

3.3.1.3 Recurrent neural networks

RNNs belong to a class of neural networks that utilizes previous outputs while having hidden states. This mechanism allows the model to take sequences of inputs and use the model's internal state or memory to process them. RNNs enjoy the advantages of processing input of possibly any length. However, the main drawback lies in its inability to access information from a long time ago. To mitigate this problem, the long short-term memory network (LSTM) was introduced to enable the learning of long-term dependencies. It utilizes a number of different gates to control information passing to cell states, while the cell states act as the "memory" of the network to carry relevant information from the earlier time steps. RNNs and LSTMS are commonly used when the inputs are sequences of values, texts, or images.

3.3.1.4 Generative adversarial networks

GANs typically have two main components, a generative network (a.k.a. a generator) and a discriminative network (a.k.a. a discriminator). The goal of the generator is to generate a candidate to fool the discriminator, while the discriminator tries to evaluate the candidate against the true data distribution. The goal of training involves finding a point of equilibrium between the two competing networks. By playing this contest game, the generator learns to generate new data with the same distribution as the training set. In medical imaging research, GAN models have been used mainly for synthetic data generation, image reconstruction, segmentation, classification, and detection (Yi et al., 2019).

3.3.2 Pretrained models and transfer learning

As a DNN goes deeper, there are more parameters to be trained and thus more data is required to train the model. However, in many cases, the desired data size is not available, especially in the medical domain. For image-based deep learning models, a common practice is to utilize pretrained models that were trained mainly based on the ImageNet dataset (Jia Deng et al., 2009). It is a publicly available natural image dataset, now with more than 21,000 subcategories and 14 million images. Well-known CNN models like VGG, ResNet, and inception, all have pretrained models available on the major deep learning platforms such as TensorFlow (n.d.) and PyTorch (n.d.).

Using one's own data to adjust the pretrained model weights is called fine-tuning. The idea behind this is the so-called transfer learning, where the model learns to adapt from a source domain data distribution to a target domain data distribution. Even though radiology images such as MRI and CT are quite different from natural images in many aspects (e.g., texture and color distributions), many studies have shown success in utilizing a pretrained model and fine-tuning it with medical data. This can be explained by the observation that what had been learned in the lower convolutional layers were basic lines and shapes (Lee et al., 2009), and therefore could be a common basis for transferring from the natural image domain to the radiology image domain.

3.3.3 Model interpretability

A major criticism against deep learning and most nonlinear machine learning models is their black-box nature (Tjoa & Guan, 2020). Model transparency and the ability to explain why a model predicts what it predicts is especially crucial in medical applications for the justification of decision-making. Different approaches have been developed to improve model interpretability or explainability. Below, we focus on the most commonly used approach, namely the saliency mask approach, for DNNs (Guidotti et al., 2018). It assigns values reflecting the importance of input components according to their contribution to the algorithm decision.

One strain of the methods computes the first-order linear approximation of the model by taking the gradient of the output with respect to the input (Simonyan et al., 2014). Many extensions have been developed since, such as DeepLIFT (Shrikumar et al., 2017), integrated gradients (Shrikumar et al., 2017), and SmoothGrad (Smilkov et al., 2017). For example, SmoothGrad was used to provide interpretability to a CNN model designed for several cardiac function predictions based on echocardiograms (Ghorbani et al., 2020). Another popular subset of methods is based on class activation mapping (CAM), where network activations are integrated into visualizations, utilizing global average pooling (GAP), and back-propagation of deep networks (Zhou et al., 2016). Since CAM has the drawback of requiring feature maps to directly precede softmax layers, Grad-CAM was introduced as a generalization of CAM without such limitation. Additionally, it is not limited to class-score outputs and has been shown to work well on CNN + LSTM models (Selvaraju et al., 2020). Later, Grad-CAM + + was proposed to further improve object localization and visual explanations (Chattopadhay et al., 2018).

The other popular subset of methods was based on an integrated attention mechanism. With such a mechanism, a network can weigh features according to their importance to a task and then use this weighting to help achieve the task. This latter portion distinguishes it from CAM type of methods, where important scores are often calculated after a model is trained (Xu et al., 2015). However, only a few prior studies on cardiac image prediction took model interpretability into consideration (Ghorbani et al., 2020).

3.4 Deep learning applications in cardiology

Below, we will look at the applications of deep learning to different modalities of cardiac data. Before going into further details, we first summarize the main learning tasks for cardiology. The main learning tasks for cardiology can be categorized as follows:

- cardiac disease detection or classification
- cardiac function prediction
- localization and annotation of specific waves in the signal
- waveform denoising
- image segmentation
- image classification
- image denoising

The first two are general tasks for all data modalities, the next two are waveform-specific tasks, and the last three are image–specific tasks.

3.4.1 Deep learning in electrocardiogram

The most common learning tasks for ECG-based deep learning models for cardiology are disease detection or classification, signal delineation, and denoising. Below, we will discuss recent work in the literature. Interested readers in earlier publications are referred to a good review paper by Hong, Zhou et al. (2020).

3.4.1.1 Detection/classification

The majority of ECG-based deep learning research has been on cardiac disease detection/classification. In recent years, most studies focused on arrhythmias (Cai et al., 2020; Mahmud et al., 2020; Parvaneh et al., 2019) and AF detection (Yan et al., 2020). For example, Mahmud introduced the design of DeepArrNet unit block consisting of multiple temporal convolutions with varying kernel dimensions to capture various temporal correlation from the ECG signals (Mahmud et al., 2020). On the other hand, Cai et al. proposed Multi-ECGNet for multilabel classification of 55 different categories of arrhythmias (Cai et al., 2020). It is a multilabel CNN model based on the 1D ResNet and a squeeze-and-excitation module. The model achieved a micro–F1–score of 0.863.

Other studies focused on the detection of myocardial ischemia (Ogrezeanu et al., 2020) and aortic stenosis (Hata et al., 2020; Kwon, Lee et al., 2020). Models used in these studies were either an out-of-the-box VGG model (Hata et al., 2020) or a relatively simple CNN design with a number of convolutional layers (Ogrezeanu et al., 2020) or a hybrid of such CNN (Kwon, Lee et al., 2020). It is worth noting that the last work was trained on 39,371 ECGs and validated on an internal dataset of 6453 ECGs and an external dataset from another hospital of 10,865 ECGs, achieving AUC of 0.884 and 0.861, respectively. Readers interested in earlier work are directed to a survey paper (Parvaneh et al., 2019).

A few of the studies were on multiple disease classification (Ribeiro et al., 2020; Sanjana et al., 2020). For example, a ResNet-based DNNs model was proposed for automatic diagnosis of six cardiac abnormalities, such as AF, sinus tachycardia, and I-AVB (Ribeiro et al., 2020). Their work was based on a dataset of 2,322,513 ECG records from 1,676,384 patients of 811 counties. The model achieved 80% F1 scores and a specificity over 99%.

3.4.1.2 Signal delineation

ECG signal delineation, sometimes called localization or annotation, is a task to identify P-waves, QRS complexes, or T-waves in ECG waveforms. Many deep learning models were developed and trained to tackle this task. For example, recently Wang et al. (2020) proposed an autoencoder model for signal delineation with the help of additional domain knowledge, while Peimankar et al. introduced a combined CNN−LSTM model and achieved sensitivity and precision of 97.95% and 95.68%, respectively (Peimankar & Puthusserypady, 2021).

3.4.1.3 Denoising

The acquisition of ECG signals can be accompanied by a large amount of noise. Commonly used models for denoising are autoencoder (Nurmaini et al., 2020) and GAN (Singh & Pradhan, 2020). For example, Nurmaini et al. (2020) designed an autoencoder for ECG denoising, achieving an sensitivity and specificity of 99.57% and 89.81%, respectively. They used the MIT−BIH Noise Stress Test Database, which added an electrode motion artifact to the original MIT−BIH Arrhythmia database. Singh et al., on the other hand, proposed a GAN-based model for ECG signal denoising (Singh & Pradhan, 2020). They used the MIT−BIH Arrhythmia database as a clean dataset and generated a noisy set from the clean distribution by adding different types of noises, such as black and white, additive white Gaussian, and electrode motion artifact noises.

3.4.2 Deep learning in PPG

PPG-based deep learning applications mainly focus on disease detection or classification and heart rate estimation, as discussed below.

3.4.2.1 Detection/classification

Many studies have been conducted for AF detection. For example, Yan et al. (2020) utilized a CNN-based model for AF detection based on facial PPG. Aschbacher et al. designed a CNN−RNN to discriminate between AF from sinus rhythm (SR) based on the wrist-worn trackers with PGG sensors. The study was based on 51 patients and the model achieved 0.983 AUC, 0.985 sensitivity, and 0.880 specificity. Kwon, Hong et al. (2020), on the other hand, evaluated the diagnostic performance of using a deep learning model based on PPG signals generated from a ring-type wearable device called CardioTracker. The study was based on 13,038 30-s PPG samples (5850 for SR and 7188 for AF), and they demonstrated that a CNN-based model outperformed support vector machine (SVM)-based machine learning models. Interested readers are directed to a survey paper on PPG-based AF detection by Pereira et al. (2020).

3.4.2.2 Heart rate estimation

To estimate heart rate based on PPG signals, a few deep learning approaches have been proposed. Reiss et al. introduced the PPG−DaLiA dataset for heart rate estimation and compared a few CNN-based models with the state-of-the-art methods. They have shown that their CNN-based model can achieve 8.82 MAE on the PPG−DaLiA dataset.

Several recent studies identified that one major challenge of accurate heart rate estimation based on PPG signals was the motion artifacts caused by user's physical activities while using wearable devices. Xu et al. (2020) proposed an RNN-based model for accurate cardiac-period segmentation on PPG signals and a stochastic model for clean PPG pulses extraction for accurate heart rate estimation. Chung et al. introduced a CNN−LSTM-based multiclass and multilabel classification model for accurate heart rate estimation based on PPG and acceleration signals. Chang, Li et al. (2021), on the other hand, tackled the problem as a denoising problem and proposed to utilize a denoising convolutional neural network (DnCNN). The DnCNN model was trained with the noisy PPG signals and their corresponding clean PPG signals generated from ECG signals. They have demonstrated that their approach outperformed the state-of-the-art methods.

3.4.2.3 Interpretability

To quantify model interpretability, Zhang, Yang et al. (2021) explored model attention of deep networks on PPG data and proposed two metrics, namely congruence and annotation classification, to measure the overlap of model identified regions of interest (ROIs) and human annotations.

3.4.3 Deep learning in CT

Deep learning has been applied to many cardiac CT analytic tasks, such as segmentation, detection, denoising, and function scoring (Litjens et al., 2019).

3.4.3.1 Segmentation

One focus has been on automatic segmentation of heart CT images to identify part of the anatomic structure of the heart. For example, models were developed for left ventricular myocardium segmentation (Koo et al., 2020; van Hamersvelt et al., 2019), left atrium or atrial appendage segmentation (Bratt et al., 2019; Chen et al., 2020), epicardial adipose tissue segmentation and quantification (Commandeur et al., 2019), epicardial and paracardial adipose tissue segmentation (Kroll et al., 2021), and whole heart segmentation (Bruns et al., 2020). For this type of work, quantity and quality of manually annotated ROIs is important for model performance. However, since a CT volume can have tens to hundreds of slices, it is very costly and time consuming to annotate ROIs manually.

To generate output segmented regions corresponding to the input image, segmentation methods generally take a pixelwise prediction approach. The output is a soft probability map of the same size as the input images, indicating the predicted probability of a given pixel belonging to the ROIs. FCN (Commandeur et al., 2019; Koo et al., 2020) and U-Net (Bratt et al., 2019; Chen et al., 2020; Kroll et al., 2021) are among the most popular model choices, although variants of standard CNNs (Bruns et al., 2020; van Hamersvelt et al., 2019) and other DNNs (Galea et al., 2021) have also been designed.

3.4.3.2 Detection

Closely related to segmentation, detecting specific characteristics of cardiac structure on CT images can assist further diagnosis. For example, for coronary arteries detection, for example, Wu et al. (2019) proposed a bidirectional tree LSTM models, while Zreik et al. (2020) employed a 3D convolutional autoencoder and a 3D variational convolutional autoencoder. Zhao et al. (2020) adapted a faster R-CNN object detection model for coronary arteries calcified detection based on OrCaScore, a public dataset of calcified plaque scores. On the other hand, Chang, Kim et al. (2021) utilized a 3D U-net model for aortic valve stenosis detection.

3.4.3.3 Denoising

To improve image quality, a number of CT image denoising models were introduced (Green et al., 2018; Hong, Park et al., 2020; Kang et al., 2019; Liu et al., 2020). For example, an FCN-based model was proposed for 3D coronary CT angiography denoising (Green et al., 2018), GAN-based models were introduced for 2D and 3D CT denoising (Kang et al., 2019; Liu et al., 2020), and a modified U-Net model was employed for CT angiography denoising (Hong, Park et al., 2020). Similar to the segmentation models, the output of a denoising model often has the same size as the input image. Typical output of a denoising model is either a denoised image or the noise. In the latter case, a denoised image is obtained by deducing the noise from the input image. The biggest challenge for a denoising model lies in the ground truth. Both the noisy and clean images of the same view are not usually available. Therefore a common practice is to select relatively clean images and generate different types of noises to add to the clean images. However, there is always a discrepancy between real-world noise and artificial noise, which limits the generalizability of the model. Often, when the model was applied to real-world data, the performance dropped.

The other type of denoising task is image artifact removal. For motion artifact removal, Lossau née Elss et al. (2020) introduced an ensemble of three CNNs for motion artifact removal in cardiac CT images. The SegmentationNets identified metal-affected line integrals directly in the projection domain and the InpaintingNets treated metal-affected values as missing data and refilled it based on surrounding line integrals. The article also provided a concise review of three main approaches for handling motion artifact removal. Jung et al. (2020), on the other hand, observed that it is medically impossible to obtain corresponding CT images devoid of motion artifacts. They proposed to use a style transfer method to generate a synthetic ground truth, based on which they trained a motion artifact reduction model.

3.4.3.4 Function scoring

Calcium scoring in cardiac and chest CT has been one research focus for cardiac function scoring. For example, de Vos et al. (2019) proposed a two-stage CNN approach for direct automatic coronary calcium scoring based on 903 cardiac CT and 1687 chest CT scans. The first CNN model performed registration to align the fields of view, while the second CNN acted as a regression model to predict calcium scores directly. Zeleznik et al. (2021) designed a U-Net-based pipeline for calcium scoring comprising localization and segmentation of the heart, coronary calcium segmentation, and calculation of calcium score. Their model generated scores showed high correlation with manual quantification. Zhang, Ding et al. (2021) proposed a multitask deep learning framework for coronary artery calcification based on 232 noncontrast cardiac-gated CT scans and found no statistical difference in the classification of Agatston scoring.

A few validation studies have been conducted. The study by van Assen et al. (2021) based on 95 patients has demonstrated that their ResNet-based calcium quantification showed good correlation compared to the traditional Agaston score on noncontrast chest CT. Another study by van Velzen et al. included various types of nonenhanced CT examinations on 7240 participants, such as coronary artery calcium scoring and screening CT, and low-dose chest CT. They have shown that their CNN-based model had good agreement with manual scoring (van Velzen et al., 2020). Martin et al. (2020) evaluated a ResNet based on calcium scoring software based on a dataset of 511 consecutive patients and came to a conclusion that the software could be integrated into routine clinical use, provided that a final quality check be conducted by a human observer.

3.4.4 Deep learning in echocardiography

Many analytic tasks on echocardiograms also adopted deep learning, including but not limited to, standard view classification, disease classification, cardiac structure segmentation and quantification, cardiac disease and mortality prediction, and automated pipeline (Litjens et al., 2019).

3.4.4.1 Standard view classification

As a standard full resting echocardiogram study can consist of images recorded from different viewing orientations, one important step for automatic echocardiogram analysis is the standard view classification.

Common standard views include A2C, A4C, ALAX, PLAX, PSAX, SC4C, and SCVC. Several real-time standard view classification models have been developed in recent years (Huang et al., 2020; Kusunose et al., 2020; Madani et al., 2018; Østvik et al., 2019; Zhang et al., 2018). For example, a CNN model with inception blocks for 7–view classification was trained using 460 echocardiogram videos and achieved 98% accuracy (Østvik et al., 2019), an U–Net and CNN model achieved an accuracy of 94.4% for 15–view classification (Madani et al., 2018), and a CNN model with five convolutional layers achieved 98.1% accuracy for 5–view classification (Kusunose et al., 2020).

3.4.4.2 Cardiac phase detection

Detecting cardiac phases, especially the end–systolic (ES) and end–diastolic (ED) frames in echocardiogram, is an important step leading to the quantification of size and function of cardiac chamber (Dezaki et al., 2019; Jahren et al., 2020). The ES frame is defined as the first frame after the aortic valve closure, representing the smallest left ventricle (LV) volume, while the ED frame is the first frame following the mitral valve, representing the largest LV volume.

Models for cardiac phase detection generally need to handle sequences of incoming frames and detect which ones are ES or ED. Therefore a common choice is an RNN type of sequence model. For example, a densely gated RNN model was introduced with a global extrema loss function for ES and ED phase detection (Dezaki et al., 2019). Another CNN + RNN model was developed for ED detection in cardiac spectral Doppler images (Jahren et al., 2020).

3.4.4.3 Segmentation and quantification

Automatic segmentation and quantification of cardiac structure and function is another important topic. A general trend is to use a variation of the U–Net model (Leclerc et al., 2019; Li et al., 2021), although there were exceptions (Ouyang et al., 2020).

For the standard echocardiography, Ouyang et al. (2020) introduced the EchoNet dataset of four-chamber echocardiograms and proposed a CNN model combining a module of spatiotemporal convolution with residual connections and a module of semantic segmentation with weak supervision. They demonstrated that the model surpassed human expert performance in segmenting LV, estimating ejection fraction and assessing cardiomyopathy. Leclerc et al. (2019) introduced the CAMUS dataset and evaluated the performance of popular models for segmentation, geometric, and clinical scoring tasks. They showed that U-Net outperformed other models on the dataset. Liu et al. (2021) introduced a deep pyramid local attention (PLA) network called PLANet for the segmentation of LV endocardium and myocardium. The PLA module was used for feature enhancement within the compact and sparse neighboring contexts. The study was based on the EchoNet and CAMUS datasets and showed promising results. As for myocardial contrast echocardiography, Li et al. (2021) designed an encoder–decoder segmentation model with a U-Net as encoder and a ConvLSTM as decoder. They demonstrated the proposed model outperformed traditional U-Net.

Instead of deriving estimation from segmentation results, several models were built to directly predict cardiac functions such as ES volume, ED volume, and ejection fraction (Ghorbani et al., 2020). Notably this last work provided additional mechanism for model interpretability.

3.4.4.4 Morbidity and mortality prediction

For echocardiography-based mortality prediction, Ulloa Cerna et al. (2021) designed four CNN models for 1-year-all-cause mortality. They are a 2D CNN + LSTM model, a 2D CNN + GAP, a 3D CNN model, and a 3D CNN + GAP model. The models were trained based on 812,278 echocardiographic videos from 34,362 individuals. They demonstrated that cardiologists' predictions of 1-year all-cause mortality had 13% improvement in sensitivity with the assistance of the model, while maintaining prediction specificity.

For cardiac disease prediction, Kim et al. (2020) introduced a hybrid CNN model for AF recurrence prediction after vein isolation based on the echocardiograms of 527 consecutive patients. The model combined a branch for echocardiography images and a branch of additional information such as patient demographics, LA diameter obtained from echocardiogram, and LA volume obtained from CT.

3.4.4.5 Automated pipeline

There were also some thorough studies done with the full pipeline of view identification, image segmentation, structure and function quantification, and disease detection based on 3D echocardiograms (Huang et al., 2020; Zhang et al., 2018). For example, a pipeline with a VGG network for view classification, a U-Net for segmentation, a speckle tracking mechanism based on the segmentation, and again a VGG network for hypertrophic cardiomyopathy, cardiac amyloid, and pulmonary arterial hypertension classification based on 14,035 2D echocardiograms (Zhang et al., 2018). To fully utilize the spatial–temporal information between the frames, a pipeline of a 3D CNN model for view classification, a 2D U-Net for frame segmentation, and a DenseNet with 3D convolutional layers was introduced for wall motion abnormality detection based on 10,638 echocardiograms (Huang et al., 2020).

3.4.5 Deep learning in MRI

Studies of applying deep learning to MRI data focus mainly on segmentation and quantification. Several validation types of studies have been conducted in recent years. Also, there were efforts done to generate synthetic data. Below we will give more details in these areas.

3.4.5.1 Segmentation and quantification

To quantify cardiac structure, such as LV and RV volumes, the first step is often cardiac structure segmentation. To date, U-Net is the most popular architecture for segmentation. Luo et al. (2020) proposed a hybrid architecture that combined a context feature extraction module to capture the MRI sequence context and U-Net like encoder–decoder network for pixelwise segmentation. Their study was based on the ACDC dataset for LV, RV, and myocardium segmentation. Vesal et al. (2020) introduced a fully automated segmentation framework based on a 3D dilated residual U-Net. It consists of an encoder–decoder architecture for localization, a gradient-weighted class activation maps to generate density map, and another encoder–decoder for final segmentation. Guo et al. (2020) designed a continuous kernel cut mechanism on top of a U-Net to refine its output probability maps for more accurate segmentation. Upendra et al. (2020), on the other hand, trained a GAN-based model called SegAN on MR images and showed that it outperformed U-Net on the 2017 ACDC segmentation challenge dataset.

3.4.5.2 Validation studies

To validate the performance of U-Net, Retson et al. conducted a validation study based on 200 clinical cardiac MRI examinations to evaluate the performance of U-Net for clinical quantification of LV and RV volume and function based on cardiac MR images against physicians' annotations. They found that the contours and ventricular volumes from the automated segmentation of the model strongly agreed with expert annotations. Böttcher et al. (2020) also evaluated a U–Net architecture implemented in a commercial software cvi42 based on MRI examinations of 5000 healthy subjects from UK Biobank dataset. They found that the correlation between the model results and expert-corrected results was strong and concluded that this deep learning approach is "feasible, time-efficient, and highly accurate."

Readers interested in earlier studies are directed to survey papers on deep-learning-based segmentation on MRI (Ado Bala & Kant, 2020; Irmawati et al., 2020). A Python package for sensitivity analysis of MRI segmentation models is also available (Ankenbrand et al., 2021).

3.4.5.3 Synthetic data generation

Limited amount of available data is always a challenge to training robust deep learning models. Diller et al. (2020) utilized a progressive GAN model for synthetic MRI data generation, and trained the model on their dataset of 303 patients. They have demonstrated that using their generated 10,000 MRI images to train a segmentation U-Net model has improved the model performance.

3.5 Challenges and limitations

Below, we summarize the key challenges deep learning technology faces in its application in cardiology.

3.5.1 Data annotation

Most deep learning tasks in cardiology are essentially supervised learning, requiring the so-called "ground truth" that we use to teach a model what to learn. The most common type of ground truth is data labels for some prediction tasks, such as diagnosis, prognosis, prescription, etc. This information can often be found in a patient's electronic health records. However, as we can see from above, segmentation and detection are important learning tasks in cardiology, but to obtain the ground truth for these types of tasks is not straight forward. In fact, it is often very costly and time consuming because it requires human experts to manually annotate the ROIs, for example, where the endocardium and epicardium of an atrium are. Such annotations are often in the form of a bounding box or a polygon. The former is defined by a four-point rectangle, which is quicker to annotate and easier for a machine to learn but less accurate. The latter is defined by finite number of points, connected by straight lines to form a closed shape, which is often preferred for its accuracy but requires more time to annotate.

There are user-friendly tools such as 3DSlicer or semiautomated tools developed to assist faster annotation. However, to mitigate this problem fundamentally is to minimize the amount of annotated data required for training. A common approach is to utilize publicly available pretrained models and apply transfer learning techniques to fine-tune the models using one's own data. Self-supervised learning is another approach that has attracted growing attention. It first pretrains a model to extract features with some "pretext" tasks that do not require ground truth labels. And then these features are used to train some downstream tasks with the annotations.

3.5.2 Generalizability

The majority of the studies reviewed in this chapter were based on data collected from a single site or medical institution. Even though the model might appear to perform well on the testing dataset, its good performance might by limited to a certain data distribution pertained to the site. However, a truly good and clinically useful model should be generalizable, that is, it should be robust enough to perform well on an unseen dataset. There are ways to increase model generalizability, for example, by adding random noise to the training data or applying data augmentation techniques such as rotation, flipping, and cropping. Additionally, it is always desirable to test one's model on a dataset collected from a different site to verify its generalizability, although it might not always be achievable.

3.5.3 Interpretability

Model interpretability is another crucial aspect for a model to be practically useful. Traditional generalized linear models, such as logistic regression, are interpretable because the linearity allows the weights learned by the model to be interpreted as the importance of the input variables. However, nonlinear machine learning models, such as SVM with nonlinear kernels, are often criticized for its black-box nature, because a nonlinear kernel projects the input data into a higher dimensional space that is almost impossible for human to visually or mentally comprehend. Deep learning models are on its surface not interpretable because structurally they contain many layers of neurons for the input information to pass through. As discussed in Section 3.3.3, a range of methods have been developed in recent years to explain the decisions a deep learning model made. However, not many studies in applying deep learning to cardiological problems have explored these techniques. It is recommended that future studies pay more attention to this aspect.

3.5.4 Multiple modalities

The majority of deep learning models focused on single-modality inputs, either texts, numerical values, or only one form of cardiac images, such as CT or MRI. However, different modalities of cardiac data capture different aspects of a person's heart condition. Therefore it is desirable to build a model that considers as many forms of cardiac data collected for a single patient as possible. The challenge lies first of all in the availability of multiple modalities in the training dataset. Second, given the availability of the data, how to carefully design a deep learning model to extract features from different data modalities and integrate them to form a single decision for prediction can be a challenge.

3.6 Summary

The dynamics of the human heart are very complex and need to be explored further. Deep-learning technology provides a new avenue for exploring the richness of different aspects of the heart expressed in cardiac waveforms and images. The ability of deep learning to extract information from complex data allows the model to consider different aspects of a person at the same time, and therefore enables more personalized decision support to the clinical professionals. In this chapter, we have given an overview of deep-learning-based personalized decision support methods. Particularly, we have introduced the major modalities of cardiac data and discussed recent developments of deep-learning algorithms for the tasks related to the data type. We also listed the challenges and limitations and offered suggestions for future studies.

We have seen some very promising results and expect more deep-learning-based CDSSs to be developed in the near future. However, there is still a gap between a lab-tested model and a model that works well in real-world clinical settings. More efforts need to be made to mitigate this gap and handle the challenges.

References

Ado Bala, S., & Kant, S. (2020). Deep learning based model architectures for cardiac MRI segmentation: A survey. *International Journal of Innovative Science Engineering and Technology, 7*, 129−135.

Afshar, P., Mohammadi, A., Plataniotis, K. N., Oikonomou, A., & Benali, H. (2019). From handcrafted to deep-learning-based cancer radiomics: Challenges and opportunities. *IEEE Signal Processing Magazine, 36*, 132−160. Available from https://doi.org/10.1109/MSP.2019.2900993.

Andreopoulos, A., & Tsotsos, J. K. (2008). Efficient and generalizable statistical models of shape and appearance for analysis of cardiac MRI. *Medical Image Analysis, 12*, 335−357. Available from https://doi.org/10.1016/j.media.2007.12.003.

Ankenbrand, M. J., Shainberg, L., Hock, M., Lohr, D., & Schreiber, L. M. (2021). Sensitivity analysis for interpretation of machine learning based segmentation models in cardiac MRI. *BMC Medical Genomics, 21*, 1−8. Available from https://doi.org/10.1186/s12880-021-00551-1.

Bernard, O., Lalande, A., Zotti, C., Cervenansky, F., Yang, X., Heng, P. A., Cetin, I., Lekadir, K., Camara, O., Gonzalez Ballester, M. A., Sanroma, G., Napel, S., Petersen, S., Tziritas, G., Grinias, E., Khened, M., Kollerathu, V. A., Krishnamurthi, G., Rohe, M. M., ... Jodoin, P. M. (2018). Deep learning techniques for automatic MRI cardiac multi-structures segmentation and diagnosis: Is the problem solved? *IEEE Transactions on Medical Imaging, 37*, 2514−2525. Available from https://doi.org/10.1109/TMI.2018.2837502.

Bizopoulos, P., & Koutsouris, D. (2018). Deep learning in cardiology. *IEEE Reviews in Biomedical Engineering, 12*, 168−193. Available from https://doi.org/10.1109/RBME.2018.2885714.

Böttcher, B., Beller, E., Busse, A., Cantré, D., Yücel, S., Öner, A., Ince, H., Weber, M. A., & Meinel, F. G. (2020). Fully automated quantification of left ventricular volumes and function in cardiac MRI: Clinical evaluation of a deep learning-based algorithm. *The International Journal of Cardiovascular Imaging, 36*, 2239−2247. Available from https://doi.org/10.1007/s10554-020-01935-0.

Bratt, A., Guenther, Z., Hahn, L. D., Kadoch, M., Adams, P. L., Leung, A. N. C., & Guo, H. H. (2019). Left atrial volume as a biomarker of atrial fibrillation at routine chest CT: Deep learning approach. *Radiology: Cardiothoracic Imaging, 1*, e190057. Available from https://doi.org/10.1148/ryct.2019190057.

Bruns, S., Wolterink, J. M., Takx, R. A. P., van Hamersvelt, R. W., Suchá, D., Viergever, M. A., Leiner, T., & Išgum, I. (2020). Deep learning from dual-energy information for whole-heart segmentation in dual-energy and single-energy non-contrast-enhanced cardiac CT. *Medical Physics, 47*, 5048−5060. Available from https://doi.org/10.1002/mp.14451.

Cai, J., Sun, W., Guan, J., & You, I. (2020). Multi-ECGNet for ECG arrythmia multi-label classification. *IEEE Access, 8*, 110848−110858. Available from https://doi.org/10.1109/ACCESS.2020.3001284.

Chang, S., Kim, H., Suh, Y. J., Choi, D. M., Kim, H., Kim, D. K., Kim, J. Y., Yoo, J. Y., & Choi, B. W. (2021). Development of a deep learning-based algorithm for the automatic detection and quantification of aortic valve calcium. *European Journal of Radiology, 137*, 109582. Available from https://doi.org/10.1016/j.ejrad.2021.109582.

Chang, X., Li, G., Xing, G., Zhu, K., & Tu, L. (2021). DeepHeart: A deep learning approach for accurate heart rate estimation from PPG signals. *ACM Transactions on Senor Networks, 17*. Available from https://doi.org/10.1145/3441626.

Chassagnon, G., Vakalopoulou, M., Paragios, N., & Revel, M. P. (2020). Artificial intelligence applications for thoracic imaging. *European Journal of Radiology, 123*. Available from https://doi.org/10.1016/j.ejrad.2019.108774.

3D Slicer image computing platform [WWW Document]. (2021). <https://www.slicer. org/>.

Chattopadhay, A., Sarkar, A., Howlader, P., & Balasubramanian, V. N. (2018). Grad-CAM++: Generalized gradient-based visual explanations for deep convolutional networks. In: *Proceedings of the 2018 IEEE winter conference on applications of computer vision* (pp. 839–847). https://doi.org/10.1109/WACV.2018.00097

Chen, H. H., Liu, C. M., Chang, S. L., Chang, P. Y. C., Chen, W. S., Pan, Y. M., Fang, S. T., Zhan, S. Q., Chuang, C. M., Lin, Y. J., Kuo, L., Wu, M. H., Chen, C. K., Chang, Y. Y., Shiu, Y. C., Chen, S. A., & Lu, H. H. S. (2020). Automated extraction of left atrial volumes from two-dimensional computer tomography images using a deep learning technique. *International Journal of Cardiology, 316*, 272–278. Available from https://doi.org/10.1016/j.ijcard.2020.03.075.

Chen, Y.-W., & Jain, L. C. (2020). *Deep learning in healthcare: Paradigms and applications.* Springer. Available from https://doi.org/10.1007/978-3-030-32606-7.

Clifford, G. D., Liu, C., Moody, B., Lehman, L. H., Silva, I., Li, Q., Johnson, A. E., & Mark, R. G. (2017). AF classification from a short single lead ECG recording: The PhysioNet/computing in cardiology challenge 2017. *Computers in Cardiology, 44*, 1–4. Available from https://doi.org/10.22489/CinC.2017.065-469.

Commandeur, F., Goeller, M., Razipour, A., Cadet, S., Hell, M. M., Kwiecinski, J., Chen, X., Chang, H.-J., Marwan, M., Achenbach, S., Berman, D. S., Slomka, P. J., Tamarappoo, B. K., & Dey, D. (2019). Fully automated CT quantification of epicardial adipose tissue by deep learning: A multicenter study. *Radiology: Artificial Intelligence, 1*, e190045. Available from https://doi.org/10.1148/ryai.2019190045.

De Roos, A., & Higgins, C. B. (2014). Cardiac radiology: Centenary review. *Radiology, 273*, S142–S159. Available from https://doi.org/10.1148/radiol.14140432.

de Vos, B. D., Wolterink, J. M., Leiner, T., de Jong, P. A., Lessmann, N., & Išgum, I. (2019). Direct automatic coronary calcium scoring in cardiac and chest CT. *IEEE Transactions on Medical Imaging, 38*, 2127–2138.

Deng, J., Dong, W., Socher, R., L-J Li, K Li, L Fei-Fei, 2009. ImageNet: A large-scale hierarchical image database 248–255. https://doi.org/10.1109/cvprw.2009.5206848.

Dezaki, F. T., Liao, Z., Luong, C., Girgis, H., Dhungel, N., Abdi, A. H., Behnami, D., Gin, K., Rohling, R., Abolmaesumi, P., & Tsang, T. (2019). Cardiac phase detection in echocardiograms with densely gated recurrent neural networks and global extrema loss. *IEEE Transactions on Medical Imaging, 38*, 1821–1832. Available from https://doi. org/10.1109/TMI.2018.2888807.

DICOM Digital Imaging and Communications in Medicine [WWW Document]. (2021). Med. Imaging Technol. Assoc. <https://www.dicomstandard.org/>.

Diller, G. P., Vahle, J., Radke, R., Vidal, M. L. B., Fischer, A. J., Bauer, U. M. M., Sarikouch, S., Berger, F., Beerbaum, P., Baumgartner, H., & Orwat, S. (2020). Utility of deep learning networks for the generation of artificial cardiac magnetic resonance images in congenital heart disease. *BMC Medical Imaging, 20*, 1–8. Available from https://doi.org/10.1186/s12880-020-00511-1.

Galea, R., Diosan, L., Andreica, A., Popa, L., Manole, S., & Bálint, Z. (2021). Region-of-interest-based cardiac image segmentation with deep learning. *Applied Sciences, 11*, 1–11.

Ghorbani, A., Ouyang, D., Abid, A., He, B., Chen, J. H., Harrington, R. A., Liang, D. H., Ashley, E. A., & Zou, J. Y. (2020). Deep learning interpretation of echocardiograms. *npj Digital Medicine, 3*, 1–10. Available from https://doi.org/10.1038/s41746-019-0216-8.

Goodfellow, I., Bengio, Y., & Courville, A., 2016. *Deep learning.* MIT press.

Green, M., Marom, E. M., Konen, E., Kiryati, N., & Mayer, A. (2018). 3-D neural denoising for low-dose coronary CT angiography (CCTA). *Computerized Medical*

Imaging and Graphics: The Official Journal of the Computerized Medical Imaging Society, 70, 185–191. Available from https://doi.org/10.1016/j.compmedimag.2018.07.004.

Guidotti, R., Monreale, A., Ruggieri, S., Turini, F., Giannotti, F., & Pedreschi, D. (2018). A survey of methods for explaining black box models. *ACM Computing Surveys, 51.* Available from https://doi.org/10.1145/3236009.

Guo, F., Ng, M., Goubran, M., Petersen, S. E., Piechnik, S. K., Neubauer, S., & Wright, G. (2020). Improving cardiac MRI convolutional neural network segmentation on small training datasets and dataset shift: A continuous kernel cut approach. *Medical Image Analysis, 61.* Available from https://doi.org/10.1016/j.media.2020.101636.

Hata, E., Seo, C., Nakayama, M., Iwasaki, K., Ohkawauchi, T., & Ohya, J. (2020). Classification of aortic stenosis using ECG by deep learning and its analysis using Grad-CAM. In: *Proceedings of the annual international conference of the ieee engineering in medicine and biology society* (pp. 1548–1551). https://doi.org/10.1109/EMBC44109.2020.9175151.

Hong, J. H., Park, E. A., Lee, W., Ahn, C., & Kim, J. H. (2020). Incremental image noise reduction in coronary CT angiography using a deep learning-based technique with iterative reconstruction. *Korean Journal of Radiology: Official Journal of the Korean Radiological Society, 21,* 1165–1177. Available from https://doi.org/10.3348/kjr.2020.0020.

Hong, S., Zhou, Y., Shang, J., Xiao, C., & Sun, J. (2020). Opportunities and challenges in deep learning methods on electrocardiogram data: A systematic review. *Computers in Biology and Medicine, 122,* 103801.

Huang, M. S., Wang, C. S., Chiang, J. H., Liu, P. Y., & Tsai, W. C. (2020). Automated recognition of regional wall motion abnormalities through deep neural network interpretation of transthoracic echocardiography. *Circulation,* 1510–1520. Available from https://doi.org/10.1161/CIRCULATIONAHA.120.047530.

Irmawati, D., Wahyunggoro, O., & Soesanti, I. (2020). Recent trends of left and right ventricle segmentation in cardiac MRI using deep learning. In: *Proceedings of the 12th international conference on information technology and electrical engineering* (pp. 380–383). Available from https://doi.org/10.1109/ICITEE49829.2020.9271750.

Jahren, T. S., Steen, E. N., Aase, S. A., & Solberg, A. H. S. (2020). Estimation of end-diastole in cardiac spectral doppler using deep learning. *IEEE Transactions on Ultrasonics, Ferroelectrics, and Frequency Control, 67,* 2605–2614. Available from https://doi.org/10.1109/TUFFC.2020.2995118.

Jung, S., Lee, S., Jeon, B., Jang, Y., & Chang, H. J. (2020). Deep learning cross-phase style transfer for motion artifact correction in coronary computed tomography angiography. *IEEE Access, 8,* 81849–81863. Available from https://doi.org/10.1109/ACCESS.2020.2991445.

Kang, E., Koo, H. J., Yang, D. H., Seo, J. B., & Ye, J. C. (2019). Cycle-consistent adversarial denoising network for multiphase coronary CT angiography. *Medical Physics, 46,* 550–562. Available from https://doi.org/10.1002/mp.13284.

Kim, J. Y., Kim, Y., Oh, G.-H., Kim, S. H., Choi, Y., Hwang, Y., Kim, T.-S., Kim, S.-H., Kim, J.-H., Jang, S.-W., Oh, Y.-S., & Lee, M. Y. (2020). A deep learning model to predict recurrence of atrial fibrillation after pulmonary vein isolation. *International Journal of Arrhythmia, 21.* Available from https://doi.org/10.1186/s42444-020-00027-3.

Koo, H. J., Lee, J. G., Ko, J. Y., Lee, G., Kang, J. W., Kim, Y. H., & Yang, D. H. (2020). Automated segmentation of left ventricular myocardium on cardiac computed tomography using deep learning. *Korean Journal of Radiology: Official Journal of the Korean Radiological Society, 21,* 660–669. Available from https://doi.org/10.3348/kjr.2019.0378.

Kroll, L., Nassenstein, K., Jochims, M., Koitka, S., & Nensa, F. (2021). Assessing the role of pericardial fat as a biomarker connected to coronary calcification—A deep learning based approach using fully automated body composition analysis. *Journal of Clinical Medicine, 10*, 356. Available from https://doi.org/10.3390/jcm10020356.

Kusunose, K., Haga, A., Inoue, M., Fukuda, D., Yamada, H., & Sata, M. (2020). Clinically feasible and accurate view classification of echocardiographic images using deep learning. *Biomolecules, 10*, 1−8. Available from https://doi.org/10.3390/biom10050665.

Kwon, J. M., Lee, S. Y., Jeon, K. H., Lee, Y., Kim, K. H., Park, J., Oh, B. H., & Lee, M. M. (2020). Deep learning-based algorithm for detecting aortic stenosis using electrocardiography. *Journal of the American Heart Association, 9*, e014717. Available from https://doi.org/10.1161/JAHA.119.014717.

Kwon, S., Hong, J., Choi, E. K., Lee, B., Baik, C., Lee, E., Jeong, E. R., Koo, B. K., Oh, S., & Yi, Y. (2020). Detection of atrial fibrillation using a ring-type wearable device (CardioTracker) and deep learning analysis of photoplethysmography signals: Prospective observational proof-of-concept study. *Journal of Medical Internet Research, 22*. Available from https://doi.org/10.2196/16443.

Laney, D., et al. (2001). 3D data management: Controlling data volume, velocity and variety. *META Group Research Note, 6*, 1.

Lateef, F., & Ruichek, Y. (2019). Survey on semantic segmentation using deep learning techniques. *Neurocomputing, 338*, 321−348. Available from https://doi.org/10.1016/j.neucom.2019.02.003.

Leclerc, S., Smistad, E., Pedrosa, J., Ostvik, A., Cervenansky, F., Espinosa, F., Espeland, T., Berg, E. A. R., Jodoin, P. M., Grenier, T., Lartizien, C., Dhooge, J., Lovstakken, L., & Bernard, O. (2019). Deep learning for segmentation using an open large-scale dataset in 2D echocardiography. *IEEE Transactions on Medical Imaging, 38*, 2198−2210. Available from https://doi.org/10.1109/TMI.2019.2900516.

Lee, H., Grosse, R., Ranganath, R., & Ng, A. Y. (2009). Convolutional deep belief networks for scalable unsupervised learning of hierarchical representations. *Proc. 26th International Conference on Machine Learning, 2009*, 609−616. Available from https://doi.org/10.1145/1553374.1553453.

Li, M., Zeng, D., Xie, Q., Xu, R., Wang, Y., Ma, D., Shi, Y., Xu, X., Huang, M., & Fei, H. (2021). A deep learning approach with temporal consistency for automatic myocardial segmentation of quantitative myocardial contrast echocardiography. *The International Journal of Cardiovascular Imaging*. Available from https://doi.org/10.1007/s10554-021-02181-8.

Litjens, G., Ciompi, F., Wolterink, J. M., de Vos, B. D., Leiner, T., Teuwen, J., & Išgum, I. (2019). State-of-the-art deep learning in cardiovascular image analysis. *JACC Cardiovascular Imaging, 12*, 1549−1565. Available from https://doi.org/10.1016/j.jcmg.2019.06.009.

Liu, F., Wang, K., Liu, D., Yang, X., & Tian, J. (2021). Deep pyramid local attention neural network for cardiac structure segmentation in two-dimensional echocardiography. *Medical Image Analysis, 67*, 101873. Available from https://doi.org/10.1016/j.media.2020.101873.

Liu, P., Wang, M., Wang, Y., Yu, M., Wang, Y., Liu, Z., Li, Y., & Jin, Z. (2020). Impact of deep learning-based optimization algorithm on image quality of low-dose coronary CT angiography with noise reduction: A prospective study. *Academic Radiology, 27*, 1241−1248. Available from https://doi.org/10.1016/j.acra.2019.11.010.

Lossau (née Elss), T., Nickisch, H., Wissel, T., Morlock, M., & Grass, M. (2020). Learning metal artifact reduction in cardiac CT images with moving pacemakers. *Medical Image Analysis, 61*, 101655. Available from https://doi.org/10.1016/j.media.2020.101655.

Lowe, D. G. (2004). Distinctive image features from scale invariant keypoints. *International Journal of Computer Vision, 60*, 91−11020042. Available from https://doi.org/10.1023/B:VISI.0000029664.99615.94.

Luo, C., Shi, C., Li, X., & Gao, D. (2020). Cardiac MR segmentation based on sequence propagation by deep learning. *PLoS One*, *15*, 1−13. Available from https://doi.org/10.1371/journal.pone.0230415.

Madani, A., Ong, J. R., Tibrewal, A., & Mofrad, M. R. K. (2018). Deep echocardiography: Data-efficient supervised and semi-supervised deep learning towards automated diagnosis of cardiac disease. *npj Digital Medicine*, *1*, 1−11. Available from https://doi.org/10.1038/s41746-018-0065-x.

Mahmud, T., Fattah, S. A., & Saquib, M. (2020). DeepArrNet: An efficient deep CNN architecture for automatic arrhythmia detection and classification from denoised ECG beats. *IEEE Access*, *8*, 104788−104800. Available from https://doi.org/10.1109/ACCESS.2020.2998788.

Malinowski, M., Rohrbach, M., & Fritz, M. (2017). Ask your neurons: A deep learning approach to visual question answering. *International Journal of Computer Vision*, *125*, 110−135. Available from https://doi.org/10.1007/s11263-017-1038-2.

Martin, S. S., van Assen, M., Rapaka, S., Hudson, H. T., Fischer, A. M., Varga-Szemes, A., Sahbaee, P., Schwemmer, C., Gulsun, M. A., Cimen, S., Sharma, P., Vogl, T. J., & Schoepf, U. J. (2020). Evaluation of a deep learning−based automated CT coronary artery calcium scoring algorithm. *JACC Cardiovascular Imaging*, *13*, 524−526. Available from https://doi.org/10.1016/j.jcmg.2019.09.015.

McDonald, R. J., Schwartz, K. M., Eckel, L. J., Diehn, F. E., Hunt, C. H., Bartholmai, B. J., Erickson, B. J., & Kallmes, D. F. (2015). The effects of changes in utilization and technological advancements of cross-sectional imaging on radiologist workload. *Academic Radiology*, *22*, 1191−1198. Available from https://doi.org/10.1016/j.acra.2015.05.007.

MICCAI Challenges [WWW Document]. (2021). The Medical Image Computing and Computer Assisted Intervention Society. <http://www.miccai.org/events/challenges/>.

Moody, B., Moody, G., Villarroel, M., Clifford, G., & Silva, I. (2020). *MIMIC-III Waveform Database (version 1.0)*. PhysioNet. https://doi.org/10.13026/c2607m.

Nurmaini, S., Darmawahyuni, A., Mukti, A. N. S., Rachmatullah, M. N., Firdaus, F., & Tutuko, B. (2020). Deep learning-based stacked denoising and autoencoder for ECG heartbeat classification. *Electronics*, *9*, 1−17. Available from https://doi.org/10.3390/electronics9010135.

Ogrezeanu, I., Stoian, D., Turcea, A., & Itu, L. M. (2020). Deep learning based myocardial ischemia detection in ECG signals. In: *Proceedings of the 24th international conference on system theory, control and computing* (pp. 250−253). Available from https://doi.org/10.1109/ICSTCC50638.2020.9259714.

Østvik, A., Smistad, E., Aase, S. A., Haugen, B. O., & Lovstakken, L. (2019). Real-time standard view classification in transthoracic echocardiography using convolutional neural networks. *Ultrasound in Medicine & Biology*, *45*, 374−384. Available from https://doi.org/10.1016/j.ultrasmedbio.2018.07.024.

Ouyang, D., He, B., Ghorbani, A., Yuan, N., Ebinger, J., Langlotz, C. P., Heidenreich, P. A., Harrington, R. A., Liang, D. H., Ashley, E. A., & Zou, J. Y. (2020). Video-based AI for beat-to-beat assessment of cardiac function. *Nature*, *580*, 252−256. Available from https://doi.org/10.1038/s41586-020-2145-8.

Parvaneh, S., Rubin, J., Babaeizadeh, S., & Xu-Wilson, M. (2019). Cardiac arrhythmia detection using deep learning: A review. *Journal of Electrocardiology*, *57*, S70−S74. Available from https://doi.org/10.1016/j.jelectrocard.2019.08.004.

Peimankar, A., & Puthusserypady, S. (2021). DENS-ECG: A deep learning approach for ECG signal delineation. *Expert System Applications*, *165*, 113911. Available from https://doi.org/10.1016/j.eswa.2020.113911.

Pereira, T., Tran, N., Gadhoumi, K., Pelter, M. M., Do, D. H., Lee, R. J., Colorado, R., Meisel, K., & Hu, X. (2020). Photoplethysmography based atrial fibrillation detection:

A review. *npj Digital Medicine*, 3. Available from https://doi.org/10.1038/s41746-019-0207-9.

Pydicom [WWW Document]. (2020). <https://pydicom.github.io/>.

PyTorch [WWW Document]. (n.d.). <https://pytorch.org/>.

Raisi-Estabragh, Z., Harvey, N. C., Neubauer, S., & Petersen, S. E. (2020). Cardiovascular magnetic resonance imaging in the UK Biobank: A major international health research resource. *European Heart Journal - Cardiovascular Imaging*, 251−258. Available from https://doi.org/10.1093/ehjci/jeaa297.

Reiss, A., Indlekofer, I., Schmidt, P., & Van Laerhoven, K. (2019). Deep PPG: Large-scale heart rate estimation with convolutional neural networks. *Sensors*, *19*, 1−27. Available from https://doi.org/10.3390/s19143079.

Ribeiro, A. H., Ribeiro, M. H., Paixão, G. M. M., Oliveira, D. M., Gomes, P. R., Canazart, J. A., Ferreira, M. P. S., Andersson, C. R., Macfarlane, P. W., Wagner, M., Schön, T. B., & Ribeiro, A. L. P. (2020). Automatic diagnosis of the 12-lead ECG using a deep neural network. *Nature Communications*, *11*, 1−9. Available from https://doi.org/10.1038/s41467-020-15432-4.

Ronneberger, O., Fischer, P., & Brox, T. (2015). U-net: Convolutional networks for biomedical image segmentation. In: *Proceedings of the international conference on medical image computing and computer assisted intervention* (pp. 234−241). Available from https://doi.org/10.1107/978-3-319-24574-4_28.

Sanjana, K., Sowmya, V., Gopalakrishnan, E. A., & Soman, K. P. (2020). Explainable artificial intelligence for heart rate variability in ECG signal. *Healthcare Technology Letters*, 7, 146−154. Available from https://doi.org/10.1049/htl.2020.0033.

Selvaraju, R. R., Cogswell, M., Das, A., Vedantam, R., Parikh, D., & Batra, D. (2020). Grad-CAM: Visual explanations from deep networks via gradient-based localization. *International Journal of Computer Vision*, *128*, 336−359. Available from https://doi.org/10.1007/s11263-019-01228-7.

Shrikumar, A., Greenside, P., & Kundaje, A. (2017). Learning important features through propagating activation differences. In: *Proceedings of the 34th international conference on machine learning* (pp. 4844−4866).

Simonyan, K., Vedaldi, A., & Zisserman, A. (2014). Deep inside convolutional networks: Visualising image classification models and saliency maps. In: *Proceedings of the 2nd international conference on learning representation*.

Singh, P., & Pradhan, G. (2020). A new ECG denoising framework using generative adversarial network. *IEEE/ACM Transactions on Computational Biology and Bioinformatics*, *5963*, 1. Available from https://doi.org/10.1109/tcbb.2020.2976981.

Smilkov, D., Thorat, N., Kim, B., Viégas, F., & Wattenberg, M. (2017). SmoothGrad: Removing noise by adding noise. In: *Workshop on visualization for deep learning*.

Sutton, R. T., Pincock, D., Baumgart, D. C., Sadowski, D. C., Fedorak, R. N., & Kroeker, K. I. (2020). An overview of clinical decision support systems: Benefits, risks, and strategies for success. *npj Digital Medicine*, *3*, 1−10. Available from https://doi.org/10.1038/s41746-020-0221-y.

TensorFlow [WWW Document]. (n.d.). <https://www.tensorflow.org/>.

Tjoa, E., & Guan, C. (2020). A survey on explainable artificial intelligence (XAI): Toward medical XAI. *IEEE Transactions on Neural Networks Learning Systems*, 1−21. Available from https://doi.org/10.1109/tnnls.2020.3027314.

Ulloa Cerna, A. E., Jing, L., Good, C. W., VanMaanen, D. P., Raghunath, S., Suever, J. D., Nevius, C. D., Wehner, G. J., Hartzel, D. N., Leader, J. B., Alsaid, A., Patel, A. A., Kirchner, H. L., Pfeifer, J. M., Carry, B. J., Pattichis, M. S., Haggerty, C. M., & Fornwalt, B. K. (2021). Deep-learning-assisted analysis of echocardiographic videos improves predictions of all-cause mortality. *Nature Biomedical Engineering*. Available from https://doi.org/10.1038/s41551-020-00667-9.

Upendra, R. R., Dangi, S., & Linte, C. A. (2020). Automated segmentation of cardiac chambers from cine cardiac MRI using an adversarial network architecture. In: *Proceedings of SPIE international society for optical engineering* (Vol. 100). Available from https://doi.org/10.1117/12.2550656.

van Assen, M., Martin, S. S., Varga-Szemes, A., Rapaka, S., Cimen, S., Sharma, P., Sahbaee, P., De Cecco, C. N., Vliegenthart, R., Leonard, T. J., Burt, J. R., & Schoepf, U. J. (2021). Automatic coronary calcium scoring in chest CT using a deep neural network in direct comparison with non-contrast cardiac CT: A validation study. *European Journal of Radiology, 134*, 109428. Available from https://doi.org/10.1016/j.ejrad.2020.109428.

van Hamersvelt, R. W., Zreik, M., Voskuil, M., Viergever, M. A., Išgum, I., & Leiner, T. (2019). Deep learning analysis of left ventricular myocardium in CT angiographic intermediate-degree coronary stenosis improves the diagnostic accuracy for identification of functionally significant stenosis. *European Radiology, 29*, 2350–2359. Available from https://doi.org/10.1007/s00330-018-5822-3.

van Velzen, S. G. M., Lessmann, N., Velthuis, B. K., Bank, I. E. M., van den Bongard, D. H. J. G., Leiner, T., de Jong, P. A., Veldhuis, W. B., Correa, A., Terry, J. G., Carr, J. J., Viergever, M. A., Verkooijen, H. M., & Išgum, I. (2020). Deep learning for automatic calcium scoring in CT: Validation using multiple cardiac CT and chest CT protocols. *Radiology, 295*, 66–79. Available from https://doi.org/10.1148/radiol.2020191621.

Vesal, S., Maier, A., & Ravikumar, N. (2020). Fully automated 3D cardiac MRI localisation and segmentation using deep neural networks. *Journal of Imaging, 6*, 1–19. Available from https://doi.org/10.3390/JIMAGING6070065.

Virani, S. S., Alonso, A., Benjamin, E. J., Bittencourt, M. S., Callaway, C. W., Carson, A. P., Chamberlain, A. M., Chang, A. R., Cheng, S., Delling, F. N., Djousse, L., Elkind, M. S. V., Ferguson, J. F., Fornage, M., Khan, S. S., Kissela, B. M., Knutson, K. L., Kwan, T. W., Lackland, D. T., . . . Heard, D. G. (2020). Heart disease and stroke statistics—2020 update: A report from the American Heart Association. *Circulation*. Available from https://doi.org/10.1161/CIR.0000000000000757.

Wagner, P., Strodthoff, N., Bousseljot, R. D., Kreiseler, D., Lunze, F. I., Samek, W., & Schaeffter, T. (2020). PTB-XL, a large publicly available electrocardiography dataset. *Scientific Data, 7*, 1–15. Available from https://doi.org/10.1038/s41597-020-0495-6.

Wang, J., Li, R., Li., & Rui, F. , B. (2020). A knowledge-based deep learning method for ECG signal delineation. *Future Generation Computer Systems, 109*, 56–66. Available from https://doi.org/10.1016/j.future.2020.02.068.

Wang, M., & Deng, W. (2018). Deep visual domain adaptation: A survey. *Neurocomputing, 312*, 135–153. Available from https://doi.org/10.1016/j.neucom.2018.05.083.

Wu, D., Wang, X., Bai, J., Xu, X., Ouyang, B., Li, Y., Zhang, H., Song, Q., Cao, K., & Yin, Y. (2019). Automated anatomical labeling of coronary arteries via bidirectional tree LSTMs. *International Journal of Computer Assisted Radiology and Surgery, 14*, 271–280. Available from https://doi.org/10.1007/s11548-018-1884-6.

Xu, K., Ba, J.L., Kiros, R., Cho, K., Courville, A., Salakhutdinov, R., Zemel, R.S., & Bengio, Y. (2015). Show, attend and tell: Neural image caption generation with visual attention. In: *Proceedings of 32nd international conference on machine learning* (Vol. 3, pp. 2048–2057).

Xu, K., Jiang, X., Lin, S., Dai, C., & Chen, W. (2020). Stochastic modeling based nonlinear Bayesian filtering for photoplethysmography denoising in wearable devices. *IEEE Transactions on Industrial Informatics, 16*, 7219–7230. Available from https://doi.org/10.1109/TII.2020.2988097.

Yan, B. P., Lai, W. H. S., Chan, C. K. Y., Au, A. C. K., Freedman, B., Poh, Y. C., & Poh, M. Z. (2020). High-throughput, contact-free detection of atrial fibrillation from video with deep learning. *JAMA Cardiology*, *5*, 105−107. Available from https://doi.org/10.1001/jamacardio.2019.4004.

Yi, X., Walia, E., & Babyn, P. (2019). Generative adversarial network in medical imaging: A review. *Medical Image Analysis*, *58*. Available from https://doi.org/10.1016/j.media.2019.101552.

Zeleznik, R., Foldyna, B., Eslami, P., Weiss, J., Alexander, I., Taron, J., Parmar, C., Alvi, R. M., Banerji, D., Uno, M., Kikuchi, Y., Karady, J., Zhang, L., Scholtz, J. E., Mayrhofer, T., Lyass, A., Mahoney, T. F., Massaro, J. M., Vasan, R. S., . . . Aerts, H. J. W. L. (2021). Deep convolutional neural networks to predict cardiovascular risk from computed tomography. *Natural Communications*, *12*. Available from https://doi.org/10.1038/s41467-021-20966-2.

Zhang, J., Gajjala, S., Agrawal, P., Tison, G. H., Hallock, L. A., Beussink-Nelson, L., Lassen, M. H., Fan, E., Aras, M. A., Jordan, C. R., Fleischmann, K. E., Melisko, M., Qasim, A., Shah, S. J., Bajcsy, R., & Deo, R. C. (2018). Fully automated echocardiogram interpretation in clinical practice: Feasibility and diagnostic accuracy. *Circulation*, *138*, 1623−1635. Available from https://doi.org/10.1161/CIRCULATIONAHA.118.034338.

Zhang, N., Yang, G., Zhang, W., Wang, W., Zhou, Z., Zhang, H., Xu, L., & Chen, Y. (2021). Fully automatic framework for comprehensive coronary artery calcium scores analysis on non-contrast cardiac-gated CT scan: Total and vessel-specific quantifications. *European Journal of Radiology*, *134*, 109420. Available from https://doi.org/10.1016/j.ejrad.2020.109420.

Zhang, O., Ding, C., Pereira, T., Xiao, R., Gadhoumi, K., Meisel, K., Lee, R., Chen, Y., & Hu, X. (2021). Explainability metrics of deep convolutional networks for photoplethysmography quality assessment. *IEEE Access*, *9*. Available from https://doi.org/10.1109/ACCESS.2021.3054613.

Zhang, Z., Pi, Z., & Liu, B. (2015). TROIKA: A general framework for heart rate monitoring using wrist-type photoplethysmographic signals during intensive physical exercise. *IEEE Transactions on Bio-Medical Engineering*, *62*, 522−531. Available from https://doi.org/10.1109/TBME.2014.2359372.

Zhao, M., Che, X., Liu, H., & Liu, Q. (2020). Medical prior knowledge guided automatic detection of coronary arteries calcified plaque with cardiac ct. *Electronics*, *9*, 1−14. Available from https://doi.org/10.3390/electronics9122122.

Zhao, Z. Q., Zheng, P., Xu, S. T., & Wu, X. (2019). Object detection with deep learning: A review. *IEEE Transactions on Neural Networks Learning Systems*, *30*, 3212−3232. Available from https://doi.org/10.1109/TNNLS.2018.2876865.

Zhou, B., Khosla, A., Lapedriza, A., Oliva, A., & Torralba, A. (2016). Learning deep features for discriminative localization. In: *Proceedings of IEEE computer society conference on computer vision pattern recognition* (pp. 2921−2929). Available from https://doi.org/10.1109/CVPR.2016.319.

Zreik, M., Van Hamersvelt, R. W., Khalili, N., Wolterink, J. M., Voskuil, M., Viergever, M. A., Leiner, T., & Isgum, I. (2020). Deep learning analysis of coronary arteries in cardiac CT angiography for detection of patients requiring invasive coronary angiography. *IEEE Transactions on Medical Imaging*, *39*, 1545−1557. Available from https://doi.org/10.1109/TMI.2019.2953054.

CHAPTER 4

Data-driven models for cuffless blood pressure estimation using ECG and PPG signals

Geerthy Thambiraj[1], Uma Gandhi[2] and Umapathy Mangalanathan[2]
[1]Centre for Healthcare Entrepreneurship, Indian Institute of Technology, Hyderabad, India
[2]Department of Instrumentation and Control Engineering, National Institute of Technology Tiruchirappalli, Tiruchirappalli, India

4.1 Introduction

Blood pressure (BP) is a crucial hemodynamic variable that is often observed continuously during surgery and critically ill patients. It is considered as one of the modifiable risk factors for cardiovascular diseases. High BP, known as hypertension (HTN), is the leading cause of a global premature mortality rate affecting 1 billion and expected to surge around 1.56 billion in 2025 (Kazemi Korayem et al., 2018), and 25.3% prevalent in India as well (Dadlani et al., 2019). HTN is considered to be a silent killer as it shows little symptoms and acts as modifiable risk factors for stroke, congestive heart failure, coronary heart disease, and chronic kidney diseases. Detection and treatment in an initial phase are very much essential for control of hypertensive population to reduce its huge adverse effect on public health. The report confirms the prevalence of HTN is 30 times higher in urban and 10 times higher in the rural community over 55 and 36 years, respectively, in India (Karmakar et al., 2018). It is also reported that the overestimation of BP by 5 mmHg would result in the subjects' inappropriate medication (Handler, 2009). Therefore there is a high necessity of ambulatory prone BP estimation in the clinical field, and this chapter dealt with the novel algorithm based on data-driven models for reliable and continuous BP measurement.

4.1.1 The conventional techniques of BP measurement

William Harvey describes the first study on blood circulation in the 17th century. Followed by Stephen hales in 1733, measured BP in the horse through arterial cannulation connecting to the tube where the pulsatile flow is identified. Later, Poiseuille invented a mercury manometer in

Edge-of-Things in Personalized Healthcare Support Systems
DOI: https://doi.org/10.1016/B978-0-323-90585-5.00017-5

1828 to measure the arterial BP. Until 1855 there was no technique to measure BP without puncturing the artery. Ludwig's kymograph and Vierordt's sphygmograph were developed in 1847 and 1854, respectively, which improved the noninvasive measurement methods. Scipione Riva-Rocci introduced the BP device, sphygmomanometer, which is in use till today. Further, in 1905, Nikolai Korotkoff devised the auscultatory method and laid the foundations to measure BP with sphygmomanometer (Booth, 1977). Table 4.1 summarizes the presently available commercial BP devices using different principles.

4.1.2 Significance of continuous BP monitoring

Currently, the noninvasive devices used for ABPM or home BP measurements are designed based on the Oscillometric principle. An inflatable cuff is attached around the arm of the subject resulting in pain, discomfort, and disturbs the sleep when it does frequently repeated measurement during the night-time. It reads BP for every 10−30 min. The cuff size also plays a vital role as it results in erroneous values, which limits its clinical applications on 24-h BP monitoring. Though the technique is used for ABPM/Home BP, it provides us only intermittent values with which dynamic interactions of BP cannot be examined clearly for better disease management. Other recent BP techniques like vascular unloading and applanation tonometry are expensive, require trained personnel for the exact placement of the sensor over the artery, and cuff-based methods increase the workload of the heart (McCarthy et al., 2011) and limit its potential benefits in clinical applications. Due to the inadequacies of the conventional BP measurement methods, continuous monitoring of BP demands an intense research Hence, pulse transit time (PTT) technique, one of the most recent methodologies that has received attention to measure BP, avoids the inherent problems of cuff-based BP methods.

4.1.3 Analytical-based BP modeling

The pulsatile nature of the blood generates pressure pulse waves in the aorta, which travel down the arterial tree with a particular speed and is said to be pulse wave velocity (PWV). It is defined as (Vlachopoulos et al., 2011)

$$PWV = \frac{L}{PTT} \tag{4.1}$$

where L is the distance between two arterial sites and PTT is the time taken for the pulse wave to travel from one arterial site to the other.

Table 4.1 Commercial conventional BP techniques.

Principle	Commercial name	Method	Need for calibration	Measurement	Manual/automatic
Intraarterial Chung et al. (2013)	Catheter line	Invasive	No	Continuous	Supervised
Auscultatory Chung et al. (2013)	Sphygmomanometer	Noninvasive/occlusive	No	Intermittent	Supervised
Oscillometric Forouzanfar et al. (2015)	Omron	Noninvasive/cuff-based	No	Intermittent	Unsupervised
Tonometric Drzewiecki et al. (1983)	SphygmoCor	Noninvasive	Yes	Continuous	Unsupervised
Penaz principle Penaz (1973)	Portapres/Finapres	Noninvasive/cuff-based	No	Continuous	Unsupervised

PWV depicts the properties of the propagating pressure pulse down the segment of the arterial tree. PWV conveys the information about the biochemical and geometric characteristics of the arterial wall. Hence, it acts as a potential physiological surrogate in various clinical applications such as arterial stiffness, cuffless BP measurement, and respiratory effort measurement. The noninvasive and easily accessible nature of PWV estimation makes it a prognostic indicator of many pathophysiological conditions.

Pulse arrival time (PAT) is the time taken for the pulse wave to travel from one peripheral arterial site to another. In cuffless BP research PTT and PAT have been used interchangeably, albeit it gives different propagation time intervals. PAT consists of PTT interval and preejection period (PEP), which is the delay between left ventricular depolarization and mechanical ejection of blood. The isovolumetric contraction also occurs during the PEP with the valves closed. PAT is given by

$$PAT = PTT + PEP. \tag{4.2}$$

PAT provides clinical information about the vessel's physical and geometrical properties and, hence, arterial stiffness, which further acts as a marker of BP. The traveling of pressure wave propagation in the arterial tree is proportionate to the BP. Ubiquitously, the extraction of PAT is obtained using one of the fiducial points of the photoplethysmography (PPG) signal and the R peak of the ECG. Some literature uses PTT and PAT for the estimation of BP, arterial stiffness, etc. In recent times globally published scientific journals have investigated the combinations of various biosignals as shown in Fig. 4.1, for extracting PAT and PTT as 24%, 15%, 9%, 7%, and 5% from China, USA, Germany, Korea, and Japan, respectively, for BP measurement.

Theoretically, there is a relationship proposed between PWV and BP, which makes the cuffless BP technique feasible. The most underlying relationship was proposed by Moens–Korteweg (M–K) and Hughes et al. (1979) in their experiment based on arterial wave propagation for the PTT-based BP assessment. M–K, from his research work, deduced the physical and geometrical properties of the arteries, determined the pressure propagation speed known as PWV, and it is given as (Tijsseling & Anderson, 2012)

$$PWV = \frac{L}{PTT} = \sqrt{\frac{Eh}{\rho d}} \tag{4.3}$$

where L is the distance between two arterial sites, E is the elasticity of the arteries, h is the thickness of the vessel wall, d is the diameter of the vessel,

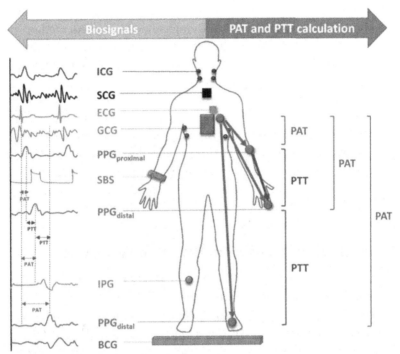

Figure 4.1 Extraction of PAT and PTT from the different biosignals. *Adapted from Welykholowa, K., Hosanee, M., Chan, G., Cooper, R., Kyriacou, P.A., Zheng, D., & Allen, J., et al. (2020). Multimodal photoplethysmography-based approaches for improved detection of hypertension. Clinical Medicine, 9, 1-20.*

and ρ is the density of the vessel. Hughes et al. (1979) proposed a relationship between arterial pressure (P) and elastic modulus of the vessel (E), and it is given by

$$E = E_o e^{\gamma P} \tag{4.4}$$

where E_o is the elastic modulus at zero arterial pressure, P is the pressure, γ is the parameter that varies with the subject, and it is related to the vessel properties.

Upon substituting Eq. (4.4) in Eq. (4.3), it gives us the relation between pressure (P) and PWV as

$$\text{PWV} = \sqrt{\frac{E_o e^{aP}}{\rho d} \frac{h}{}}. \tag{4.5}$$

From Eq. (4.5), it is evident that PWV varies in direct proportion to BP, and PTT varies inversely to BP. Many authors investigated the BP estimation using PTT grounded on the M−K relationship, which assumes blood to be a Newtonian fluid. Although the PTT-based method provides a continuous BP, there are other regulating BP parameters that should be considered in the analytical model to attain improvement in accuracy.

Nevertheless, the current PTT-based BP models face several challenges, such as recalibration of model coefficients and incorporation of various physiological parameters that are conquered by data-driven models. These data-driven models based on the machine learning (ML), such as regression techniques, extract the key features from physiological signals in time-domain, and deep learning (DL) approaches could help us to attain a calibration-free cuffless continuous BP estimation.

4.2 Review of data-driven models in the literature

To an extent, the current PTT-based methods exhibit the progress to attain continuous cuffless BP monitoring. Yet, it requires substantial research to mitigate the errors and challenges encountered during the BP measurement. Nevertheless, with developments in the ML and DL, various immanent features were extracted to predict output that varies linearly/nonlinearly with inputs becoming easier and simpler. They are capable of handling large data with great ease and are less susceptible to noise than the mathematical models, thereby resulting in better performance. Thus data-driven models for BP assessment either using raw biosignals or features extracted from ECG and PPG signals became widespread. Table 4.2 summaries the recent studies on PTT-based data-driven models for cuffless BP estimation.

Few things were observed after conducting the literature review which are as follows: There is always a trade-off between the accuracy of the PTT-based BP estimation and calibration interval frequency. The calibration is performed in either of these ways: subject-specific, population-based in the analytical model, and it uses another BP measurement device for an initial calibration, which limits its potential application in continuous monitoring. Hence, data-driven models come into a picture to overcome the constraints allied with PTT-based BP estimation. Various handcrafted time-domain features of the ECG and PPG signals are considered along with hemodynamic parameters such as the Womersley number is extracted noninvasively from the signal and further all these features given into different regression models. Not all features contribute toward

Table 4.2 Outline of the research studies based on PTT and machine learning for the cuffless continuous BP assessment.

Author, year	Algorithm	Bio-signals	BP Indicators	Reference method and calibration	Subjects	Estimated accuracy					
						SBP		DBP		MBP	
						Bias and STD (mmHg)	Corr.	Bias and STD (mmHg)	Corr.	Bias and STD (mmHg)	Corr.
Teng and Zhang (2003)	Linear regression	PPG	Width of ½ and 2/3 of pulse amplitude, systolic and diastolic time	Oscillometric BP device, individual calibration	15	0.21 ± 7.32	—	0.025 ± 4.39	—	—	—
Fung et al. (2004)	$BP = (a/PTT^2) + b$	ECG, PPG	PAT	Cuff-based subject-specific (constant, b)	22	-0.079 ± 11.3	—	—	—	—	—
Muehlsteff et al. (2006)	$SBP = A*\ln PAT + BSBP = A*L/PAT + BSBP = A*(L/PAT)^2 + B$	ECG, PPG, IPG, IMP	PAT, PEP, PTT, HR are extracted.	Cuff-based BP, BP measurement at rest for calibration	18	RMSE for 3 models: 7.56.97.3	—	—	—	—	—
Payne et al. (2006)	$SBP = a*PAT + b$ $DBP = a*PTT + b$	ECG, PPG	—	Intraarterial radial BP, (with hemodynamic drugs)	12 healthy	95% LoA: 17.0 mmHg for SBP&PATBHS standards: values within 5, 10, 15 mmHg are 44, 66, and 73%, respectively	—	95% LoA:17.3 mmHg for DBP&PTT BHS standards: values within 5, 10, 15 mmHg are 42, 64, and 72% respectively	—		

(Continued)

Table 4.2 (Continued)

Author, year	Algorithm	Bio-signals	BP Indicators	Reference method and calibration	Subjects	Estimated accuracy SBP Bias and STD (mmHg)	SBP Corr.	DBP Bias and STD (mmHg)	DBP Corr.	MBP Bias and STD (mmHg)	MBP Corr.
Chua and Heneghan (2006)	Multiple linear regression	ECG, PPG	PPG amplitude, T-wave amplitude, PAT, HR	Portapres BP measurement Individual calibration	8 healthy	0.7 ± 5.9	0.59 ± 0.16	0.3 ± 3.6	0.59 ± 0.18	—	—
Wong and Zhang (2008)	SBP = a*PAT + b SBP = a*PTT + b	ECG, PPG, ICG	PAT, PTT	Oscillometric BP (physical exercise), Oscillometric BP	11	SD: 5.6 SD: 8.6	—	—	—	—	—
Kim et al. (2008)	Multiple regression	ECG, PPG	PTT, Time and amplitude features of the PPG	Finometer Pro BP device, Intermittent calibration	10	—	0.86	—	0.83	—	—
Yoon et al. (2009)	SBP = a1* PTT + b1 DBP = a2* diastolic time + b	ECG, PPG	PTT using max slope as a fiducial point	Oscillometric BP (step-climbing)	5 healthy	Absolute mean difference (AME): 4 −11% for individual regression line	0.71	Absolute mean difference (AME): 4 −10% for individual regression line	0.76	—	—
Wong et al. (2009)	SBP = a1*PTT + b1 DBP = a2*PTT + b2	ECG, PPG	PTT	Oscillometric BP (exercise)/half-year oscillometric	41 healthy (14 for half year)	0.0 ± 5.3 With same coefficients after half-year: 1.4 ± 10.2	First test: −0.92 Repeatability test: −0.87	0.0 ± 2.9 with same coefficients after half-year: 2.1 ± 7.3	First test: −0.38 Repeatability test: −0.30	-	-

Reference	Method	Signals	Features	Validation	Subjects/Dataset						
Cattivelli and Garudadri (2009)	SBP = a1*PAT + b1* HR + c1DBP = a2*PAT + b2*HR + c2	ECG, PPG	PAT, HR	Arterial noninvasive BP Adaptive recalibration	34 subjects MIMIC database	−0.41 ± 7.77	—	−0.07 ± 4.96	—	—	—
Gesche et al. (2011)	$BPPTT= P1 \times PWV \times e$ $(P3 \times PWV) + P2 \times PWVP4-$ $(BPPTT, cal-BPcal)$	ECG, PPG	PTT	Cuff-based BP device, one-point calibration	63	—	0.83	—	—	—	—
Monte-Moreno (2011)	Different ML approaches	PPG	PPG features	Aneroid sphygmo-manometer, calibration-free	410	—	0.91	—	0.089	—	—
Ruiz-Rodriguez et al. (2013)	Deep Belief Network Restricted Boltzmann Machine (DBNRBM)	PPG	—	Invasive arterial BP, calibration-free	707	−2.98 ± 19.35	—	−3.65 ± 8.69	—	−3.38±10.35	—
Wibmer et al. (2014)	Non-linear regression $BP = a + (d/(PTT − c))^2$	ECG, PPG	PTT	Oscillometric BP (cuff-based BP during exercise)	20 (Diseased subjects)	−10.9–10.9 mmHg	—	−8.9–8.9 mmHg	—	—	—
Yoon et al. (2018)	Multiregression model	ECG, PPG	PAT, 14 features from PPG	Invasive BP	23	8.7 ± 3.2	—	4.4 ± 1.6	—	—	—
Kachuee et al. (2017)	Regression methods	ECG, PPG	PTT, whole-based PPG features	Noninvasive BP, Calibration, and Calibration-free	942	MAE: 11.7	0.59	MAE: 5.35	0.48	MAE: 5.92	0.59

(Continued)

Table 4.2 (Continued)

Author, year	Algorithm	Bio-signals	BP Indicators	Reference method and calibration	Subjects	Estimated accuracy					
						SBP		DBP		MBP	
						Bias and STD (mmHg)	Corr.	Bias and STD (mmHg)	Corr.	Bias and STD (mmHg)	Corr.
Escobar-Restrepo et al. (2018)	Linear regression	ECG, PPG	PAT	Invasive arterial BP, regression parameters are obtained using individual BP values	11 ICU patients	MAE: 3.9	−0.5	MAE: 7.6	−0.42	−	−
Khalid et al. (2018)	MLR, SVM, regression tree	PPG	pulse area, pulse rising time, and width 25%	Noninvasive BP Calibration less	32	−0.1 ± 6.5	−	−0.6 ± 5.2	−	−	−
Song et al. (2019)	Stacked DNN	ECG, PPG	PTT foot, peak, max slope, systolic and diastolic time, 2/3 width of the pulse, PWV	Mercury sphygmo-manometer, calibration-free	110	−3.0 ± 6	−	−1.5 ± 6	−	−	−

accuracy; hence, feature selection to identify the most significant features is the third objective of the work. However, to reduce the computational cost and to eliminate the handcrafted feature engineering involved in regression techniques. Hence, it demands the deep network to be executed for the successful implementation of continuous BP monitoring.

This motivates us to focus our research on BP analytical models with various prime BP regulators and data-driven models with the implementation of the optimization technique for calibration-free cuffless BP estimation.

4.3 Methodology

The entire approach of the proposed work is represented in Fig. 4.2. The dataset chosen for this work was UCI dataset by Physionet organization. Using UCI dataset, the data-driven models such as regression and deep network were modeled and validated the model with the Advancement of Medical Instrumentation (AAMI) standards.

4.3.1 Dataset

We considered the ML repository database from UCI, center for ML and intelligent systems for cuffless BP estimation provided by Physionet Organization (Goldberger et al., 2000). The processing on the MIMIC II dataset has been discussed in Kachuee et al. (2015). The database has the ECG signal, which is obtained using the lead II configuration, PPG signal from the plethysmograph placed in the finger, and IABP (intraarterial BP)

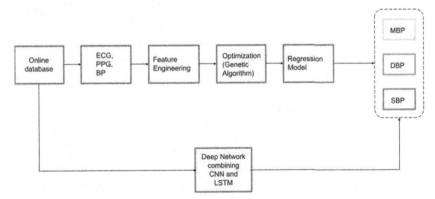

Figure 4.2 The overall proposed block diagram for Cuffless BP estimation using data-driven models.

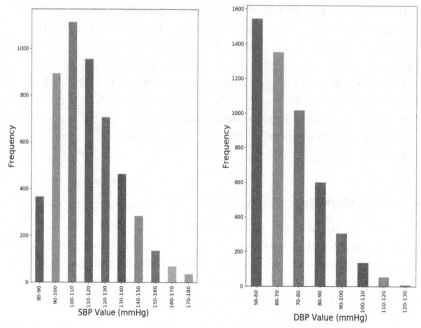

Figure 4.3 Distribution of SBP and DBP in the UCI dataset.

signals which act as the ground truth to calculate systolic BP (SBP) and diastolic BP (DBP). It has 12,000 instances of data sampled at 125 Hz. Subjects who had too high or low BP values were removed, and filtering process was performed. Every instance in the dataset contains different recording duration, therefore exact visualization of the first 1000 samples are considered. Overlapping of the data in the dataset was avoided using a unique ID for each instance. Preprocessed signal is fed as an input to the ML model. A total of 80%−20% dataset split was implemented for training and testing. Fig. 4.3 shows BP distribution in the dataset.

4.3.2 Dataset preprocessing

Removing the undesired artifacts is the significant step before extracting the physiological parameters from biosignals in estimating BP. Therefore a wavelet denoising method and zero-phase FIR filter was implemented on the raw ECG and PPG signals to eliminate the baseline wandering as shown in Fig. 4.4. The Daubechies wavelet was decomposed to the level L and then reconstructed back to get the noiseless signals. Fig. 4.5

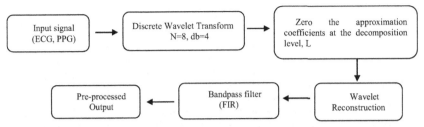

Figure 4.4 Preprocessing approach on raw ECG and PPG signals.

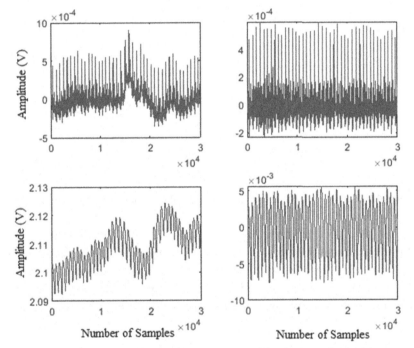

Figure 4.5 Baseline wandering suppression using DWT: (A) ECG and (B) PPG.

demonstrates the removal of motion artifacts before and after applying wavelet transform. This preprocessing step is applied for the public UCI dataset.

4.3.3 Data-driven models

Fig. 4.6 explains the block diagram to assess BP without using a cuff with ML model using an optimization method. The following steps were employed to estimate BP: (1) biosignals such as ECG, PPG, and IABP

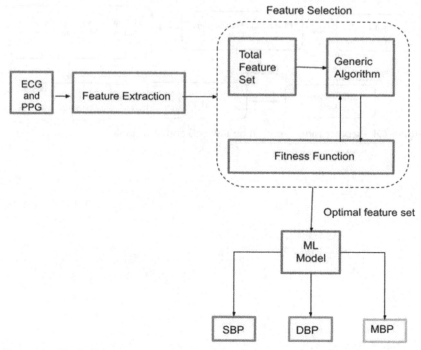

Figure 4.6 Block diagram representation of the BP estimation using ML methods.

signals were obtained from the dataset; (2) signal preprocessing; (3) the physiological signals such as ECG and PPG were used to extract the features on time-domain; (4) features extracted were given to different regression models an input; and (5) GA was implemented to augment the BP estimation. The following sections discuss the above process.

4.3.3.1 Feature extraction

Features from the literature and the proposed features such as the Womersley number are presented in this study; a total of 43 features were considered from the finger PPG, ECG, and a combination of the ECG and finger PPG using the MATLAB® software. The following subsection elucidates the extraction of features in the time-domain from ECG and PPG in an extensive way:

1. **PTT:** Cross-correlation between the ECG and sparsified PPG signals were performed to obtain the PTT and sample record is shown in Fig. 4.7. Sparsification of the pulse signal is acquired using the moving window maximum on PPG, where only the peaks are preserved,

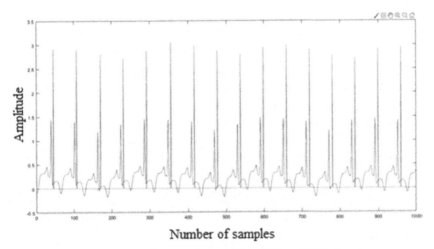

Figure 4.7 Representation of a sample sparsified PPG and ECG signals to measure PTT.

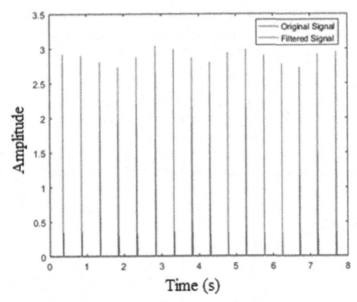

Figure 4.8 Sparsification of PPG signals before and after filtering.

which results in the sparsified signal as shown in Fig. 4.8. It is also observed that there is no delay introduced after filtering. PTT can be extracted as the time interval between R wave of the ECG wave and any one of the fiducial points of the PPG signal. In this work PTTs (time between ECG R wave and slope of the PPG), PTTmin (time between ECG R wave and minimum of the PPG), and PTTmax (time between ECG R wave and maximum of the PPG) were considered.

2. **IH**: IH represents the highest value of PPG which represents the maximum intensity recorded. This feature is extracted by applying the findpeaks function of MATLAB on the signal.

3. **IL**: IL represents the lowest value of PPG which represents the minimum intensity recorded. IL value is extracted by negating the PPG signal and then using the findpeaks function in MATLAB to find the location of the peak.

4. **Photoplethysmogram intensity ratio (PIR):** PIR is the highest value of PPG divided by the lowest value of PPG (Ding & Zhang, 2015):

$$PIR = \frac{IH}{IL}. \tag{4.6}$$

5. **Extraction of Womersley number, α**

The estimation of the Womersley number from the pulse signal is challenging. However, an effort is made to obtain the approximation of the Womersley number noninvasively from the pulse signal. It is defined as the ratio of pulsatile effects to the viscous flow, and Womersley deduces it as, $\alpha = R\sqrt{\frac{\rho\omega}{\mu}}$, where μ is the viscosity of blood, ω is the angular frequency, R is the radius of the blood vessel, and ρ is the blood density. The radius of the vessel is approximated to be the valley of the pulse signal during systole based on the Beer Lambert's law (Ding & Zhang, 2015). It is also demonstrated in the literature that intensity of the pulse signal is positively correlated with ultrasound-based diameter measurements. However, to obtain μ values, two noninvasive ways are attempted to extract from the PPG signal and its derivatives as follows:

1. *Womersley number* (α_1): Wei (2011) formulated the noninvasive method of estimating the spring constant, which reflects the peripheral arterial elasticity using the pulsating signal measured with the piezoelectric transducer on the radial artery. It was obtained using an arterial wave propagation equation as follows:

$$\frac{d^2P(z,t)}{dt^2} + b\frac{dP(z,t)}{dt} + KP(z,t) = -V_\infty^2 \frac{d^2P(z,t)}{dz^2} \tag{4.7}$$

where $P(z,t)$ is the pressure term, b is the damping coefficient, which is related to the fluid viscosity, k is the spring constant, and V_∞^2 is the force due to the Windkessel effect.

Considering the fixed position and pressure variation to be small, the arterial diameter change [representing the displacement of the wall, $x(t)$] to the pressure variation [$P(t)$] can be approximated to be linear; therefore the equation becomes (Wei, 2011)

$$\frac{d^2x(t)}{dt^2} + b\frac{dx(t)}{dt} + Kx(t) = -V_\infty^2 \frac{d^2x(t)}{dz^2} \qquad (4.8)$$

where $x(t)$ is the displacement of the arterial wall. However, the right-hand side term is due to the Windkessel effect which decreases at the maximum slope of the pulse signal, therefore ignoring the Windkessel effect and adding the unit impulse response (Wang et al., 1997) to estimate the damping coefficient, the above equation becomes

$$\frac{d^2x(t)}{dt^2} + b\frac{dx(t)}{dt} + Kx(t) = 1. \qquad (4.9)$$

The spring constant K is estimated as mentioned by Wei, in his work as, $K = -\dfrac{\frac{d^2x(t_{\text{peak}})}{dt^2}}{x(t_{\text{peak}})}$ and substituted in the above Eq. (4.9) and the b term is given as (Thambiraj et al., 2020)

$$b = \frac{1 - Kx(t_{\text{peak}}) - \frac{d^2x(t_{\text{peak}})}{dt^2}}{\frac{dx(t_{\text{peak}})}{dt}} = \frac{1}{\frac{dx(t_{\text{peak}})}{dt}}. \qquad (4.10)$$

$x(t_{\text{peak}})$ is the peak of an arterial pulse, $\frac{d^2x(t_{\text{peak}})}{dt^2}$ is the peak of the arterial pulse's second-order derivative, and $\frac{dx(t_{\text{peak}})}{dt}$ is the first derivative of the pulse signal. The friction coefficient, b, is assumed to be proportional to the fluid viscosity based on the fanning friction formula (White, 1999) and therefore the equation can be written as

$$\alpha_1 = R\sqrt{\frac{\omega\rho}{b}}. \qquad (4.11)$$

2. **Womersley number (α_2):** PPG can be utilized for the evaluation of cardiovascular indicators such as BP and blood viscosity. Unprecedentedly, PPG and its effect on the changes in fluid viscosity have not been explored since Njoum and Kyriacou (2017) unraveled their relationship and suggest that PPG provides a plausible solution to assess blood viscosity in a noninvasive way. It is also reasoned the experimental relation of blood rheology with the AC and DC components of the PPG signal experimentally. Therefore the AC component

of the pulsatile signal was considered as a surrogate for blood viscosity [$\mu$is the pulse signal's AC amplitude, R is approximated to be the minimum of the pulse signal, and ω is the frequency of heart rate (HR)]. Also, Meu (μ) and ω are considered as separate features.

$$\alpha_2 = R\sqrt{\frac{\omega\rho}{\mu}}. \tag{4.12}$$

6. Features based on the ECG signal

The features acquired from the ECG signal are exemplified in Fig. 4.9. The time-domain features like QRS complex, PR interval, the amplitude of P, Q, R, S, T, HR, RR interval, QR interval, QT interval, QT interval corrected for HR, TQ interval, systolic–diastolic interval (SDI), and SDI corrected for HR were extracted from the ECG signal (Hall, 1995).

1. **PR interval:** The time interval between the P wave and the start of the QRS complex is said to be the PR interval. The normal PR interval is about 0.16 s.

2. **QT interval:** The time interval between the start of Q wave to the end of T wave is called the QT interval and it is about 0.35 s.

3. **RR interval:** The time interval between two successive heart beats is said to be RR interval.

4. **P:** The P wave is caused by the atrial depolarization and the duration is around 0.08–0.10 s.

5. **QRS:** The QRS complex is caused due to the ventricular depolarization. It lasts around 0.10–0.12 s.

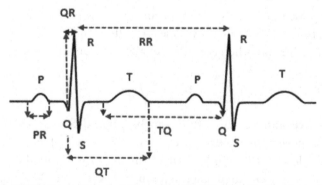

Figure 4.9 ECG features. *Adapted from Thambiraj G., Gandhi, U., Mangalnathan U., Jeya Maria J., and Anand M. (2020). Investigation on the effect of Womersley number, ECG and PPG features for cuff less blood pressure estimation using machine learning,* Biomedical Signal Processing Control 60, 101942.

6. **T wave:** The T wave happens when the ventricle recovers from the depolarization known as a repolarization wave. It occurs after 0.25—0.35 s of depolarization.

7. **HR:** The HR is the reciprocal of the time interval between two successive heartbeats. The interval between two consecutive QRS complexes in the subject is about 0.83 s.

8. **QTc interval:**

    The QTc interval is correlated with MBP (Satpathy et al., 2018), therefore Bazzet's formula is implemented to obtain QTc, and it is given as

$$QT_c = \frac{QT}{\sqrt{RR\ interval}}. \tag{4.13}$$

9. **SDI:**

    SDI features were extracted based on two methods proposed in the literature (Fossa et al., 2007; Imam et al., 2015). Fossa et al. (2007) define SDI as a ratio of QT and TQ interval within each cardiac cycle. Imam et al (2015). describe the ratio of QT and RR interval, as SDI which is mentioned as SDIr, in this study. Besides, SDIc is defined as the corrected SDI for HR.

7. **PPG features**

    The PPG signal features (20 features) were obtained based on Liu et al. (2017) suggestions. To name a few, systolic upstroke time (ST), diastolic upstroke time (DT), and the ratio of ST and DT.

1. **ST**: It is the time interval between the start of the pulse and the peak value of the pulse. It is calculated by finding the time interval between the max peak value location and min peak value location.

2. **DT**: It is the time interval between the peak and end of a pulse. The time interval between adjacent max and min peak gives this value.

3. **ST (10, 25, 33, 50, 66, 75)**: It is the time interval in the systolic range where the amplitude of PPG is 10%, 25%, 33%, 50%, 66%, and 75% of peak value. The calculation was performed as follows:

    First, the peak value of the signal is calculated. Then, all PPG values lying in between the min and max peak (systolic range) are copied onto a temporary array. From peak, 10% of peak is found out, and the temporary array containing PPG values is checked for exact presence of 10% value or a value which is closest and slightly lower than the required value.

4. **DT (10, 25, 33, 50, 66, 75)**: It is the time interval in the diastolic range where the amplitude of PPG is 10%, 25%, 33%, 50%, 66%, and 75% of peak value. It is calculated as the systolic time @ different

values. The PPG signal features suggested by Liu et al. were extracted and some of the features were not considered due to its values which were approximately zero.

Hence, the features considered to be systolic width @ 10% of the finger PPG magnitude, addition of ST and DT @ 10% of the PPG magnitude, division of ST, DT @ 10% of the PPG magnitude, addition of ST and DT @ 10% of the PPG magnitude, division of ST and DT @ 10% of the PPG magnitude, addition of ST and DT @ 33% of the PPG magnitude, division of ST and DT @ 33% of the PPG magnitude, addition of ST and DT @ 50% of the PPG magnitude, division of ST and DT @ 50% of the PPG magnitude, ST, and DT.

The features with physiological significance were obtained from pulse signals are shown in Fig. 4.10. The features extracted were based on Kachuee et al. (2017).

5. **AI:** Augmentation index represents the wave reflection on the arteries. It is defined as the ratio between the maximum peak of the finger PPG and the first inflection point and it given as

$$AI = a/b. \tag{4.14}$$

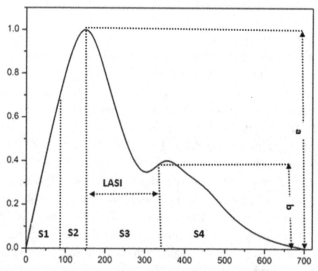

Figure 4.10 Physiological features. *Adapted from Thambiraj G., Gandhi, U., Mangalnathan U., Jeya Maria J., and Anand M. (2020). Investigation on the effect of Womersley number, ECG and PPG features for cuff less blood pressure estimation using machine learning,* Biomedical Signal Processing Control 60, 101942.

6. **Large artery stiffness index (LASI)**: LASI is an index of the arterial stiffness, and it is inversely related to the time interval between the systolic peak and the point where the curvature is zero (first inflection point) (Fig. 4.10).
7. **Inflection Point Area (IPA)**: IPA is defined as the areas under the PPG curve between selected points, denoted by S1, S2, S3, and S4. IPA provides information about the total systemic resistance. The points chosen in this case are the inflection points, that is, the points where the second derivative of PPG (SDPPG) are zero.

4.3.3.2 Regression methods

The extracted features were given to various ML models to analyze the possible relation between features and BP. All 43 features were trained and tested using various regression techniques. The separate models were trained for the targets to estimate such as SBP, DBP, and MBP. The following regression methods such as linear, ridge, support vector, adaboost, and random forest (RF) regression were considered.

4.3.3.3 Feature selection

All 43 handcrafted features given to the RF model provided the better MAE than other regression methods. Yet, there can be an irrelevant feature which does not contribute much to the model's accuracy. Hence, to identify such features in the proposed feature set, binary genetic algorithm (GA) was implemented which further reduces the computational cost. It returns the fittest individual in the population.

The ML algorithm that results in the least MAE is considered to combine with a GA to obtain the feature subset separately for SBP, DBP, and MBP (mean BP). In this work the feature exploration using binary GA is implemented to find the best combination of features by removing the least significant features and effectively constructing a model for better estimation of BP. Initially, the population was randomly generated and updated at every iteration. The length of the chromosome is 43, which consists of 0 and 1 indicates its absence and presence of features, respectively. The probability rate of single point crossover and mutation is 1 and 0.25, respectively. The workstation used here to train and test the model was the Nvidia V-100 consisting of Quadro GV100 GPU with 125 GB RAM. The fitness function of GA is considered to be the inverse of MAE.

4.3.4 Deep network

DL is the buzzword in a recent technological community. The neural networks with more than two layers are termed as deep networks which is a specific subset of ML. Implications of DL are widely adopted in the field of radiology, pathology, drug discovery, medical informatics, natural language processing, image classification, and object detection. The most commonly used deep network for spatial features-based classification is convolutional networks; similarly for sequence prediction, LSTM networks are widely used.

The visualization of the proposed DL is shown in Fig. 4.11. We propose a novel neural network architecture for prediction of SBP and DBP. The input is a 2*1000 data consisting of 1000 samples of ECG and PPG signals as explained in the previous section. The proposed architecture consists of four parts: (1) convolutional block; (2) LSTM block; (3) aggregation of features; and (4) regression layer. Blocks 1 and 2 contain a 2D convolutional layer, with a set of 2×2 convolutional filters, stride of 1, and a padding of 2 across each direction. The number of channels is set as 32 in the first block and 64 in the second block. The convolutional layer in each block is followed by a batch normalization and ReLU activation. The output of the convolutional block consists of 32,256 extracted

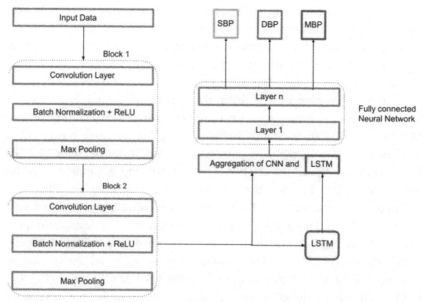

Figure 4.11 Block diagram of the proposed DL network (CNN and LSTM combined).

features. In the LSTM block the number of features in the hidden state is set as 100. A dropout layer with a dropout probability of 0.3 is also introduced to prevent over fitting. Then the CNN and LSTM block combined together to aggregate the spatial and temporal features which is further given to the fully connected network for regression. The dataset is divided into 80% for training, 15% for validation, and 5% for testing. Ten-fold cross-validation is performed on the training data to enhance the training accuracy. The splitting of the dataset has been carried out meticulously to avoid overlapping of data.

4.4 Results

This section of the chapter discusses the results of the regression and DL model in each of the following subsections.

4.4.1 Results of regression models

Table 4.3 exemplifies the various ML models output in terms of MAE. Different feature set combinations were investigated to demonstrate improved performance due to the proposed features on various regression methods. It is evident that PTT, ECG features, and alpha provide better BP estimation which is validated by the least MAE of the RF regression model.

4.4.1.1 Results of genetic algorithm

The GA is implemented for the dataset of size 100*43, a feature subset was acquired, and the results were compared. The Bland−Altman plot is shown in Fig. 4.12A−F for both before and after applying GA to the feature set, signifying the mean difference with the limits of agreement along with an ICC estimates. Fig. 4.13A−C demonstrates that before applying the GA's optimal feature set the mean difference and limits of agreement are too high, reflecting its unsuitability in clinical BP measurements. However, after applying GA's optimal feature set, the mean difference and limits of agreement are reduced twice as that of before GA is shown in Fig. 4.13D−F. The limits of agreement and the mean difference were within the clinically acceptable range with the optimal feature set. Fig. 4.13 explains the convergence plot of GA algorithm.

Table 4.4 illustrates the different optimal feature set obtained using GA for SBP, DBP, and MBP. The size of the dataset was reduced to 100*43 due to the computational cost and fed into GA with the RF

Table 4.3 ML models performance on different feature set.

BP	Feature set	Machine learning techniques				
		Linear regression	Ridge regression	Support vector regression	Adaboost regression	Random forest regression
		MAE	MAE	MAE	MAE	MAE
SBP	ALL	14.05	14.23	14.85	14.20	10.83
	PTT + PPG	15.27	15.19	15.14	15.63	14.14
	Physiological	15.13	15.10	14.70	16.90	11.89
	PTT + ECG + alpha	15.06	15.08	14.94	16.18	11.16
DBP	ALL	10.76	10.94	10.98	11.99	8.43
	PTT + PPG	11.37	11.41	11.17	12.52	10.58
	Physiological	11.06	11.40	11.08	11.81	9.38
	PTT + ECG + alpha	11.35	11.21	11.05	12.11	8.55
MBP	ALL	10.44	10.59	10.70	10.95	8.35
	PTT + PPG	11.19	11.14	10.95	12.06	10.22
	Physiological	10.86	11.09	10.77	11.27	9.11
	PTT + ECG + alpha	11.08	10.95	10.75	11.96	8.45

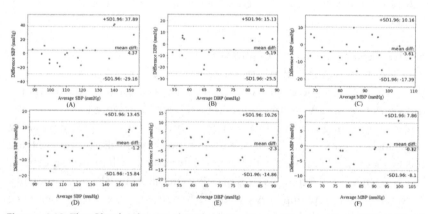

Figure 4.12 The Bland–Altman plot showing mean difference and 95% limit of agreement for before applying GA: (A) SBP, (B) DBP, (C) MBP and after applying GA (D) SBP, (E) DBP, and (F) MBP.

regression to obtain the optimal feature set, which results in many ECG features and the proposed feature, alpha. It was also seen in the reduction in MAE of SBP, DBP, and MBP when providing an optimal feature set

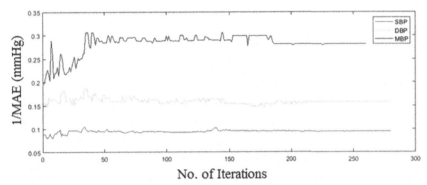

No. of Iterations

Figure 4.13 Convergence plot of optimization technique.

Table 4.4 Genetic algorithm's optimal feature set (Thambiraj et al., 2020).

	Genetic algorithm's optimal feature set
SBP	PTTs, I_H,Systolic width @10%,@ 10% ST + DT,@ 10% DT/ST,@ 25% ST + DT,@25% DT/ST,@ 50% DT/ST, AI, LASI, S1, S2, S4, PATmin, Meun, Alpha1, SDIr, QT, QRS, P, S, T, HR
DBP	I_H, I_LSystolic width @10%,@ 10% ST + DT,@ 10% DT/ST,@ 25% ST + DT,@25% DT/ST,AI, Meun, Omega, Alpha1, SDI, QTc, QRS, Q, R, T
MBP	PIR, I_H, I_LSystolic width @10%,@ 25% ST + DT,@25% DT/ST, @33% DT/ST,@ 50% DT/ST,Diastolic time, AI, S1, PATs, PATmin, PATmax, Omega, Alpha2, SDIr, QTc, QRS, P

Table 4.5 Comparison of all the features with GA's optimal feature set (Thambiraj et al., 2020).

RF model	43 features		GA's optimal subset	
	MAE	ICC and 95% CI	MAE	ICC and 95% CI
SBP	13.53	0.78 (0.46−0.91)	9.54	0.89 (0.72−0.95)
DBP	9.51	0.72 (0.30−0.88)	5.48	0.90 (0.76−0.96)
MBP	6.70	0.91 (0.75−0.96)	3.27	0.97 (0.93−0.98)

as an input compared with a complete feature set. The GA algorithm results further conclude that some of the features from the ECG signal were strongly associated with BP.

It is apparent from Table 4.5 that the feature set before and after GA exhibit the perceptible difference in metrics like MAE, ICC estimates and

95% CI. ICC has demonstrated moderate reliability with every feature extracted from the signal, whereas, after the optimal feature set from GA, ICC indicates excellent reliability.

4.4.2 Deep learning model

In this work, the first 1000 samples of ECG and PPG signals are concatenated to form a 1000*2 matrix instead of using the ECG or PPG signals separately. An ablation study is performed for the proposed network, and it is tabulated in Table 4.6. It is observable that there is a vast improvement in the performance when the dataset is fed as a combination of ECG and PPG signals.

4.4.2.1 Comparison with other existing networks

BP-Net is compared with some of the widely used network architecture in DL, such as ResNet -18 (He et al. 2016), VGGNet-16 (Simonyan & Zisserman, 2015), and GoogleNet (Szegedy et al., 2015). ImageNet Large Scale Visual Recognition Challenge (ILSVRC) is a software contest where different architectures are proposed to detect and classify images precisely. GoogleNet and VGGNet is proposed in the ILSVRC contest, 2014. GoogleNet architecture has 22 layers, with 4 million parameters achieving the detection error of 6.67%, while VGGNet achieved the error rate of 7.3%. ResNet proposed by He et al., in the ILSVRC contest 2015, is the CNN-based architecture with skip connections, and this network trains the neural network with 152 layers, achieving an error rate of 3.57%. Table 4.7 shows the performance comparison of BP-Net with these existing networks.

An assessment and comparison of results obtained using the BP-Net model have been made with the AAMI criterion. Table 4.8 illustrates the tabulation of the same. From Table 4.8, it is evident that our results are within the range specified by AAMI for DBP and MBP prediction. However, the STD of SBP is slightly out of the acceptable limit of AAMI standards. This may be due to the variability associated with the SBP. Also, the number of a test subject is around 251, which is much greater than the required criterion, thereby ensuring the strong reliability of our proposed work.

4.5 Discussion

The prevailing BP methods for continuous monitoring in clinical practice is based on cuff-based methods. However, it causes discomfort and provides intermittent values which negates the dynamic interaction involved in its

Table 4.6 Ablation study.

Input	Learning method	Method (input)	SBP		DBP		MBP	
			MAE	RMSE	MAE	RMSE	MAE	RMSE
ECG	MTL	CNN	9.63	12.08	6.74	8.56	6.68	8.31
PPG		CNN	37.69	38.94	21.70	22.91	25.18	26.17
ECG + PPG	MTL	CNN	9.14	10.90	7.58	9.59	7.06	8.75
ECG + PPG		LSTM	9.71	11.21	7.65	10.32	7.58	10.01
ECG + PPG	MTL	CNN + LSTM	10.71	12.73	7.41	8.91	7.76	9.36
ECG + PPG	STL	CNN + LSTM + BN	9.20	10.94	6.96	8.61	8.03	9.95
ECG + PPG	MTL	BP-Net	8.14	10.54	6.47	8.25	6.26	7.85

Table 4.7 Performance comparison of BP-Net with ResNet-18, VGGNet-16, and Google Net.

Network	SBP		DBP		MBP	
	MAE (mmHg)	SD	MAE (mmHg)	SD	MAE (mmHg)	SD
VGGNet-16	10.12	8.05	6.82	4.96	6.67	4.52
GoogleNet	8.51	5.98	6.80	4.96	6.50	4.39
Res Net-18	9.02	6.65	7.23	4.79	7.45	5.10
BP-Net (the proposed network)	8.14	5.90	6.47	4.34	6.26	4.10

Table 4.8 Proposed network's validation using AAMI standards.

		Mean difference (mmHg)	STD (mmHg)	Subjects
Proposed results	Systolic blood pressure	− 0.96	10.02	251
	Diastolic blood pressure	− 1.18	7.71	251
	Mean blood pressure	− 0.41	7.48	251
AAMI	SBP and DBP	≤ 5	≤ 8	≥ 85

regulation. Therefore in recent years there is a paradigm shift in estimating BP continuously using the PTT principle that resolves the challenges posed by the cuff-based methods. Though the analytical model based on PTT provides us the continuous assessment, it is still overshadowed by its limitations such as frequent recalibration and incorporation of the intrinsic features implying the BP's physiological regulation was not feasible. Unlike analytical driven models, ML which is a subset of artificial intelligence learns a complex feature involving physiological regulation from biosignals when it is provided with good training. Yet, feature extraction is the tedious process and cost computational power using regression methods, hence DL comes into picture, which could process the raw bio signals retaining the calibration-free, continuous,s and accurate BP assessment.

In this work, the data collected from the subjects in SMC and UCI online dataset was also considered to estimate BP using data-driven methods. Initially, PTT and HR from both the dataset was extracted and fed into the simple analytical model. The result demonstrates that PTT with HR model provides least MAE rather than the model with PTT alone in both the dataset. This exemplifies the significance of other regulating factors to be included in the model for the reliable tracking of BP. It is also obvious that availability of lesser training data increases the MAE on SMC dataset. Hence, for further analysis using regression and DL models SMC dataset was not considered as it needs huge data for training to make the model robust and generalized one.

So far in the literature, only PPG features were considered as it has better correlation with BP. However, in this work, an effort was taken to analyze the relation between ECG time-domain features and BP. It is observed that both ECG and PPG features were contributing to the improvement in BP's accuracy. To further demonstrate the role of ECG features in BP estimation, we implemented GA as a feature selection technique. The results revealed that apart from PPG features such as S1, S2, and S4, 25% ST + DT time, PTT, PIR, etc., ECG features were also present. The features such as QRS, P, and T amplitude, QT interval proved to have a strong association with the hypertensive heart diseases. Apart from this, the proposed feature Womersley number was also considered a significant factor in estimating SBP, DBP, and MBP, showing its strong relationship with BP. The features chosen by GA to estimate SBP, DBP, and MBP were not the same is also in accord with the literature.

Tables 4.9 and 4.10 describe the proposed work comparison with few other studies in the literature. Compared with other studies in the literature, proposed features to ML methods and DL model has a profound impact in improving BP estimation.

Though the regression methods were quite promising to estimate a calibration-free BP, it requires a feature extraction and frequency-domain, complexity features were also not considered in analyzing its relation with BP in this work. Nonetheless, it was overshadowed by the DL model which considered raw bio signals to learn all the hidden features involved in governing BP. In the proposed work, we combined both CNN and LSTM to obtain spatial and temporal information.

From the ablation study it was obvious that using combined CNN and LSTM results in the least MAE rather than using either CNN or LSTM network. The proposed work also uses multitasking which helps to learn the model in a better way to improve results can also be perceived from Table. The model's performance was also abided by AAMI standards.

Table 4.9 Comparison of the proposed work with studies in the literature.

Work	Method	SBP				DBP				MBP			
		MAE	RMSE	r	Bias & 95% CI	MAE	RMSE	r	Bias & 95% CI	MAE	RMSE	r	Bias & 95% CI
Kurylyak et al. (2013)	SVR	13.6	11.6	—	—	7.7	7.9	—	—	—	—	—	—
Kachuee et al. (2015)	SVR	12.38	—	—	—	6.34	—	—	—	—	—	—	—
Kachuee et al., 2017	Random forest	11.17	—	0.59	—	5.35	—	0.48	—	5.92	—	0.56	—
Liu et al. (2017)	SVR	8.54	10.9	—	—	4.34	5.8	—	—	—	—	—	—
Attarpour et al. (2019)	ANN	4.94	5.87	0.94	—	4.03	5.50	0.84	—	—	—	—	—
Proposed work	Random forest	9.54	13.83	0.85	−1.2 & (9.35, 5.26)	5.48	6.80	0.84	−2.30 & (0.77, −5.37)	3.27	4.0	0.94	−0.11 & (1.83, −2.07)

Table 4.10 Comparison with other DL works in the literature.

Work	Physiological signals	Deep network	Subjects	MAE SBP	MAE DBP	MAE MBP	RMSE SBP	RMSE DBP	RMSE MBP
Yan et al. (2019)	ECG and PPG (calibration free)	CNN	MIMIC II 604 subjects	12.49	8.03	–	–	–	–
Slapni Č Ar et al. (2019)	PPG, 1dPPG, and 2dPPG (without personalization)	Spectro temporal ResNet	MIMIC III 510 subjects	15.41	12.38	–	–	–	–
Ghosh et al. (2018)	ECG, PPG, PTT, Accelerometric data, Gyroscopic data	LSTM	50	–	–	–	4.83	3.95	–
Su et al. (2018)	Features from ECG and PPG are given to the network (PTT approach)	LSTM	84	–	–	–	3.90	2.66	–
Xing and Sun (2016)	Features from PPG using FFT	ANN	MIMIC III 69 subjects and 23	–	–	–	0.06	0.01	–
Tanveer and Hasan (2019)	Features from ECG and PPG	ANN + LSTM	MIMIC I (39 subjects)	0.92	0.51	–	1.26	0.72	–
Baek (2019)	ECG and PGG (raw signals & calibration free)	CNN	MIMIC II (379)	9.30	5.12	–	–	–	–
BP-Net (ours)	ECG and PPG signals (raw signals)	CNN + LSTM	MIMIC II (251)	8.14	6.47	6.26	10.54	8.25	7.85

4.6 Limitations

This study has its own limitations. Concerning the features from bio signals, only the time-domain features are used in this thesis. However, it was apparent from the literature that frequency-domain and complexity

---

BOX 4.1 Technological opportunities

Technological opportunities

Exploring the adversarial networks further and combine with the clinical data such as height and weight for a reliable and continuous assessment of BP.

Drawbacks

Dataset employed do not contain much information about the subjects and hence the optimal feature set by the GA cannot be made as a generalized feature set for healthy and diseased conditions.

The technique is validated on the fewer dataset.

Better tuning of hyperparameters in the deep neural network.

Challenges

Computational cost involved in choosing the optimal feature set using the genetic algorithm

Benefits

Reduce the computational cost as DL model has the inherent ability to learn from the raw signals to estimate BP.

BOX 4.2 Impacts (economic and environmental assessment)
Table 4.11

Table 4.11 Environmental/economic/technical indicators.

Category	Criteria	Indicator
Environmental	—	—
Economic	Mitigate the nation's burden on the cost of HTN and its diagnosis and treatment	—
Technical	Improvement in accuracy	MAE (mmHg)
Social	Awareness	Ambulatory BP monitoring in practice
	Affordable healthcare	

features can also contribute to the estimated BP using regression techniques. Optimization technique must be employed on the larger dataset encompassing to obtain the reliable feature sets for BP assessment. Exploration of various other deep networks might improve BP accuracy. The technological oppurtunities, drawbacks of the present work and impact of this research are explained in the Box 4.1 and Box 4.2 respectively.

4.7 Conclusions

Calibration, one of the drawbacks involved in analytical models, is dealt with using data-driven models such as regression and deep network. Various time-domain features of the ECG and PPG signals (an online database) were extracted and fed to the feature selection algorithms. The Womersley number (α) with other PPG and ECG features in the time-domain was obtained, and its relevance with BP was also studied. The GA was applied and it gave the most important features that improved the model's performance in estimating BP compared with all the features given to the model. Separate optimal feature sets were obtained for SBP, DBP, and MBP, which were further fed into the RF regression model to estimate the continuous BP. The manual feature extraction was performed in the regression models to estimate the BP continuously. However, there are many more immanent factors of the physiological signals that were not considered in the regression models, which may contribute to accuracy. Therefore raw signals were provided to the deep CNN and LSTM network, tracking the BP with the least error and computation time. The estimated SBP, DBP, and MBP also complied with the AAMI standards. To reduce the mortality rate that arises from HTN comorbid conditions, a cuffless and noninvasive BP device must be designed with the proposed algorithm to give a continuous, reliable, and accurate measurement.

References

Attarpour, A., Mahnam, A., Aminitabar, A., & Samani, H. (2019). *Cuff-less continuous measurement of blood pressure using wrist and fingertip photo-plethysmograms: Evaluation and feature analysis. Biomedical Signal Processing Control* (p. 49).

Baek, S. (2019). End-to-end blood pressure prediction via fully convolutional networks. *IEEE Access*, 7, 185458–185468.

Booth, J. (1977). A short history of blood pressure measurement. *Proceedings of the Royal Society of Medicine*, 70(11), 793–799.

Cattivelli, F.S. & Garudadri, H. (2009). Noninvasive cuffless estimation of blood pressure from pulse arrival time and heart rate with adaptive calibration. In: *Proceedings 2009 6th international workshop on wearable and implantable body sensor networks* (Vol. 1, pp. 114–119).

Christian Szegedy, Liu, W., Jia, Y., Sermanet, P., Reed, S., Anguelov, D., ... Rabinovich, et al. (2015). Going deeper with convolutions. In: *Proceedings of the IEEE conference on computer vision and pattern recognition.*

Chua, C.P. & Heneghan, C. (2006). Continuous blood pressure monitoring using ECG and finger photoplethysmogram. In: *Proceedings of the annual international conference of the IEEE engineering in medicine and biology* (Vol. 1, pp. 5117–5120).

Chung, E., Chen, G., Alexander, B., & Cannesson, M. (2013). Non-invasive continuous blood pressure monitoring: A review of current applications. *Frontiers of Medicine in China,* 7(1), 91–101.

Dadlani, A., Madan, K., & Sawhney, J. P. S. (2019). Ambulatory blood pressure monitoring in clinical practice. *Indian Heart Journal,* 71(1), 91–97.

Ding, X.R. & Zhang, Y.T. (2015). Photoplethysmogram intensity ratio: A potential indicator for improving the accuracy of PTT-based cuffless blood pressure estimation. In: *Proceedings of the annual international conference of the IEEE engineering in medicine and biology society* (pp. 398–401).

Drzewiecki, G. M., Melbin, J., & Noordergraaf, A. (1983). Arterial tonometry: Review and analysis. *Journal of Biomechanics,* 16, 141–152.

Escobar-Restrepo, B., Torres-Villa, R., Kyriacou, P. A., Zheng, D., Chen, F., Mukkamala, R., & Pan, F. (2018). Evaluation of the linear relationship between pulse arrival time and blood pressure in ICU patients: Potential and limitations. *Frontiers in Physiology,* 1–9.

Forouzanfar, M., Dajani, H. R., Groza, V. Z., Bolic, M., Rajan, S., & Batkin, I. (2015). Oscillometric blood pressure estimation: Past, present, and future. *IEEE Reviews in Biomedical Engineering,* 8, 44–63.

Fossa, A. A., Wisialowski, T., Crimin, K., Ph, D., Wolfgang, E., Couderc, J., et al. (2007). Analyses of dynamic beat-to-beat QT − TQ interval (ECG restitution) changes in humans under normal sinus rhythm and prior to an event of Torsades de Pointes during QT prolongation caused by sotalol. *Annals of Noninvasive Electrocardiology,* 12(4), 338–348.

Fung, P., Dumont, G., Ries, C., Mott, C. & Ansermino, M. (2004). Continuous noninvasive blood pressure measurement by pulse transit time. In: *Proceedings of the annual international conference of the IEEE engineering in medicine and biology* (Vol. 26 I(4), pp. 738–741).

Gesche, H., Grosskurth, D., Küchler, G., & Patzak, A. (2011). Continuous blood pressure measurement by using the pulse transit time: Comparison to a cuff-based method. *European Journal of Applied Physiology,* 112(1), 309–315.

Ghosh, S., Banerjee, A., Ray, N., Wood, P.W., Boulanger, P. & Padwal, R. (2018). Using accelerometric and gyroscopic data to improve blood pressure prediction from pulse transit time using recurrent neural network. In: *Proceedings of the IEEE international conference on acoustics, speech and signal processing* (pp. 935–939).

Goldberger, A. L., Amaral, L. A. N., Glass, L., Hausdorff, J. M., et al. (2000). PhysioBank, PhysioToolkit and PhysioNet components of a new research resource for complex physiologic signals ary. *Circulation,* 215–220.

Hall, J. (1995). *Guyon and Hall Textbook of Medical Physiology.* Elsevier.

Handler, J. (2009). The importance of accurate blood pressure measurement. *The Permanente Journal,* 13(3), 51–54.

He, K., Zhang, X., Ren, S. and Sun, J. "Deep residual learning for image recognition," 2016 IEEE Conference on Computer Vision and Pattern Recognition (CVPR), 2016, pp. 770–778, https://doi.org/10.1109/CVPR.2016.90.

Hughes, D. J., Babbs., Charles, F., Geddes, L. A., & Bourland, J. D. (1979). *Measurements of Young's modulus of elasticity of the canine aorta with ultrasound.* Weldon School of Biomedical Engineering Faculty Publications.

Imam, M. H., Karmakar, C. K., Jelinek, H. F., Palaniswami, M., & Khandoker, A. H. (2015). Analyzing systolic—Diastolic interval interaction characteristics in diabetic cardiac autonomic neuropathy progression. *IEEE Journal of Translational Engineering in Health, 3.*

Kachuee, M., Kiani, M.M., Mohammadzade, H. & Shabany, M. (2015) Cuff-less high-accuracy calibration-free blood pressure estimation using pulse transit time. In: *Proceedings of the IEEE international symposium on circuits and systems* (pp. 1006-1009).

Kachuee, M., Kiani, M. M., Mohammadzade, H., & Shabany, M. (2017). Cuffless blood pressure estimation algorithms for continuous health-care monitoring. *IEEE Transactions on Biomedical Engineering, 64*(4), 859—869.

Karmakar, N., Nag, K., Saha, I., Parthasarathi, R., Patra, M., & Sinha, R. (2018). Awareness, treatment, and control of hypertension among adult population in a rural community of Singur block, Hooghly District, West Bengal. *Journal of Education and Health Promotion, 7,* 1—6.

Kazemi Korayem, A., Ghamami, S., & Bahrami, Z. (2018). Fractal properties and morphological investigation of nano hydrochlorothiazide is used to treat hypertension. *BMC Pharmacology and Toxicology, 19*(1), 1—9.

Khalid, S. G., Zhang, J., Chen, F., & Zheng, D. (2018). Blood pressure estimation using photoplethysmography only: Comparison between different machine learning approaches. *Hindawi,* 2018.

Kim, J. S., Kim, K. K., Baek, H. J., & Park, K. S. (2008). Effect of confounding factors on blood pressure estimation using pulse arrival time. *Physiological Measurement, 29*(5), 615—624.

Kurylyak, Y., Lamonaca, F. & Grimaldi, D. (2013). A neural network-based method for continuous blood pressure estimation from a PPG signal. In: *Proceedings of IEEE instrumentation and measurement technology conference* (pp. 280—283).

Liu, M., Po, L.-M., & Fu, H. (2017). Cuffless blood pressure estimation based on photoplethysmography signal and its second derivative. *International Journal of Computer Theory and Engineering, 9*(3), 202—206.

McCarthy, B. M., O'Flynn, B., & Mathewson, A. (2011). An investigation of pulse transit time as a non-invasive blood pressure measurement method. *Journal of Physics Conference Series, 307*(1), 1—5.

Monte-Moreno, E. (2011). Non-invasive estimate of blood glucose and blood pressure from a photoplethysmograph by means of machine learning techniques. *Artificial Intelligence in Medicine, 53*(2), 127—138.

Muehlsteff, J., Aubert, X.L. & Schuett, M. (2006). Cuffless estimation of systolic blood pressure for short effort bicycle tests: The prominent role of the pre-ejection period. In: *Proceedings of the annual international conference of the IEEE engineering in medicine and biology* (pp. 5088—5092).

Njoum, H., & Kyriacou, P. A. (2017). Photoplethysmography for the assessment of hemorheology. *Scientific Reports, 7*(1), 1—11.

Payne, R. A., Symeonides, C. N., Webb, D. J., & Maxwell, S. R. J. (2006). Pulse transit time measured from the ECG: An unreliable marker of beat-to-beat blood pressure. *Journal of Applied Physiology, 100*(1), 136—141.

Penaz J. (1973). Photoelectric measurement of blood pressure, volume and flow in the finger. In: *Digest of the 10th international conference on medical and biological engineering.*

Ruiz-Rodríguez, J. C., Ruiz-Sanmartín, A., Ribas, V., Caballero, J., García-Roche, A., Riera, J., Nuvials, X., et al. (2013). Innovative continuous non-invasive cuffless blood pressure monitoring based on photoplethysmography technology. *Intensive Care Medicine, 39*(9), 1618—1625.

Satpathy, S., Satpathy, S., & Nayak, P. K. (2018). Correlation of blood pressure and QT interval. *National Journal of Physiology, Pharmacy and Pharmacology, 8*(2), 207−210.

Simonyan, K. & Zisserman, A. (2015). Very deep convolutional networks for large-scale image recognition. In: *3rd international conference on learning representations* (pp. 1−14).

Slapni Č Ar, G., Mlakar, N., & Luštrek, M. (2019). Blood pressure estimation from photoplethysmogram using a spectro-temporal deep neural network. *Sensors, 19*(15).

Song, K., Chung, K., & Chang, J.-H. (2019). Cuff-less deep learning-based blood pressure estimation for smart wristwatches. *IEEE Transactions on Instrumentation and Measurement, 69*(7), 4292−4302.

Su, P., Ding, X.R., Zhang, Y.T., Liu, J., Miao, F. & Zhao, N. (2018). Long-term blood pressure prediction with deep recurrent neural networks. In: *Proceedings of the IEEE EMBS international conference on biomedical and health informatics* (pp. 323−328).

Tanveer, M. S., & Hasan, M. K. (2019). Cuffless blood pressure estimation from electrocardiogram and photoplethysmogram using waveform-based ANN-LSTM network. *Biomedical Signal Processing and Control., 51*, 382−392.

Teng, X.F. & Zhang, Y.T. (2003). Continuous and noninvasive estimation of arterial blood pressure using a photoplethysmographic approach. In: *Proceedings of the annual international conference of the IEEE engineering in medicine and biology* (Vol. 4, pp. 3153−3156).

Thambiraj, G., Gandhi, U., Mangalnathan, U., Jeya Maria, J., & Anand, M. (2020). Investigation on the effect of Womersley number, ECG and PPG features for cuff less blood pressure estimation using machine learning. *Biomedical Signal Processing and Control., 60*, 101942.

Tijsseling, A.S. & Anderson, A. (2012). A. Isebree Moens and D.J. Korteweg: On the speed of propagation of waves in elastic tubes. In: *Proceedings of 11th international conferences on pressure surges* (pp. 227−245).

Vlachopoulos, C., O'Rourke, M., & Nichols, W. W. (2011). *McDonald's blood flow in arteries: Theoretical, experimental and clinical principles*. Boca Raton, FL, USA: CRC Press.

Wang, Y.-Y. L., Chang, C. C., Chen, J. C., Hsiu, H., & Wang, W. K. (1997). Pressure wave propagation in arteries. A model with radial dilatation for simulating the behavior of a real artery. *IEEE Engineering in Medicine and Biology Magazine, 16* (1), 51−54.

Wei, C. C. (2011). An innovative method to measure the peripheral arterial elasticity: Spring constant modeling based on the arterial pressure wave with radial vibration. *Annals of Biomedical Engineering, 39*(11), 2695−2705.

Welykholowa, K., Hosanee, M., Chan, G., Cooper, R., Kyriacou, P. A., Zheng, D., Allen, J., et al. (2020). Multimodal photoplethysmography-based approaches for improved detection of hypertension. *Clinical Medicine, 9*, 1−20.

White, F. M. (1999). *Fluid mechanics*. Boston: Mass: WCB/McGraw-Hill.

Wibmer, T., Doering, K., Kropf-Sanchen, C., Rüdiger, S., Blanta, I., Stoiber, K. M., Rottbauer, W., et al. (2014). Pulse transit time and blood pressure during cardiopulmonary exercise tests. *Physiological Research, 63*(3), 287−296.

Wong, M.Y.M. & Zhang, Y.T. (2008). The effects of pre-ejection period on the blood pressure estimation using pulse transit time. In: *Proceedings of 5th international workshop on wearable and implantable body sensor networks* (pp. 254−255).

Wong, M. Y. M., Poon, C. C. Y., & Zhang, Y. T. (2009). An evaluation of the cuffless blood pressure estimation based on pulse transit time technique: A half year study on normotensive subjects. *Cardiovascular Engineering, 9*(1), 32−38.

Xing, X., & Sun, M. (2016). Optical blood pressure estimation with photoplethysmography and FFT-based neural networks. *Biomedical Optics Express, 7*(8), 3007.

Yan, C., Li, Z., Zhao, W., Hu, J., Jia, D., Wang, H. & You, T. (2019) Novel deep convolutional neural network for cuff-less blood pressure measurement using ECG and PPG signals. In: *Proceedings of the 41st annual international conference of the IEEE engineering in medicine and biology society* (pp. 1917−1920).

Yoon, Y. Z., Kang, J. M., Kwon, Y., Park, S., Noh, S., Kim, Y., Park, J., & Hwang, S. W. (2018). Cuff-less blood pressure estimation using pulse waveform analysis and pulse arrival time. *IEEE Journal of Biomedical Health Informatics*, *22*(4), 1068−1074.

Yoon, Y., Cho, J. H., & Yoon, G. (2009). Non-constrained blood pressure monitoring using ECG and PPG for personal healthcare. *Journal of Medical Systems*, *33*(4), 261−266.

CHAPTER 5

A recommendation system for the prediction of drug−target associations

Simone Contini and Simona Ester Rombo
Department of Mathematics and Computer Science, University of Palermo, Palermo, Italy

5.1 Introduction

One of the most important challenges in the field of drug discovery is the prediction of new targets for existing drugs (Chong & Sullivan, 2007). Indeed, drug discovery is a costly and time-consuming task, and new drugs have to pass through several steps before being approved. The repurposal of old drugs in the context of new therapies has shown to be a successful approach, allowing conspicuous money and time saving. In the last few years computational techniques have started to be proposed to this aim [see Li et al. (2015) for a good review on the topic and more recent approaches such as Keiser et al. (2009); Olayan et al. (2017); Luo et al. (2018); Wu et al. (2016); Cheng et al. (2018); and Che et al. (2020)].

Keiser et al. (2009) compare both approved and investigational drugs against hundreds of targets, and define each target by its ligands. In particular, they search for chemical similarities between drugs and ligand sets, predicting this way thousands of candidate novel associations. As a final step of the research is presented in Keiser et al. (2009), the authors test experimentally the supposed new drug−target associations, and 23 of them have been confirmed. Interestingly, they point out that the physiological relevance of one, the drug N,N-dimethyltryptamine on serotonergic receptors, has been confirmed in a knockout mouse. Moreover, according to their study such a chemical similarity approach has resulted to be systematic and comprehensive, and may suggest side-effects and new indications for many drugs.

Olayan et al. (2017) try to solve the problem that the computational prediction of drug−target interactions often suffers for high false positive

Edge-of-Things in Personalized Healthcare Support Systems
DOI: https://doi.org/10.1016/B978-0-323-90585-5.00004-7
115

prediction rate. They use a heterogeneous graph that contains known drug—target interactions with multiple similarities between drugs and multiple similarities between target proteins. They apply nonlinear similarity fusion in order to combine different similarities. Before fusion, they heuristically select a subset of similarities and obtain an optimized combination of similarities. Then a random forest model is applied, which uses different graph-based features extracted from the Drug-Target Interactions (DTI) heterogeneous graph. Olayan et al. (2017) show that their approach significantly reduces the false positive prediction rate in novel predictions.

In Luo et al. (2018), the drug repositioning problem is modeled as a recommendation system that recommends novel treatments based on known drug—disease associations. In particular, a heterogeneous drug—disease interaction network is built by integrating drug—drug, disease—disease, and drug—disease networks, obtained from existing curated datasets. Such a network is represented by a large drug—disease adjacency matrix, whose entries include drug pairs, disease pairs, and both known and unknown drug—disease pairs. A fast Singular Value Thresholding algorithm is applied in order to complete the drug—disease adjacency matrix with predicted scores for unknown drug—disease pairs. The authors show the advantages of this approach when compared against its competitors, and they also illustrate how it can be used on specific case studies for several selected drugs.

Wu et al. (2016) present their research for prioritizing potential targets for old drugs, failed drugs, and new chemical entities. The approach proposed incorporates network and chemoinformatics and has shown good performance on four benchmark datasets, covering G protein-coupled receptors, kinases, ion channels, and nuclear receptors. Moreover, the authors discuss how to apply their approach to identify novel anticancer indications for nonsteroidal antiinflammatory drugs by inhibiting AKR1C3, CA9, or CA12.

Cheng et al. (2018) integrate protein—protein interaction (PPI) and large-scale patient-level longitudinal data. They quantify the network proximity of disease genes and drug targets and search for causal relationships. Among the validated predictions, they have found carbamazepine associated with an increased risk of coronary artery disease (CAD) and hydroxychloroquine associated with a decreased risk of CAD.

Che et al. (2020) use existing drug classifications based on their target proteins and apply a multilabel classification model to assign drugs into correct target groups. For each drug property, they build a network and extract suitable drug features from these networks based on which several machine learning algorithms are used to build the classification model.

From the analysis of the literature, the effectiveness of approaches able to integrate different data and model them as networks is evident. This is due to the fact that the way cellular components interact each other often influences the effect of drugs on specific targets.

In this chapter, we show how a recommendation system can be built, relying on a PPI network associations between drugs and targets. The main aim of the chapter is to guide the reader through modeling the problem of drug repurposing as a recommendation system problem. We have downloaded the PPI network from the *Intact* database (Hermjakob et al., 2004) (https://www.ebi.ac.uk/intact/) and the associations between drugs and targets from the *DrugBank* database (Wishart et al., 2006) (https://go.drugbank.com/). The main assumption is that, depending on how proteins are connected on the PPI network, given an input drug new targets can be suggested for further investigation.

A secondary goal of this chapter is showing to the reader how Big Data technologies can be applied in the context of the considered problem, such that the number of available data involving drug–target associations is notably increasing. The framework adopted here is Apache Spark (Zaharia et al., 2010), useful for loading, managing, and manipulating data by means of appropriate resilient distributed datasets (Zaharia et al., 2012), and for providing suitable libraries for the implementation of the alternating least square (ALS) machine learning algorithm, a matrix factorization algorithm for distributed and parallel computing (Ricci et al., 2010).

Finally, how to provide analysis of the obtained results by the creation of interactive graphs that intuitively show interactions between the predicted proteins and the proteins that have a direct relationship with the drug chosen for the recommendation system is also shown.

5.2 Approach

The main idea at the basis of the proposed approach is that novel targets for existing drugs may be searched for among "neighbors" of their known targets in the associated PPI network. It is worth to recall that a PPI network associated with an organism (e.g., human) is a network where nodes represent the proteins of that organism, and there is an edge between two nodes if a physical interaction between the corresponding two proteins has been discovered and experimentally demonstrated in the laboratory

(Panni & Rombo, 2013). Given a node p in the network, we define as its *neighborhood* the set of nodes linked to p by an edge in the network.

For the purposes explained in this chapter, it is reasonable to observe that not every protein that physically interacts with one of the targets of a drug has to be a possible target in its turn. However, if a set of existing drug—target associations is used to train the recommendation system, then only those proteins which effectively present a behavior similar to that of the real targets can be identified. To this aim, as already explained, a parallel version of the ALS machine learning algorithm (Ricci et al., 2010) is used here and recalled below, for the convenience of the reader.

Finally, a visual inspection of the predicted new targets in the network is also provided by the system. The proposed pipeline consists of the four steps summarized in the following and shown in Fig. 5.1:

1. creation, cleaning, filtering, and analysis of the model integrating PPI network and known drug—target associations
2. creation of the ALS model and data fitting
3. implementation and evaluation of the main core recommendation system
4. graphs generation and visual overview

5.2.1 Alternating least square algorithm: matrix factorization

Recommender systems (Aggarwal, 2016) are systems providing a preference rating for a given user with references to a specific item. They are often based on collaborative filters, that is, methods to provide automatic predictions based on the concept of "collaboration" between users. In more detail, if a user U has shown in the past to have common interests with the user V, this behavior can be assumed to be conserved also in the

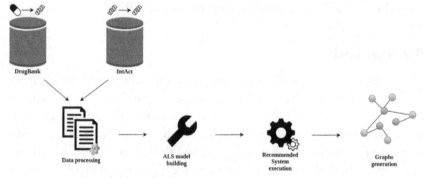

Figure 5.1 Pipeline of the proposed recommendation system.

future, therefore if V will choose an item, the same item can be suggested to U as well. This reasoning is generalized for large sets of interacting users, and the resulting problem may be modeled by *matrix factorization*, working by considering a user—item interaction matrix which is decomposed into the product of two lower dimensionality rectangular matrices. The ALS algorithm provides a way to factorize a given matrix M into two factors U (user matrix) and I (item matrix) such that $M \approx U^T I$. The unknown row dimension is given as a parameter to the algorithm and it is called *latent factors*. An optimization problem is then formulated in order to find the matrices U and I (Ricci et al., 2010).

5.3 Implementation

Python was used as a programming language, and Jupyter Notebook as IDE. The exploited libraries are the following:
- PySpark
- Pandas
- Numpy
- NetworkX
- PyVis

The main steps implementing the proposed pipeline are described in detail below. Attention will be paid to the more technical aspects, in order to guide the reader through the practical use of the adopted Big Data technologies.

5.3.1 Creation, cleaning, filtering, and analysis of PPI and drug—target data

Two datasets are used for this work as follows:
- Intact database (https://www.ebi.ac.uk/intact/) for the human PPI network, containing 15 features [ID(s) interactor A, ID(s) interactor B, Alt. ID(s) interactor A, Alt. ID(s) interactor B, Alias(es) interactor A, Alias(es) interactor B, Interaction detection method(s), Publication 1st author(s), Publication Identifier(s), Taxid interactor A, Taxid interactor B, Interaction type(s), Source database(s), Interaction identifier(s), and Confidence value(s)] and 1,054,920 records
- DrugBank database (https://go.drugbank.com/) for direct associations between drug and target, containing five features (DrugBank ID, Name, Type, UniProt ID, and UniProt Name) and 20,941 records.

Both files have been downloaded in csv format and loaded in the form of DataFrame (*ppiDF* for *Intact* and *drugBankDF* for *DrugBank*), using specific functions of the PySpark library in order to be processed. Since many features, in both DataFrames, are not useful for the purpose of this work, only the necessary ones have been selected. Figs. 5.2 and 5.3 show the files considered as input, while Figs. 5.4 and 5.5 show the features of *ppiDF* and *drugBankDF* after some further step of cleaning and renaming.

Finally, a third DataFrame has been created, *drugTargetsDF* (Fig. 5.6), such that, for each drug, there is a list of associated proteins. This will be

```
ppiDF = spark.read.csv("DATA/species_13.csv", header=True, inferSchema=True, sep='\t')
ppiDF.show()
+----------------+----------------+----------------+----------------+----------------+----
------+----------------+----------------+----------------+----------------+----------------+----
------+
|#ID(s) interactor A|ID(s) interactor B|Alt. ID(s) interactor A|Alt. ID(s) interactor B|Alias(es) interactor A|Ali
as(es) interactor B|Interaction detection method(s)|Publication 1st author(s)|Publication Identifier(s)|  Taxid in
teractor A|  Taxid interactor B|  Interaction type(s)|  Source database(s)|Interaction identifier(s)|Confidence val
ue(s)|
+----------------+----------------+----------------+----------------+----------------+----
----+----------------+----------------+----------------+----------------+----------------+----
----+
|  uniprotkb:P38764|  uniprotkb:P40016|  intact:EBI-15913|...|  intact:EBI-15927|...|  psi-mi:rpn1_yeast...|  p
si-mi:rpn3_yeast...|  psi-mi:"MI:0676"(...|  Krogan et al. (2006)|  pubmed:16554755|i...|taxid:5592
92(yeas...|taxid:559292(yeas...|psi-mi:"MI:0915"(...|psi-mi:"MI:0471"(...|  intact:EBI-694186...|intact-miscor
e:0.76|
|  uniprotkb:Q01939|  uniprotkb:P40016|  intact:EBI-13914|...|  intact:EBI-15927|...|  psi-mi:prs8_yeast...|  p
si-mi:rpn3_yeast...|  psi-mi:"MI:0676"(...|  Krogan et al. (2006)|  pubmed:16554755|i...|taxid:5592
92(yeas...|taxid:559292(yeas...|psi-mi:"MI:0915"(...|psi-mi:"MI:0471"(...|  intact:EBI-694187...|intact-miscor
e:0.40|
|  uniprotkb:P33299|  uniprotkb:P40016|  intact:EBI-13910|...|  intact:EBI-15927|...|  psi-mi:prs7_yeast...|  p
si-mi:rpn3_yeast...|  psi-mi:"MI:0676"(...|  Krogan et al. (2006)|  pubmed:16554755|i...|taxid:5592
92(yeas...|taxid:559292(yeas...|psi-mi:"MI:0915"(...|psi-mi:"MI:0471"(...|  intact:EBI-694190...|intact-miscor
e:0.69|
|  uniprotkb:Q06103|  uniprotkb:P40016|  intact:EBI-15940|...|  intact:EBI-15927|...|  psi-mi:rpn7_yeast...|  p
si-mi:rpn3_yeast...|  psi-mi:"MI:0676"(...|  Krogan et al. (2006)|  pubmed:16554755|i...|taxid:5592
92(yeas...|taxid:559292(yeas...|psi-mi:"MI:0915"(...|psi-mi:"MI:0471"(...|  intact:EBI-694189...|intact-miscor
e:0.81|
```

Figure 5.2 Top four rows of *ppiDF* after csv loading.

```
drugBankDF = spark.read.csv("DATA/uniprot links.csv", header=True, inferSchema=True)
drugBankDF.show()
+----------+-------------------+-----------+----------+-------------------+
|DrugBank ID|               Name|Type|UniProt ID|       UniProt Name|
+----------+-------------------+-----------+----------+-------------------+
|   DB00001|          Lepirudin|BiotechDrug|   P00734|        Prothrombin|
|   DB00002|          Cetuximab|BiotechDrug|   P00533|Epidermal growth ...|
|   DB00002|          Cetuximab|BiotechDrug|   O75015|Low affinity immu...|
|   DB00002|          Cetuximab|BiotechDrug|   P00736|Complement C1r su...|
|   DB00002|          Cetuximab|BiotechDrug|   P02745|Complement C1q su...|
|   DB00002|          Cetuximab|BiotechDrug|   P02746|Complement C1q su...|
|   DB00002|          Cetuximab|BiotechDrug|   P02747|Complement C1q su...|
|   DB00002|          Cetuximab|BiotechDrug|   P08637|Low affinity immu...|
|   DB00002|          Cetuximab|BiotechDrug|   P09871|Complement C1s su...|
|   DB00002|          Cetuximab|BiotechDrug|   P12314|High affinity imm...|
|   DB00002|          Cetuximab|BiotechDrug|   P12318|Low affinity immu...|
|   DB00002|          Cetuximab|BiotechDrug|   P31994|Low affinity immu...|
|   DB00002|          Cetuximab|BiotechDrug|   P31995|Low affinity immu...|
|   DB00004|Denileukin diftitox|BiotechDrug|   P61589|Interleukin-2 rec...|
|   DB00004|Denileukin diftitox|BiotechDrug|   P14784|Interleukin-2 rec...|
|   DB00004|Denileukin diftitox|BiotechDrug|   P31785|Cytokine receptor...|
|   DB00005|          Etanercept|BiotechDrug|   P01375|Tumor necrosis fa...|
|   DB00005|          Etanercept|BiotechDrug|   P20333|Tumor necrosis fa...|
|   DB00005|          Etanercept|BiotechDrug|   P12314|High affinity imm...|
|   DB00005|          Etanercept|BiotechDrug|   P08637|Low affinity immu...|
+----------+-------------------+-----------+----------+-------------------+
only showing top 20 rows
```

Figure 5.3 Top 20 rows of *drugBankDF* after csv loading.

```
ppiDF = ppiDF.withColumnRenamed("#ID(s) interactor A", "Interactor_A")
ppiDF = ppiDF.withColumnRenamed("ID(s) interactor B", "Interactor_B")
ppiDF = ppiDF.withColumnRenamed("Confidence value(s)", "Confidence")

ppiDF.show()

+-----------+-----------+----------+
|Interactor_A|Interactor_B|Confidence|
+-----------+-----------+----------+
|     P38764|     P40016|      0.76|
|     Q01939|     P40016|      0.40|
|     P33299|     P40016|      0.69|
|     Q06103|     P40016|      0.81|
|     P38764|     P40016|      0.76|
|     P40016|     P38764|      0.76|
|     P53549|     P40016|      0.55|
|     P40016|     Q08723|      0.70|
|     Q08723|     P40016|      0.70|
|     P40016|     P38886|      0.76|
|     P40016|     P43588|      0.76|
|     P40016|     P38764|      0.76|
|     P40016|     Q06103|      0.81|
|     P53008|     P40016|      0.40|
|     P32565|     P40016|      0.40|
|     P40016|     P32565|      0.40|
|     P38764|     P40016|      0.76|
|     P53549|     P40016|      0.55|
|     P40016|     P38764|      0.76|
|     Q08723|     P40016|      0.70|
+-----------+-----------+----------+
only showing top 20 rows
```

Figure 5.4 Top 20 rows of *ppiDF* after cleaning and renaming.

```
drugBankDF = drugBankDF.withColumnRenamed("DrugBank ID", "ID_DrugBank")
drugBankDF = drugBankDF.withColumnRenamed("UniProt ID", "ID_UniProt")

drugBankDF.show()

+-----------+----------+
|ID_DrugBank|ID_UniProt|
+-----------+----------+
|    DB00001|    P00734|
|    DB00002|    P00533|
|    DB00002|    O75015|
|    DB00002|    P00736|
|    DB00002|    P02745|
|    DB00002|    P02746|
|    DB00002|    P02747|
|    DB00002|    P08637|
|    DB00002|    P09871|
|    DB00002|    P12314|
|    DB00002|    P12318|
|    DB00002|    P31994|
|    DB00002|    P31995|
|    DB00004|    P01589|
|    DB00004|    P14784|
|    DB00004|    P31785|
|    DB00005|    P01375|
|    DB00005|    P20333|
|    DB00005|    P12314|
|    DB00005|    P08637|
+-----------+----------+
only showing top 20 rows
```

Figure 5.5 Top 20 rows of *drugBankDF* after cleaning and renaming.

useful for the system to recommend, for each drug, proteins that are not on this list for a given drug, since the system will have to recommend proteins that are not already targets for that drug.

5.3.2 Creation of the DataFrame to be used for the recommendation system

In this section it is shown how the DataFrame used by the recommendation system is generated and refined (Fig. 5.7). In this regard the

```
ppiDF = ppiDF.withColumn("Confidence", ppiDF["Confidence"].cast(FloatType()))
ppiDF.printSchema()

root
 |-- Interactor_A: string (nullable = true)
 |-- Interactor_B: string (nullable = true)
 |-- Confidence: float (nullable = true)

drugTargetsDF = drugBankDF.groupBy("ID_DrugBank").agg(collect_list("ID_UniProt").alias("Proteins")) \
                          .orderBy('ID_DrugBank')
drugTargetsDF.show()

+-----------+--------------------+
|ID_DrugBank|            Proteins|
+-----------+--------------------+
|   DB00001|            [P00734]|
|   DB00002|[P00533, O75015, ...|
|   DB00004|[P01589, P14784, ...|
|   DB00005|[P01375, P20333, ...|
|   DB00006|            [P00734]|
|   DB00007|            [P30968]|
|   DB00008|   [P48551, P17181]|
|   DB00009|[P00747, P02671, ...|
|   DB00010|            [Q02643]|
|   DB00011|   [P48551, P17181]|
|   DB00012|            [P19235]|
|   DB00013|[P00747, Q03405, ...|
|   DB00014|   [P22888, P30968]|
|   DB00015|[P00747, P02671, ...|
|   DB00016|            [P19235]|
|   DB00017|            [P30988]|
|   DB00018|   [P48551, P17181]|
|   DB00019|            [Q99062]|
|   DB00020|[P15509, P26951, ...|
|   DB00022|   [P48551, P17181]|
+-----------+--------------------+
only showing top 20 rows
```

Figure 5.6 Top 20 rows of *drugTargetsDF*.

```
joinedDF = drugBankDF.join(ppiDF, (drugBankDF.ID_UniProt == ppiDF.Interactor_A) & (ppiDF.Confidence >= 0.5))
joinedDF.orderBy('ID_DrugBank').show()

+-----------+----------+----------+----------+----------+
|ID_DrugBank|ID_UniProt|Interactor_A|Interactor_B|Confidence|
+-----------+----------+----------+----------+----------+
|   DB00001|   P00734|   P00734| EBI-941456|     0.56|
|   DB00001|   P00734|   P00734|    Q846V4|     0.73|
|   DB00001|   P00734|   P00734|    Q846V4|     0.73|
|   DB00001|   P00734|   P00734|    Q846V4|     0.73|
|   DB00001|   P00734|   P00734| EBI-941456|     0.56|
|   DB00002|   P00533|   P00533|    P00533|     0.98|
|   DB00002|   P12314|   P12314|  Q9BXN2-6|     0.56|
|   DB00002|   P00533|   P00533|    P62994|     0.56|
|   DB00002|   P31994|   P31994|    P01857|     0.56|
|   DB00002|   P31994|   P31994|    P01857|     0.56|
|   DB00002|   P12314|   P12314|  Q8N6F1-2|     0.56|
|   DB00002|   P31994|   P31994|    P01857|     0.56|
|   DB00002|   P00533|   P00533|    P07948|     0.82|
|   DB00002|   P31994|   P31994|    P01857|     0.56|
|   DB00002|   P00533|   P00533|  P07948-1|     0.54|
|   DB00002|   P12318|   P12318|    Q9UMX0|     0.56|
|   DB00002|   P12314|   P12314|    P07306|     0.56|
|   DB00002|   P00533|   P00533|    P00533|     0.98|
|   DB00002|   P31994|   P31994|    P01857|     0.56|
|   DB00002|   P12314|   P12314|  Q6UX27-3|     0.56|
+-----------+----------+----------+----------+----------+
only showing top 20 rows
```

Figure 5.7 Top 20 rows of *joinedDF*.

DataFrame *joinedDF* has been created, resulting as the join between *ID_UniProt* values of *drugBankDF* and *Interactor_A* values of *ppiDF*, and only those *ppiDF* proteins that have a *Confidence* value $> = 0.5$ have been selected, in order to avoid unreliable associations.

Each *joinedDF* record contains a protein such that its *ID_UniProt* and *Interactor_A* fields have the same value, which is directly associated with the drug *ID_DrugBank*, and another protein identified by the *Interactor_B* field which has an interaction with the first protein and may potentially be a recommended protein for the drug identified by the *ID_DrugBank* field, if the corresponding *Confidence* value is sufficiently high. Therefore the feature *Proteins* of *drugTargetsDF*, containing the list of proteins directly associated with the drug identified by the *ID_DrugBank* field, is taken into account (Fig. 5.8). Then the *Confidence* feature is removed from the selection, as it is no longer useful (Fig. 5.9).

```
joinedDF = joinedDF.withColumnRenamed("ID_DrugBank", "ID_Drug")
joinedDF = joinedDF.join(drugTargetsDF, drugTargetsDF.ID_DrugBank == joinedDF.ID_Drug)
joinedDF = joinedDF.select(joinedDF['ID_Drug'], joinedDF['ID_UniProt'], joinedDF['Interactor_A'],
                           joinedDF['Interactor_B'], joinedDF['Proteins'])

joinedDF.orderBy('ID_Drug').show()

+--------+----------+------------+------------+--------------------+
|ID_Drug|ID_UniProt|Interactor_A|Interactor_B|            Proteins|
+--------+----------+------------+------------+--------------------+
|DB00001|    P00734|      P00734| EBI-941456|            [P00734]|
|DB00001|    P00734|      P00734|     Q846V4|            [P00734]|
|DB00001|    P00734|      P00734|     Q846V4|            [P00734]|
|DB00001|    P00734|      P00734|     Q846V4|            [P00734]|
|DB00001|    P00734|      P00734| EBI-941456|            [P00734]|
|DB00002|    P00533|      P00533|     P62993|[P00533, O75015, ...|
|DB00002|    P00533|      P00533|     P06493|[P00533, O75015, ...|
|DB00002|    P00533|      P00533|     P38646|[P00533, O75015, ...|
|DB00002|    P00533|      P00533|     P11142|[P00533, O75015, ...|
|DB00002|    P00533|      P00533|     P22681|[P00533, O75015, ...|
|DB00002|    P09871|      P09871|     P00736|[P00533, O75015, ...|
|DB00002|    P00533|      P00533|     P31943|[P00533, O75015, ...|
|DB00002|    P00533|      P00533|     P62993|[P00533, O75015, ...|
|DB00002|    P00533|      P00533|     Q06124|[P00533, O75015, ...|
|DB00002|    P00533|      P00533|     Q05397|[P00533, O75015, ...|
|DB00002|    P00533|      P00533|     P29353|[P00533, O75015, ...|
|DB00002|    P00533|      P00533|     Q71U36|[P00533, O75015, ...|
|DB00002|    P00533|      P00533|     P62993|[P00533, O75015, ...|
|DB00002|    P00533|      P00533|     P10599|[P00533, O75015, ...|
|DB00002|    P00533|      P00533|     P63104|[P00533, O75015, ...|
+--------+----------+------------+------------+--------------------+
only showing top 20 rows
```

Figure 5.8 Top 20 rows of *joinedDF* after adding *Proteins* feature.

```
joinedDF = joinedDF.withColumn("Interaction", lit(1))
joinedDF.orderBy('ID_Drug').show()

+--------+----------+------------+------------+--------------------+-----------+
|ID_Drug|ID_UniProt|Interactor_A|Interactor_B|            Proteins|Interaction|
+--------+----------+------------+------------+--------------------+-----------+
|DB00001|    P00734|      P00734| EBI-941456|            [P00734]|          1|
|DB00001|    P00734|      P00734|     Q846V4|            [P00734]|          1|
|DB00001|    P00734|      P00734|     Q846V4|            [P00734]|          1|
|DB00001|    P00734|      P00734|     Q846V4|            [P00734]|          1|
|DB00001|    P00734|      P00734| EBI-941456|            [P00734]|          1|
|DB00002|    P00533|      P00533|     P62993|[P00533, O75015, ...|          1|
|DB00002|    P00533|      P00533|     P06493|[P00533, O75015, ...|          1|
|DB00002|    P00533|      P00533|     P38646|[P00533, O75015, ...|          1|
|DB00002|    P00533|      P00533|     P11142|[P00533, O75015, ...|          1|
|DB00002|    P00533|      P00533|     P22681|[P00533, O75015, ...|          1|
|DB00002|    P09871|      P09871|     P00736|[P00533, O75015, ...|          1|
|DB00002|    P00533|      P00533|     P31943|[P00533, O75015, ...|          1|
|DB00002|    P00533|      P00533|     P62993|[P00533, O75015, ...|          1|
|DB00002|    P00533|      P00533|     Q06124|[P00533, O75015, ...|          1|
|DB00002|    P00533|      P00533|     Q05397|[P00533, O75015, ...|          1|
|DB00002|    P00533|      P00533|     P29353|[P00533, O75015, ...|          1|
|DB00002|    P00533|      P00533|     Q71U36|[P00533, O75015, ...|          1|
|DB00002|    P00533|      P00533|     P62993|[P00533, O75015, ...|          1|
|DB00002|    P00533|      P00533|     P10599|[P00533, O75015, ...|          1|
|DB00002|    P00533|      P00533|     P63104|[P00533, O75015, ...|          1|
+--------+----------+------------+------------+--------------------+-----------+
only showing top 20 rows
```

Figure 5.9 Top 20 rows of *joinedDF* after adding the *Interaction* feature.

In each record if a protein associated with the *Interactor_B* field occurs in the *Proteins* list, which contains proteins directly associated with the drug identified by *ID_Drug*, then it is discarded. Then a further feature, *Interaction*, is added to *joinedDF*, with all values set to 1, useful later to establish the recommendation criterion. A new boolean feature, *Interactor_drug_target*, has been added to *joinedDF*, with *true* value if *Interactor_B* protein of the considered record is present in *Proteins* list (and, therefore has a direct association with the drug *ID_Drug*), *false* otherwise (Fig. 5.10).

Finally, to establish the criterion to be adopted for the recommendation system, a grouping of all those records with the same *ID_Drug* and *Interactor_B* and *Interactor_drug_target* set to *false* is carried out. The feature

```
joinedDF = joinedDF.withColumn("Interactor_drug_target", expr("array_contains(Proteins, Interactor_B)"))
joinedDF.orderBy('ID_DrugBank', 'Interactor_B').show(50)
+--------+----------+------------+------------+-------------------------+-----------+----------------------+
|ID_Drug|ID_UniProt|Interactor_A|Interactor_B|         Proteins|Interaction|Interactor_drug_target|
+--------+----------+------------+------------+-------------------------+-----------+----------------------+
|DB00001|   P00734|      P00734|  EBI-941456|                [P00734]|         1|                 false|
|DB00001|   P00734|      P00734|      Q846V4|                [P00734]|         1|                 false|
|DB00002|   P00533|      P00533|      A4FU49|[P00533, O75015, ...]|         1|                 false|
|DB00002|   P00533|      P00533|EBI-4399559|[P00533, O75015, ...]|         1|                 false|
|DB00002|   P00533|      P00533|  NP_059022|[P00533, O75015, ...]|         1|                 false|
|DB00002|   P00533|      P00533|      000170|[P00533, O75015, ...]|         1|                 false|
|DB00002|   P00533|      P00533|      000401|[P00533, O75015, ...]|         1|                 false|
|DB00002|   P00533|      P00533|      000459|[P00533, O75015, ...]|         1|                 false|
|DB00002|   P12314|      P12314|      000526|[P00533, O75015, ...]|         1|                 false|
|DB00002|   P00533|      P00533|      000750|[P00533, O75015, ...]|         1|                 false|
|DB00002|   P00533|      P00533|      014543|[P00533, O75015, ...]|         1|                 false|
|DB00002|   P00533|      P00533|      014544|[P00533, O75015, ...]|         1|                 false|
|DB00002|   P00533|      P00533|      014818|[P00533, O75015, ...]|         1|                 false|
|DB00002|   P00533|      P00533|      014944|[P00533, O75015, ...]|         1|                 false|
|DB00002|   P00533|      P00533|      014965|[P00533, O75015, ...]|         1|                 false|
|DB00002|   P00533|      P00533|      015511|[P00533, O75015, ...]|         1|                 false|
|DB00002|   P12314|      P12314|      043491|[P00533, O75015, ...]|         1|                 false|
|DB00002|   P00533|      P00533|      043561|[P00533, O75015, ...]|         1|                 false|
|DB00002|   P00533|      P00533|      043639|[P00533, O75015, ...]|         1|                 false|
|DB00002|   P00533|      P00533|      043707|[P00533, O75015, ...]|         1|                 false|
|DB00002|   P12314|      P12314|      043736|[P00533, O75015, ...]|         1|                 false|
|DB00002|   P00533|      P00533|      060603|[P00533, O75015, ...]|         1|                 false|
|DB00002|   P00533|      P00533|      060716|[P00533, O75015, ...]|         1|                 false|
|DB00002|   P00533|      P00533|      060884|[P00533, O75015, ...]|         1|                 false|
|DB00002|   P00533|      P00533|      075095|[P00533, O75015, ...]|         1|                 false|
|DB00002|   P00533|      P00533|      075368|[P00533, O75015, ...]|         1|                 false|
|DB00002|   P00533|      P00533|      075674|[P00533, O75015, ...]|         1|                 false|
|DB00002|   P00533|      P00533|      075791|[P00533, O75015, ...]|         1|                 false|
|DB00002|   P00533|      P00533|      094875|[P00533, O75015, ...]|         1|                 false|
|DB00002|   P12314|      P12314|      095393|[P00533, O75015, ...]|         1|                 false|
|DB00002|   P00533|      P00533|      095433|[P00533, O75015, ...]|         1|                 false|
|DB00002|   P00533|      P00533|      095782|[P00533, O75015, ...]|         1|                 false|
|DB00002|   P00533|      P00533|      P00533|[P00533, O75015, ...]|         1|                  true|
|DB00002|   P09871|      P09871|      P00736|[P00533, O75015, ...]|         1|                  true|
|DB00002|   P00736|      P00736|      P00736|[P00533, O75015, ...]|         1|                  true|
|DB00002|   P00533|      P00533|      P01133|[P00533, O75015, ...]|         1|                 false|
|DB00002|   P00533|      P00533|      P01135|[P00533, O75015, ...]|         1|                 false|
|DB00002|   P12314|      P12314|      P01857|[P00533, O75015, ...]|         1|                 false|
|DB00002|   P31994|      P31994|      P01857|[P00533, O75015, ...]|         1|                 false|
|DB00002|   P02747|      P02747|      P02745|[P00533, O75015, ...]|         1|                  true|
|DB00002|   P02746|      P02746|      P02745|[P00533, O75015, ...]|         1|                  true|
|DB00002|   P02745|      P02745|      P02746|[P00533, O75015, ...]|         1|                  true|
|DB00002|   P02745|      P02745|      P02747|[P00533, O75015, ...]|         1|                  true|
|DB00002|   P00533|      P00533|    P03372-4|[P00533, O75015, ...]|         1|                 false|
|DB00002|   P00533|      P00533|      P04083|[P00533, O75015, ...]|         1|                 false|
|DB00002|   P00533|      P00533|      P04406|[P00533, O75015, ...]|         1|                 false|
|DB00002|   P00533|      P00533|      P04626|[P00533, O75015, ...]|         1|                 false|
|DB00002|   P00533|      P00533|      P04792|[P00533, O75015, ...]|         1|                 false|
|DB00002|   P00533|      P00533|      P05067|[P00533, O75015, ...]|         1|                 false|
|DB00002|   P00533|      P00533|      P06493|[P00533, O75015, ...]|         1|                 false|
+--------+----------+------------+------------+-------------------------+-----------+----------------------+
only showing top 50 rows
```

Figure 5.10 Top 50 rows of *joinedDF* after adding *Interactor_drug_target* feature.

Interactions contains the counts for each of these groupings. This feature is used as a rating for the recommendation system.

Therefore for each record, *Interactions* represent the number of proteins that are directly associated with *ID_Drug* which have an interaction with the *Interactor_B* protein (not directly associated with *ID_Drug* and, therefore a potential recommended).

Since the values to be used to train the recommendation model can only be Integer or Float type (in this case, *ID_Drug* and *Interactor_B* are String type), *StringIndexer* function from PySpark was used to encode all String type values. Since this function can be used one column at a time, and since indexes have to be able to be distinguished in all columns, the *Pipelane* function from PySpark has been used to overcome this problem. Therefore *indexedDF* DataFrame is created in which the features *ID_Drug_Index* and *Interactor_B_Index* are added (Fig. 5.11).

5.3.3 Creation of the ALS model and data fitting

In this section it is explained how the model used for the recommended system is created and evaluated. The implemented algorithm is the *ALS*, a type of matrix factorization for "collaborative filtering," as already explained above. Several hyperparameters are set to find the smallest RMSE value, used to test the accuracy of model predictions.

```
drugIndexer = StringIndexer(inputCol='ID_Drug', outputCol='ID_Drug_index').fit(joinedDF)
proteinIndexer = StringIndexer(inputCol='Interactor_B', outputCol='Interactor_B_index').fit(joinedDF)

pipeline = Pipeline(stages=[drugIndexer, proteinIndexer])

indexedDF = pipeline.fit(joinedDF).transform(joinedDF)

indexedDF.show()

+--------+------------+------------+-------------+------------------+
|ID_Drug |Interactor_B|Interactions|ID_Drug_index|Interactor_B_index|
+--------+------------+------------+-------------+------------------+
|DB11712 |     P12956 |          1 |       329.0 |            551.0 |
|DB00114 |     Q9NVD7 |          1 |       200.0 |           4935.0 |
|DB04988 |     P04406 |          2 |        41.0 |            256.0 |
|DB07529 |     O75496 |          1 |       525.0 |            102.0 |
|DB00823 |     P78362 |          1 |       150.0 |            147.0 |
|DB12010 |     Q9GZT8 |          2 |         0.0 |           3232.0 |
|DB06624 |     Q15084 |          1 |        81.0 |            122.0 |
|DB00074 |     O43491 |          1 |      1220.0 |           1123.0 |
|DB14487 |   O75182-2 |          1 |         1.0 |           1540.0 |
|DB00907 |     P30825 |          1 |       214.0 |            736.0 |
|DB06870 |   Q9BY66-3 |          1 |       619.0 |            317.0 |
|DB04573 |     Q9BQ39 |          1 |       102.0 |            200.0 |
|DB01412 |     O15354 |          1 |      1605.0 |            263.0 |
|DB07126 |     Q00526 |          2 |       377.0 |             74.0 |
|DB12267 |     P0DOD2 |          1 |        17.0 |           1363.0 |
|DB06454 |     P47211 |          1 |      2544.0 |            266.0 |
|DB01169 |     Q92793 |          2 |        15.0 |              7.0 |
|DB01108 |EBI-11177835|          1 |       101.0 |            463.0 |
|DB06492 |     O95967 |          1 |      2399.0 |           1488.0 |
|DB00527 |     O43303 |          1 |       864.0 |            852.0 |
+--------+------------+------------+-------------+------------------+
only showing top 20 rows
```

Figure 5.11 Top 20 rows of *indexedDF*.

For training, first *indexedDF* is randomly split (80% for training and 20% for testing), then the *ALS* function from PySpark was used, whose implemented parameters are as follows:

- *maxIter:* maximum number of iterations to run (default: 10)
- *regParam:* specifies the regularization parameter in ALS (default: 1.0)
- *rank:* number of latent factors in the model (default: 10)
- *alpha:* parameter applicable to the implicit feedback variant of ALS that governs the baseline confidence in preference observations (default: 1.0)
- *coldStartStrategy:* strategy to manage new or unknown items/users. Setting it to "drop" excludes these items from the results (default: nan)

Finally, the *RegressionEvaluator* function allows us to evaluate the RMSE value for the chosen *regParam, rank,* and *alpha* parameters.

Fig. 5.12 illustrates the abovementioned procedure.

```
(training,test) = indexedDF.randomSplit([0.8, 0.2])

regParams = [0.01, 0.1]
ranks = [25]
alphas = [10.0, 20.0, 40.0, 60.0, 80.0]

aus_regParam = 0.0
aus_rank = 0
aus_alpha = 0.0
aus_rmse = 0.0

print('Creating ALS model ...')
for regParam in regParams:
    for rank in ranks:
        for alpha in alphas:
            aus_als = ALS(maxIter=10, regParam=regParam, rank=rank, alpha=alpha, userCol='ID_Drug_index',
                          itemCol="Interactor_B_index", ratingCol="Interactions", coldStartStrategy="drop")
            aus_model = aus_als.fit(training)
            predictions = aus_model.transform(test)
            evaluator = RegressionEvaluator(metricName="rmse", labelCol="Interactions", predictionCol="prediction")
            rmse = evaluator.evaluate(predictions)

            if(aus_rmse == 0.0 or rmse < aus_rmse):
                aus_regParam = regParam
                aus_rank = rank
                aus_alpha = alpha
                aus_rmse = rmse
                model = aus_model

            print("For regParam: {0}, rank: {1}, alpha: {2}, RMSE: {3}".format(regParam, rank, alpha, rmse))

print('Chosen parameters: regParam = {0}, rank = {1}, alpha = {2}'.format(aus_regParam, aus_rank, aus_alpha))
Creating ALS model ...
For regParam: 0.01, rank: 25, alpha: 10.0, RMSE: 0.2846906755205193
For regParam: 0.01, rank: 25, alpha: 20.0, RMSE: 0.2846906755205193
For regParam: 0.01, rank: 25, alpha: 40.0, RMSE: 0.2846906755205193
For regParam: 0.01, rank: 25, alpha: 60.0, RMSE: 0.2846906755205193
For regParam: 0.01, rank: 25, alpha: 80.0, RMSE: 0.2846906755205193
For regParam: 0.1, rank: 25, alpha: 10.0, RMSE: 0.2689031685224229
For regParam: 0.1, rank: 25, alpha: 20.0, RMSE: 0.2689031685224229
For regParam: 0.1, rank: 25, alpha: 40.0, RMSE: 0.26890316852224296
For regParam: 0.1, rank: 25, alpha: 60.0, RMSE: 0.2689031685224229
For regParam: 0.1, rank: 25, alpha: 80.0, RMSE: 0.26890316852224296
Chosen parameters: regParam = 0.1, rank = 25, alpha = 10.0
```

Figure 5.12 Creation of the ALS model according to the parameters that minimize the RMSE value.

The model created with a combination of *regParam*, *rank*, and *alpha* that generates the lowest RMSE value is that chosen for the recommendation system.

5.3.4 Implementation of the recommendation system

Once the model has been generated, the *recommendForAllUsers* function from PySpark is used which takes as a parameter the number *n* of proteins that we want the system to recommend for each drug, and returns, for each *ID_Drug_index*, *n* recommended *[Interactor_B_index, rating]* pairs (Fig. 5.13).

Therefore the *recommendations* column is divided into the two columns *Interactor_B_index* and *rating* (Fig. 5.14).

Finally, since it is not intuitive to establish to which *ID_Drug* a value of *ID_Drug_index* refers (respectively, to which *Interactor_B* a value of *Interactor_B_index* refers), the *IndexToString* and *Pipeline* functions from PySpark were used to convert indexes into their source strings (Fig. 5.15).

5.3.5 Evaluation of the recommendation system

Given an *ID_Drug*, the *n* proteins recommended for that specific *ID_Drug* are considered and the *rec.csv* file containing these records is

```
print("Insert a number of recommendations per drug:")
n = int(input())

protein_recs = model.recommendForAllUsers(n)

protein_recs.show()

Insert a number of recommendations per drug:
5
+------------+--------------------+
|ID_Drug_index|     recommendations|
+------------+--------------------+
|        1580|[[5263, 3.0869117...|
|         471|[[5263, 3.0935702...|
|        1591|[[3543, 3.6073778...|
|        4101|[[5263, 3.1222749...|
|        1342|[[5263, 3.0735283...|
|        2122|[[5263, 3.1680398...|
|        2142|[[5263, 3.194394]...|
|         463|[[5263, 2.9468124...|
|         833|[[5263, 3.0409713...|
|        3794|[[5263, 2.6746826...|
|        1645|[[5263, 3.204765]...|
|        3175|[[5263, 2.935325]...|
|         496|[[3543, 2.74789],...|
|        2366|[[5263, 3.0821824...|
|        2866|[[5263, 3.0592232...|
|        3997|[[3543, 2.924361]...|
|         148|[[5263, 2.983145]...|
|        1088|[[5263, 2.9785137...|
|        1238|[[5263, 3.1817076...|
|        3918|[[5263, 3.1363788...|
+------------+--------------------+
only showing top 20 rows
```

Figure 5.13 Top 20 rows of *protein_recs* DataFrame showing *[Interactor_B_index, rating]* pairs for each *ID_Drug_index* chosen by recommendation system.

```
flatDrugRecs = protein_recs.withColumn('proteinAndRating', explode(protein_recs.recommendations))\
                           .select('ID_Drug_index', 'proteinAndRating.*')

flatDrugRecs.show()

+-------------+-----------------+---------+
|ID_Drug_index|Interactor_B_index|   rating|
+-------------+-----------------+---------+
|         1580|             5263|3.0869117|
|         1580|             6119|3.0869117|
|         1580|             3543|2.8562443|
|         1580|             6280|2.3151839|
|         1580|             1137|1.9725999|
|          471|             5263|3.0935702|
|          471|             6119|3.0935702|
|          471|             3543| 2.849777|
|          471|             6280|2.3201783|
|          471|             1137|2.0504289|
|         1591|             3543|3.6073778|
|         1591|             5263|3.2085395|
|         1591|             6119|3.2085395|
|         1591|             6280|2.4064047|
|         1591|             3937|2.0296333|
|         4101|             5263|3.1222749|
|         4101|             6119|3.1222749|
|         4101|             3543|2.8859322|
|         4101|             6280|2.3417063|
|         4101|             1137|1.9761636|
+-------------+-----------------+---------+
only showing top 20 rows
```

Figure 5.14 Top 20 rows of *flatDrugRecs*.

```
drugString = IndexToString(inputCol='ID_Drug_index', outputCol='ID_Drug', labels=drugIndexer.labels)
proteinString = IndexToString(inputCol='Interactor_B_index', outputCol='ID_Protein', labels=proteinIndexer.labels)

convertedDrugRecs = Pipeline(stages=[drugString, proteinString]).fit(indexedDF).transform(flatDrugRecs)
convertedDrugRecs = convertedDrugRecs.select(convertedDrugRecs['ID_Drug'], convertedDrugRecs['ID_Protein'],
                                             convertedDrugRecs['rating'])

convertedDrugRecs.select('ID_Drug', 'ID_Protein', 'rating').orderBy('rating', ascending=False).show()

+-------+----------+---------+
|ID_Drug|ID_Protein|   rating|
+-------+----------+---------+
|DB12010|    P08238|12.920674|
|DB12010|    P63104| 8.531653|
|DB12010|    P61981| 7.272068|
|DB12010|    Q16543|6.8025374|
|DB12010|    Q04917|6.7631745|
|DB08236|    Q9BPX5| 4.915586|
|DB08235|    Q9BPX5|4.8028083|
|DB08515|    Q9Y244|4.7891808|
|DB07728|    Q9Y375| 4.620285|
|DB07728|    Q9P032| 4.620285|
|DB04160|    Q9P032| 4.534795|
|DB04160|    Q9Y375| 4.534795|
|DB08236|    Q9P032| 4.426154|
|DB08236|    Q9Y375| 4.426154|
|DB08358|    Q9P032|4.2863016|
|DB08358|    Q9Y375|4.2863016|
|DB12695|    Q9UQL6|  4.26186|
|DB12695|    P30305|  4.26186|
|DB07080|    Q9Y244|4.1399984|
|DB00162|    Q9P032|4.0918837|
+-------+----------+---------+
only showing top 20 rows
```

Figure 5.15 Top 20 rows of *convertedDrugRecs*.

created, useful later for the generation and visualization of the associated graphs. In addition, the *joinedDF* view filtered with *ID_Drug* equal to the *ID_Drug* inserted, in descending order of *Interactions*, is also flanked. It is possible this way to establish whether the predicted proteins are actually those with the highest *Interactions* values (Fig. 5.16).

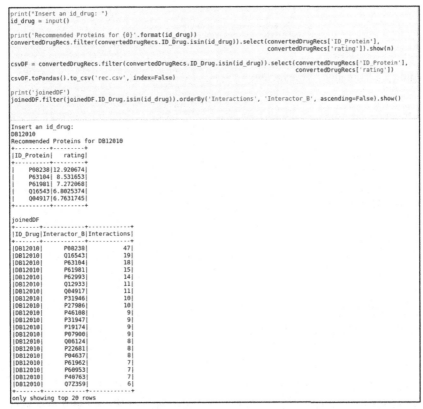

```
print("Insert an id_drug: ")
id_drug = input()

print('Recommended Proteins for {0}'.format(id_drug))
convertedDrugRecs.filter(convertedDrugRecs.ID_Drug.isin(id_drug)).select(convertedDrugRecs['ID_Protein'],
                                                             convertedDrugRecs['rating']).show(n)

csvDF = convertedDrugRecs.filter(convertedDrugRecs.ID_Drug.isin(id_drug)).select(convertedDrugRecs['ID_Protein'],
                                                             convertedDrugRecs['rating'])
csvDF.toPandas().to_csv('rec.csv', index=False)

print('joinedDF')
joinedDF.filter(joinedDF.ID_Drug.isin(id_drug)).orderBy('Interactions', 'Interactor_B', ascending=False).show()
```

```
Insert an id_drug:
DB12010
Recommended Proteins for DB12010
+---------+---------+
|ID_Protein|   rating|
+---------+---------+
|   P08238|12.920674|
|   P63104| 8.531653|
|   P61981| 7.272068|
|   Q16543|6.8025374|
|   Q04917|6.7631745|
+---------+---------+

joinedDF
+-------+-----------+------------+
|ID_Drug|Interactor_B|Interactions|
+-------+-----------+------------+
|DB12010|     P08238|          47|
|DB12010|     Q16543|          19|
|DB12010|     P63104|          18|
|DB12010|     P61981|          15|
|DB12010|     P62993|          14|
|DB12010|     Q12933|          11|
|DB12010|     Q04917|          11|
|DB12010|     P31946|          10|
|DB12010|     P27986|          10|
|DB12010|     P46108|           9|
|DB12010|     P31947|           9|
|DB12010|     P19174|           9|
|DB12010|     P07900|           9|
|DB12010|     Q06124|           8|
|DB12010|     P22681|           8|
|DB12010|     P04637|           8|
|DB12010|     P61962|           7|
|DB12010|     P60953|           7|
|DB12010|     P40763|           7|
|DB12010|     O7Z359|           6|
+-------+-----------+------------+
only showing top 20 rows
```

Figure 5.16 Five recommended proteins for drug *DB12010* chosen by the recommended system. The flanked *joinedDF* DataFrame shows how the proteins chosen are actually those with the highest *Interactions* value.

5.3.6 Graphs generation and visual overview

Graphs are generated for a more accurate and visually intuitive analysis of the results. First, from the previously created *rec.csv* file containing proteins recommended by the system for a specific drug, a *rec_list* is generated containing these proteins (Fig. 5.17).

Similarly, the *dir_list* is created containing the proteins that interact directly with the drug inserted for the recommendation system (Fig. 5.18).

By exploiting the *NetworkX* library, the graph G was generated by *DiGraph* and *add_edge* functions, containing all the PPIs taken from the previously created file *ppi.csv* (Fig. 5.19).

```
rec_list = []

rec_prot_DF = spark.read.csv("rec.csv", header=True, inferSchema=True)
rec_prot_array = np.array(rec_prot_DF.select("ID_Protein").collect())

for rec in rec_prot_array:
    rec_list.append(rec[0])
```

Figure 5.17 Creation of the *rec_list*.

```
dir_list = []

dir_prot_DF = drugBankDF.filter(drugBankDF.ID_DrugBank.isin(id_drug))
dir_prot_array = np.array(dir_prot_DF.select("ID_UniProt").collect())

for dir in dir_prot_array:
    dir_list.append(dir[0])
```

Figure 5.18 Creation of *dir_list*.

```
G = nx.DiGraph()

f = open("ppi.csv", "r")

for line in f:
    node1, node2, weight = line.split(",")
    G.add_edge(node1, node2, weight=float(weight))
```

Figure 5.19 Creation of the *G* graph.

The graph is then filtered in such a way as to contain only the nodes connected with a specific protein (recommended by the system) at a certain distance in the network, that is, linked to that protein by a pathway of at most a chosen number of edges. To this aim, the *ego_graph* function from *NetworkX* library is used, which takes as parameters the reference graph, the central node, and the radius. Then a new list is created, *nodes_list*, containing these nodes with their relative weight (Fig. 5.20).

For the representation of interactive graphs, appropriate functions from *PyVis* library have been used. In particular, for a given recommended protein, two types of graphs can be generated: a complete one (Figs. 5.22 and 5.24), containing all the interactions of this protein, and a partial one (Figs. 5.21 and 5.23), containing only the interactions of this protein with proteins that are directly associated with the reference drug, and any interactions with other recommended proteins. This is due to the fact that in the PPI network each protein could have many interactions, and often a complete representation of the graph, albeit interactive, due to such a high number of interactions would not be very intuitive (Fig. 5.22, Fig. 5.23, and Fig. 5.24).

```
print("Insert the center node ")
node = input()
print("Insert the radius ")
radius = input()
G = nx.generators.ego_graph(G, node, radius=int(radius))

nodes_list = []

for line in G.edges():
    nodes_list.append([line[0], line[1], G.edges[line[0], line[1]]["weight"]])

Insert the center node
P61981
Insert the radius
1
```

Figure 5.20 Creation of the *nodes_list*.

```
def partial_graph(nodes_list):
    net = Network(height="100%", width="100%", bgcolor="#222222", font_color="white")

    for i in range(len(nodes_list)):
        node1 = nodes_list[i][0]
        node2 = nodes_list[i][1]
        w = float(nodes_list[i][2])

        if node1 in rec_list:
            if node2 in rec_list:
                net.add_node(node1, color="#ff4d4d")
                net.add_node(node2, color="#ff4d4d")
                net.add_edge(node1, node2, value=w, title=w, color="#ff3300", width=float(w))
            elif node2 in dir_list:
                net.add_node(node1, color="#ff4d4d")
                net.add_node(node2, color="#80ff80")
                net.add_edge(node1, node2, title=w, color="#ffcc66", width=float(w))
            else:
                continue
        elif node1 in dir_list:
            if node2 in rec_list:
                net.add_node(node1, color="#80ff80")
                net.add_node(node2, color="#ff4d4d")
                net.add_edge(node1, node2, value=w, title=w, color="#ffcc66", width=float(w))
            elif node2 in dir_list:
                net.add_node(node1, color="#80ff80")
                net.add_node(node2, color="#80ff80")
                net.add_edge(node1, node2, value=w, title=w, color="#66ff66", width=float(w))
            else:
                continue
        else:
            continue

    #net.show_buttons(filter_=['physics'])
    #net.show_buttons(filter_=['nodes'])

    net.set_options(
    """
    var options = {
        "physics": {
            "forceAtlas2Based": {
                "gravitationalConstant": -268,
                "centralGravity": 0.025,
                "springLength": 265,
                "springConstant": 0.14,
                "damping": 0.17
            },
            "maxVelocity": 0,
            "minVelocity": 0.01,
            "solver": "forceAtlas2Based",
            "timestep": 0.01
        },
        "nodes": {
            "borderWidthSelected": 4
        }
    }
    """
    )

    net.show("{0}_partial_network_for_{1}_IDdrug.html".format(node ,id_drug))
```

Figure 5.21 Creation of the partial graph.

It is worth pointing out that, in the graphs considered here, the weights on the edges represent the confidence values of the corresponding PPI association. These values are used only as a validation measure to

```
def complete_graph(nodes_list):
    net = Network(height="100%", width="100%", bgcolor="#222222", font_color="white")

    for i in range(len(nodes_list)):
        node1 = nodes_list[i][0]
        node2 = nodes_list[i][1]
        w = float(nodes_list[i][2])

        if node1 in rec_list:
            if node2 in rec_list:
                net.add_node(node1, color="#ff4d4d")
                net.add_node(node2, color="#ff4d4d")
                net.add_edge(node1, node2, value=w, title=w, color="#ff3300", width=float(w))
            elif node2 in dir_list:
                net.add_node(node1, color="#ff4d4d")
                net.add_node(node2, color="#80ff80")
                net.add_edge(node1, node2, title=w, color="#ffcc66", width=float(w))
            else:
                net.add_node(node1, color="#ff4d4d")
                net.add_node(node2)
                net.add_edge(node1, node2, value=w, title=w, color='black', width=float(w))
        elif node1 in dir_list:
            if node2 in rec_list:
                net.add_node(node1, color="#80ff80")
                net.add_node(node2, color="#ff4d4d")
                net.add_edge(node1, node2, value=w, title=w, color="#ffcc66", width=float(w))
            elif node2 in dir_list:
                net.add_node(node1, color="#80ff80")
                net.add_node(node2, color="#80ff80")
                net.add_edge(node1, node2, value=w, title=w, color="#66ff66", width=float(w))
            else:
                net.add_node(node1, color="#80ff80")
                net.add_node(node2)
                net.add_edge(node1, node2, value=w, title=w, color='black', width=float(w))
        else:
            if node2 in rec_list:
                net.add_node(node1)
                net.add_node(node2, color="#ff4d4d")
                net.add_edge(node1, node2, value=w, title=w, color='black', width=float(w))
            elif node2 in dir_list:
                net.add_node(node1)
                net.add_node(node2, color="#80ff80")
                net.add_edge(node1, node2, value=w, title=w, color='black', width=float(w))
            else:
                net.add_node(node1)
                net.add_node(node2)
                net.add_edge(node1, node2, value=w, title=w, color='black', width=float(w))

    #net.show_buttons(filter_=['physics'])
    #net.show_buttons(filter_=['nodes'])

    net.set_options(
    """
    var options = {
      "physics": {
        "forceAtlas2Based": {
          "gravitationalConstant": -268,
          "centralGravity": 0.025,
          "springLength": 265,
          "springConstant": 0.14,
          "damping": 0.17
        },
        "maxVelocity": 0,
        "minVelocity": 0.01,
        "solver": "forceAtlas2Based",
        "timestep": 0.01
      },
      "nodes": {
        "borderWidthSelected": 4
      }
    }
    """
    )

    net.show("{0}_complete_network_for_{1}_IDdrug.html".format(node ,id_drug))
```

Figure 5.22 Creation of the complete graph.

highlight the fact that proteins suggested by the model for a given drug have high confidence value with the target proteins of that drug.

As shown in Fig. 5.23, the three recommended proteins are linked together. In detail, P61981 is linked with Q04917 (confidence = 0.64) and with P63104 (confidence = 0.64); and Q04917 is linked with P63104 (confidence = 0.73). Since three proteins recommended by the system

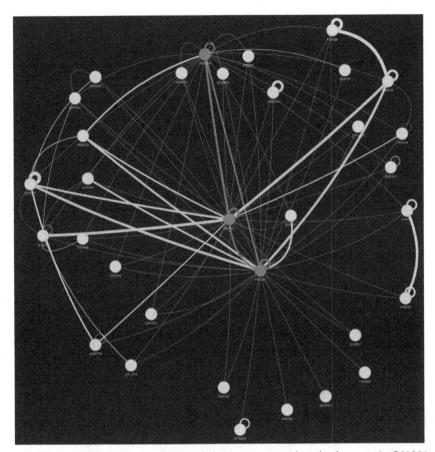

Figure 5.23 Example of a partial graph having as central node the protein P61981 recommended by the system for the drug DB12010, with radius set to 1. Proteins recommended by the system for the DB12010 drug are shown in red (from this example, in fact, we can see how the P61981 protein has an interaction with two other proteins recommended by the system), while in green the proteins that have a direct association with the drug DB12010 and which are linked at the same time with the P61981 protein. The size of the edges refers to the confidence value between the two proteins: the larger this value, the greater the size of the edge. The red-colored edges are those edges that link two recommended proteins; the orange-colored edges link recommended proteins with proteins that have a direct association with the drug; and green-colored edges link two associated proteins directly with the drug.

interact with each other in the PPI network with high confidence values, and since they interact in common (also with high confidence values) with numerous proteins directly associated with the drug, it is possible to

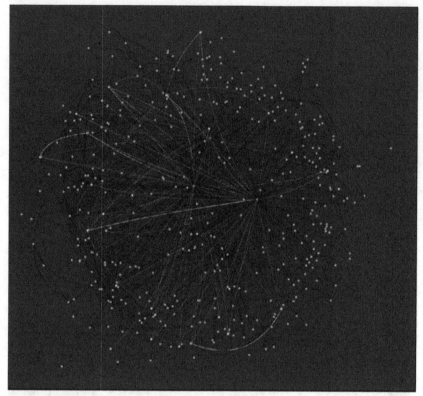

Figure 5.24 Example of a complete graph having as central node the protein P61981 recommended by the system for the drug DB12010, with radius set to 1. In this graph all the proteins of the PPI network that have a relationship with the central protein inserted, and any other relationships, are represented in blue. Their interactions are represented by black edges.

conclude that these recommended proteins are indeed excellent candidates for the drug DB12010.

5.4 Conclusions

Identifying new targets for existing drugs (drug repositioning or repurposing) is an important task in drug discovery. Indeed, the purpose of drug repositioning is the detection of new clinical uses for existing drugs. Since in vitro experiments are extremely expensive and time-consuming, and considering that known drug–target interactions based on wet-lab experiments are limited to a very small number, computational prediction has become relevant in the last few years.

In this chapter we have first summarized some of the most recent research in drug–target associations automatic prediction, and then we have explained how a recommendation system can be designed relying on Big Data technologies in order to perform this task. In carrying out this work we have built a human PPI network from available PPIs and we have used it, together with known drug–target associations, in order to train the system. The main idea of the approach is that, depending on how proteins are connected inside the PPI network, and on the behavior of already known targets for a specific drug, new targets can be predicted for that drug.

Technical details on how implementing such a recommendation system in Apache Spark are provided, and visual inspection of the obtained results, including subnetworks involving the predicted targets, are described.

References

Aggarwal, C. C. (2016). An introduction to recommender systems. *Recommender Systems: The Textbook*, 1−28.

Che, J., Chen, L., Guo, Z., Wang, S., & Aorigele. (2020). Drug target group prediction with multiple drug networks. *Combinatorial Chemistry & High Throughput Screening, 23* (4), 274−284.

Cheng, F., Desai, R. J., Handy, D. E., Wang, R., Schneeweiss, S., Barabási, A., & Loscalzo, J. (2018). Network-based approach to prediction and population-based validation of in silico drug repurposing. *Nature Communications, 9*, Article number: 2691.

Chong, C. R., & Sullivan, D. J. (2007). New uses for old drugs. *Nature, 448*, 646.

Hermjakob, H., Montecchi-Palazzi, L., Lewington, C., Mudali, S., Kerrien, S., Orchard, S., Vingron, M., Roechert, B., Roepstorff, P., Valencia, A., Margalit, H., Armstrong, J., Bairoch, A., Cesareni, G., Sherman, D., & Apweiler, R. (2004). IntAct: An open source molecular interaction database. *Nucleic Acids Research*, D452−D455, 2004 Jan 1; 32(Database issue).

Keiser, M. J., Setola, V., Irwin, J. J., Laggner, C., Abbas, A. I., Hufeisen, S. J., Jensen, N. H., Kuijer, M. B., Matos, R. C., Tran, T. B., Whaley, R., Glennon, R. A., Hert, J., Thomas, K. L. H., Edwards, D. D., Shoichet, B. K., & Roth, B. L. (2009). Predicting new molecular targets for known drugs. *Nature, 462*, 175−181.

Li, J., Zheng, S., Chen, B., Butte, A. J., Swamidass, S. J., & Lu, Z. (2015). A survey of current trends in computational drug repositioning. *Briefing in Bioinformatics, 17*(1), 2−12.

Luo, H., Li, M., Wang, S., Liu, Q., Li, Y., & Wang, J. (2018). Computational drug repositioning using low-rank matrix approximation and randomized algorithms. *Bioinformatics, 34*(11), 1904−1912.

Olayan, R. S., Ashoor, H., & Bajic, V. B. (2017). DDR: Efficient computational method to predict drug-target interactions using graph mining and machine learning approaches. *Bioinformatics, 34*(7), 1164−1173.

Panni, S., & Rombo, S. E. (2013). Searching for repetitions in biological networks: Methods, resources and tools. *Briefings in Bioinformatics*, *16*(1), 118−136.

Ricci, F., Rokach, L., & Shapira, B. (2010). Introduction to recommender systems handbook. *Recommender Systems Handbook*, 1−35.

Wishart, D. S., Knox, C., Guo, A. C., Shrivastava, S., Hassanali, M., Stothard, P., Chang, Z., & Woolsey, J. (2006). Drugbank: A comprehensive resource for in silico drug discovery and exploration. *Nucleic Acids Research*, 2006 Jan 1;34 (Database issue):D668-72. 16381955.

Wu, Z., Cheng, F., Li, J., Liu, G., & Tang, Y. (2016). SDTNBI: An integrated network and chemoinformatics tool for systematic prediction of drug-target interactions and drug repositioning. *Briefings in Bioinformatics*, *18*(2), 333−347.

Zaharia, M., Chowdhury, M., Das, T., Dave, A., Ma, J., McCauley, M., Franklin, M. J., Shenker, S., & Stoica, I. (2012). *Resilient distributed datasets: A fault-tolerant abstraction for in-memory cluster computing*. Berkeley: University of California.

Zaharia, M., Chowdhury, M., Franklin, M. J., Shenker, S., & Stoica, I. (2010). *Spark: cluster computing with working sets*. Berkeley: University of California.

CHAPTER 6

Towards building an efficient deep neural network based on YOLO detector for fetal head localization from ultrasound images

M. Ramla, S. Sangeetha and S. Nickolas
Department of Computer Applications, National Institute of Technology Tiruchirappalli, Tiruchirappalli, India

6.1 Introduction

High-risk pregnancy complications include ectopic pregnancy, spontaneous abortion, fetal chromosomal abnormalities, fetal growth restriction, congenital anomalies, placenta previa and abruption, gestational diabetes, preeclampsia, and cesarean delivery. All these problems require increased monitoring or surveillance. Particularly, fetal growth restriction is one of the major risk concerns of high-risk pregnancy problems. In the field of Obstetrics, measuring fetal biometry through ultrasound has become a regular practice (Zhang et al., 2020). To ascertain the well-being of the fetal and safe pregnancy, measuring the head circumference (HC) of fetal at different gestational age is of primary importance. HC measurement is one of the basic biometric parameters used to ascertain the fetal growth and identify cerebral anomalies. This, in turn, is extrapolated to approximate the fetal age and estimated due date. But, ultrasound images are idealistic and naïve for most pregnant women as it requires trained radiologists or sonographers to interpret the images (Van Den Heuvel et al., 2018a). Considering the scarcity of trained sonographers, there is an absolute necessity for automated HC estimation from ultrasound images. Of late, the arena of computer vision (CV) has monumental success in visual object detection and recognition. CV applications can render a trustworthy help for physicians and radiologists by improving the accuracy of screening. Inspired by the tremendous

Edge-of-Things in Personalized Healthcare Support Systems
DOI: https://doi.org/10.1016/B978-0-323-90585-5.00005-9

advancements in the area of CV, automatic HC detection is proposed in this chapter employing the efficient state-of-the-art object detection algorithm "You Only Look Once" (YOLO). YOLO is a very adept convolutional neural network (CNN) that runs in real time for object detection. Object detection is rather more intricate than classification as it not only identifies the objects but also indicates where the location of the object in the image. YOLO detectors are giving a stunning performance in terms of speed, accuracy, and learning capabilities (Redmon & Farhadi, 2018). YOLO applies a single feed forward propagation through the NN speeding up the process of object recognition. Ultimately, the aim of this study revolves on actively pushing the envelope of this technology and to build a model that precisely locates the head of the fetal from US images and consequently measures the fetal HC.

6.2 Related works

Myriad research works have been carried out for measuring the fetal head from ultrasound images. Sobhaninia, Rafiei, et al. (2019) proposed a multiscale light CNN for automatically measuring the HC and it is efficient in terms of time and works with lesser parameters (Van Den Heuvel et al., 2018a). In 2019 Cahya Perbawa Aji, Muhammad Hilman Fatoni, and Tri Arief Sardjono proposed a CNN model to semantically segment fetal head from maternal tissues (Aji et al., 2019). Nadiyah et al. (2019) experimented using Haar Cascade and Fit Ellipse method to extract the curve shape of the fetal head (Nadiyah et al., 2019). Li et al. (2018) incorporated the prior knowledge about the gestational weeks and ultrasound scanning into a very powerful RF Tree classifier to localize the fetal head (Li et al., 2018). Hanene Sahli et al. described a statistical study of main biometric parameters for fetal head anomaly diagnostics (Ni et al., 2013). Sinclair et al. (2018) proposed a fully connected CNN to determine the measurements of fetal HC that performed similar to a human expert (Rajinikanth et al., 2019; Satwika et al., 2014) described an elliptical detection approach to spot and estimate the head measurement of the fetal automatically using the Hough transform method and optimized using Particle Swarm Optimization (PSO) (Sahli et al., 2018). Ni et al. (2013) exploited the AdaBoost learning algorithm by using Haar-like features to train the classifier and performed a sliding-window-based head detection (Satwika et al., 2014). Sinclair et al. (2018) explored the performance of fully CNN model for automatic head biometrics in US images

(Sinclair et al., 2018). Sobhaninia et al. proposed a light CNN for automatic HC measurement that takes fewer parameters and less training time (Sobhaninia, Emami, et al., 2019). Moving ahead in this direction to isolate the fetal head from US images and estimate its circumference is the ultimate motive of this chapter.

6.3 The dataset

The HC Grand Challenge Dataset comprises 1334 2D US images taken at the standard plane position which aids in measuring the HC (Van Den Heuvel et al., 2018b). This dataset is publicly available and it facilitates a platform to share and compare the different methodologies and algorithms developed for estimating the fetal HC from 2D US images (Van Den Heuvel et al., 2018b).

Out of the total 1334 images, 999 images are taken as training images and the left over 335 images are set aside for the phase of testing. Each of the ultrasound images is of size 800*540 pixels with a pixel size spanning from 0.052 to 0.326 mm in size. The comma separated value files (CSV) contain the pixel size for each image. Manual annotation of the HC for each of the training set images is also provided. Out of the 999 images present in the training set, few files are not considered as they have similar appearances (Fig. 6.1).

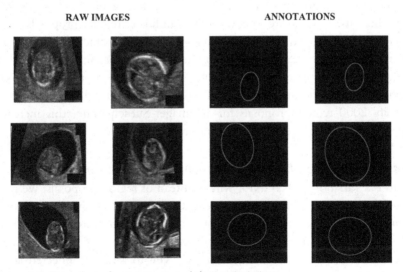

Figure 6.1 Snap shot of raw images and their annotations.

6.4 Object detection

Unlike classification, the main aim of object detection is to locate the object via bounding boxes (BBs) and also give the classification of the located objects. In Image classification problem, the assumption is that there is only one main target object and the focus is to identify or classify the target. In Object detection approach, the aim is not just to classify but to get the position of the target on the image. There are twofold steps in object detection algorithm:

- Localize the object by drawing a BB around each object.
- Classify the bounded Region Of Interest (ROI).

The output of these object detection will be the box coordinates and the class probabilities. There are many object detection algorithms like faster R-CNN (region-based CNN) family of networks, SSD and YOLO family of networks. These are anchor-based network structures. Among them, YOLO has a very clear advantage of running speed and better performance in detecting smaller and objects that are farther.

6.4.1 Object detection algorithms

6.4.1.1 Region-based convolutional neural network

The R-CNN utilizes a region proposal algorithm with the sole aim to overcome the difficulty of selecting huge number of regions. The selective search algorithm uses the greedy principle to generate candidate regions, by recursively combining identical areas into bigger ones. These regions are then used to create the ultimate candidate region proposals. The CNN undertakes the responsibility of feature extraction and these extracted features are provided as input to a classification algorithm for detecting the presence of objects within the region proposals. R-CNN spends a large amount of time in the network training phase as it has to classify 2000 region proposals for an image. Subsequently, this increases the time taken to classify the images in real environment.

6.4.1.2 Fast region-based convolutional neural network

This is a variation of R-CNN, where instead of region proposals; we provide the original source image to the ConvNet to produce a convolutional feature map. We identify the region of proposals from these extracted feature maps and wrap them using ROI pooling layer, reshaping them to predetermined size that can be given to a fully connected layer. Softmax layer is used to classify the proposed region and also the offset values for the BB. This alternative way of CNN is faster as the convolution dot operation is performed just only once for every image for feature extraction.

6.4.1.3 Faster region-based convolutional neural network

Selective search is a slow-moving and extensive process that affects the network performance. This faster R-CNN eliminated the selective search process and allows the network to learn the region proposals. The image is provided as input to a CNN that produces feature map. A dedicated network is used to predict the region proposals. ROI Pooling layer is added to reshape the predicted region proposals. Finally, the image contained in the proposed region is classified along with the predicted offset values of the BBs are given as outcome.

6.4.1.4 How YOLO works?

Almost all the above-discussed algorithms use regions for object localizations. They do not scan the complete image, but they speculate only on those sections of the image that is anticipated to have greater probabilities of containing the object. In YOLO a single CNN predicts the BBs and confidence scores for these boxes. The algorithm begins by taking a source image and splits it into S*S cells, within each of the grid we take "n" BBs. The network outputs confidence score of the classes for each of the BBs. If the probability is above an upper limit value, the BB can be is utilized to pinpoint the object within the image. Putting it all together, YOLO is faster and efficient than other object detection algorithms.

6.5 YOLO family of algorithms

YOLO is a very intelligible real-time object detection CNN. It runs with a single NN applied to the entire image, dividing the image into regions, and predicts the BBs and class probabilities of each region. It takes only one forward pass to make the prediction. A single CNN simultaneously predicts multiple BBs and class probabilities for each of those boxes. Furthermore, as YOLO is trained on full images, it directly optimizes the detection performance. This realistic model has significant advantages and can give a competitive edge over the conventional approaches for object detection.

6.5.1 YOLOv1

First, YOLO partitions the input image into S*S grid, say 7*7. Each 7*7 detects B BBs B = 2. If the objects center lies on a grid cell, then that particular grid cell is liable for detecting the object. Classification and localization are run on each grid cell simultaneously, which makes YOLO a very fast model. The model outputs a confidence score, C. Any BB with the higher intersection over union (IOU) with the ground truth will bound equal object. Higher the IOU, higher is the probability of detection. YOLO was trained with 20 output classes, BBs = 2, grid cell = 7*7. 98 boxes in total are predicted and we drop of the boxes with very less confidence score. The offset predictions are arranged as S*S* (B*5 + number of classes) tensors (Fig. 6.2).

The network uses a sum of squared error as the loss function as optimization becomes easier. With 24 convolutional layers and 2 fully connected layers, trained on ImageNet 1000 class dataset, the final output is 7*7*30 tensors. Occasionally most of the grid cells may not hold any object. Hence, we predict the class probability, pc, which decides to get

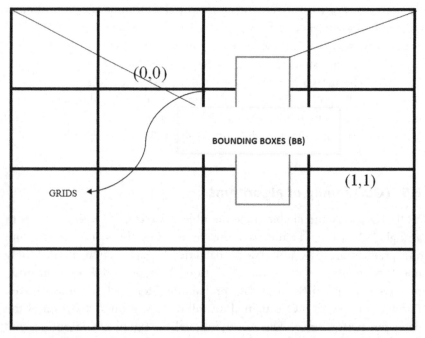

Figure 6.2 Grids and bounding boxes—components of YOLO. *YOLO,* You Only Look Once.

rid of the boxes with low confidence scores and the BBs with the high overlapping area using a technique called Non-Max Suppression (NMS).

6.5.2 YOLOv2

YOLOv1 can detect only 49 objects and there is significant localization error. Batch normalization and high-resolution classifier network were the crucial factors in YOLOv2. The convergence of the model was achieved by adding batch normalization. YOLOV2 trains on higher resolution images (448*448) and apparently it provides the network the adequate time to regulate the filters necessary for higher resolution. Also, multiple object detection per grid cell was introduced. Intuitively, rather than predicting the box relative to the image, anchor boxes are defined to predict offsets and confidences for anchor boxes. YOLOV2 which is a high-resolution classifier was introduced with the sole aim of improving the accuracy. YOLOv2 employed accuracy improvements techniques like batch normalization and convolutional with anchor boxes. Anchor boxes also called prior boxes are estimated rather than using arbitrary BBs. These anchor boxes will make the initial training stable as we focus on specific shape. Using the anchor box approach we move prediction from cell level to boundary box level and the probability of detecting all the ground-truth objects increases. Darknet-19 is the central foundation for YOLOv2. This architecture has lower processing requirements than other architectures. 10 convolutional layers and 5 maxpooling layers are present in Darknet-19. For pascal visual object classes format, we predict 5 boxes with 5 coordinates [tx, ty, tw, th, to (objectness score)] each with *20 classes* per box. So the total number of filters is 125. To eliminate choosing priors manually, k-means clustering is run on the training set BBs to precisely find the prior boxes that lead to good IOU. K = 5 strikes a perfect balance between the complexity of model and recall (Fig. 6.3).

6.5.3 YOLOv3

To boost the accuracy, the complexity of the underlying architecture of Darknet was increased. This gave rise to a new hybrid network. Apart from borrowing the inspiration from ResNet and feature-pyramid network architecture, YOLOv3 uses DarkNet-53, where 53 layer networks were trained on ImageNet and 53 additional layers were stacked for the task of object detection. YOLOv3 incorporates residual skip connections and upsampling. YOLOv3 performs detection at three levels of scales by

Type	Filters	Size/Stride	Output
Convolutional	32	3*3	224*224
Maxpool		2*2 / 2	112*112
Convolutional	64	3*3	112*112
MaxPool		2*2/2	56*56
Convolutional	128	3*3	56*56
Convolutional	64	1*1	56*56
Convolutional	128	3*3	56*56
MaxPool		2*2/2	28*28
Convolutional	256	3*3	28*28
Convolutional	128	1*1	28*28
Convolutional	256	3*3	28*28
MaxPool		2*2/2	14*14
Convolutional	512	3*3	14*14
Convolutional	256	1*1	14*14
Convolutional	512	3*3	14*14
Convolutional	256	1*1	14*14
Convolutional	512	3*3	14*14
MaxPool		2*2/2	7*7
Convolutional	1024	3*3	7*7
Convolutional	512	1*1	7*7
Convolutional	1024	3*3	7*7
Convolutional	512	1*1	7*7
Convolutional	1024	3*3	7*7
Convolutional	1000	1*1	7*7
AvgPool		Global	1000
Softmax			

Figure 6.3 Darknet-19 architecture.

reducing the spatial dimensions of the input image by 32, 16 and 8, respectively. FAIR (FB AI Research) developed a tropical arrangement known as "feature–pyramid" in 2017. This feature–pyramid allows feature map to gradually decrease in spatial dimension, but later it increases and it is combined with previous feature maps with corresponding sizes. The process is recursively done and each of the combined feature maps is branched out to a separate detection path. Each of the cell in the output layer's feature map detects 3 boxes for YOLOV3 and 5 boxes for

YOLOV2—one box per anchor. The 4 coordinates for each predicted BB contains box center offsets, box size scales, objectness score, and the class score values. Logistic regression is used for predicting the objectness score with the loss function of binary cross entropy for the prediction of the class.

6.6 Proposed methodology

This proposed work aims at employing the YOLO framework for object localization to decide "where it is?" and "what it is?" YOLO looks into the full image in one swoop and forecasts the BB parameters and the probabilities of the classes for the boxes. YOLO is incredibly faster and it can process at a dramatic speed understanding the generalized object representation. The proposed work takes up an input image and converts the gray-scale images into RGB images. The images are then resized to 224*224*3 color channels. The essential step in a supervised image classification task is to perform data labeling to mark rectangular ROI labels. The Image Labeller application is used for this purpose by performing the same on each of the training set images and to label the ground truth manually. The labeled ground truth is exported and used for training. The BBs are rectangular boxes that define the location of target object. They are determined by the coordinate positions (x,y) of the upper-left and bottom-right corner of the enclosing box.

- Divides the input image into grids (S*S) and image localization and classification are applied on each grid. We choose S = 5
- The labeled data is passed to the model to train it.
- The grid cell that holds the objects center is responsible for detecting the object and this continues for each grid cells.
- YOLO detects the BB parameters and their respective confidence score which is determined by the vector as given in (6.1).

$$Y = \{cs, boxx, boxy, boxh, boxw, class1, class2, class3\} \qquad (6.1)$$

wherecs—confidence or score, takes 0 or 1boxx, boxy, boxh, boxw—denote the BB coordinate values, if the object is presentclass1, class2, class3—denote the different classes, here only one class, c1 = "Head."

- IOU and NMS are applied on the predicted boxes to get highly accurate BBs for localizing the ROIs.

The major idea behind YOLO is to construct a CNN to reduce the spatial dimension. The prediction of YOLO has a shape of (5*5*6*N) with the estimated number of anchor boxes. The training process takes an input 2D image of any shape and maps it with the target shape of 5*5*6*N. This varies based on the size of grid, number of anchors, and number of classes. All the grid cells simultaneously predict the BBs and YOLO is dramatically faster. In spite of the challenging nature of the US images such as the low light, presence of artifacts, signal-to-noise ratio, and noise interference, YOLO was able to localize the fetal head on the US images (Fig. 6.4).

6.6.1 Representation of bounding boxes

A BB is a rectangle that encloses an object. Intuitively, a ground-truth BB can be represented as [x,y,w,h] where x, y are coordinate points of the BBs, w and h are the width of the BB. Coordinates of the BBs corners are calculated with reference to the upper-left corner of the image with coordinates (0,0). The different formats of BBs are pascal-voc, albumentations, coco, yolo format. In Pascal_voc format, the BBs are encoded with 4 values in pixels [min_x, min_y. max_x, max_y]. Albumentations format is similar to pascal_voc but they use normalized values. COCO (common objects in context) format uses BB defined by the four values in pixels [x-min,y_min, width, height]. This is the centroid representation of the BBs. There are other ways of representing the BBs like corner representation

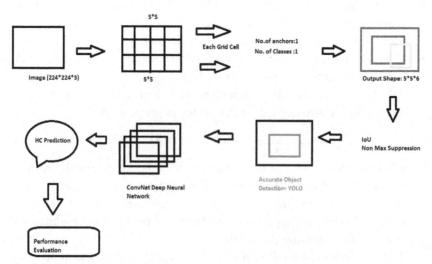

Figure 6.4 Process flow of the proposed work.

and min–max representation. In this work the centroid representation of BB is used to locate the fetal head.

6.6.2 Encoding anchor boxes

Anchor boxes, also called prior boxes, are predefined fix-sized BBs on image input or feature map. Anchor boxes can be configured to create predictors that detect objects on the grid. The BB regressor predicts the offset of the manually encoded ground-truth box and the predicted BB to the anchor box rather than predicting the BB location on the image, Undoubtedly, choosing the anchor box size is crucial for the network to detect the object. This is one of the important parameters to be tuned for greater performance on the dataset. These anchor boxes represent the ideal size, shape, and location and size of the object they specialize in predicting. Fig 6.6 represents the plot between mean IOU and number of anchors. High IOU is obtained when number of anchors increases (Figs. 6.5 and 6.6).

6.6.3 Intersection over union

For each of the anchor box, calculate the IOU by dividing the overlapping area by the union area. The area of union is the area occupied by the predicted BB and the ground-truth BB. If the IOU is >50%, detect the object that gave the highest IOU. True detection is ambiguous, if

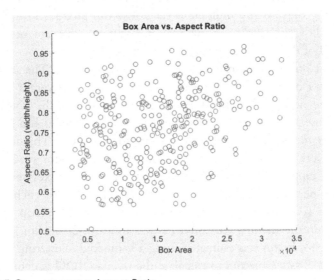

Figure 6.5 Box area versus Aspect Ratio.

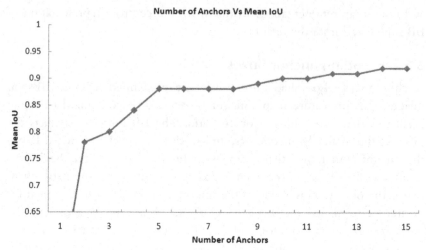

Figure 6.6 Number of anchors versus mean IOU. *IOU*, Intersection over union.

IOU > 40 and if the highest IOU < 40% then there is no object. The IOU is approximately 1, if the IOU lies between the predicted BB and ground-truth box overlaps perfectly.

6.6.4 Estimating the anchor box size

The term "anchor box" is defined as a set of predefined BBs of specific height and weight. They help in detecting objects of different scales, overlapping objects, and multiple objects. We can produce several anchor boxes with varied shapes (horizontal and vertical) that are centered on each pixel of the image (Fig. 6.7).

Anchor boxes are crucial parameters of deep learning. The shape, size, and number of anchor boxes greatly influence the effectiveness and accuracy of the detector. Anchor boxes are estimated from the training data and they use IOU distance metrics. The number of anchors to be used is a hyperparameter that demands careful empirical analysis. The numbers of anchors were analyzed by ranging from 1 to maximum of 20 in this work. An empirical analysis of the number of anchor boxes versus mean IOU was performed to estimate the required number of anchor boxes. The mean IOU > 0.5 ascertains that the anchor boxes overlap well with the boxes in the training data. With a maximum of 20 anchor boxes, the size of anchor boxes was estimated using a k-medoids clustering algorithm with IOU distance metrics to calculate the overlap. IOU is defined as the ratio between the intersection area and the union area of two BBs as

Anchor Box 1   Anchor Box 2

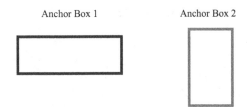

Figure 6.7 Anchor boxes in varied shapes.

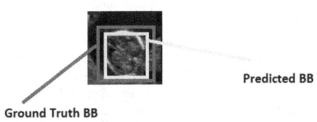

Predicted BB

Ground Truth BB

Figure 6.8 Visual example of ground-truth BB versus predicted BB with good IOU. *BB*, bounding box; *IOU*, intersection over union.

given by (6.2). The IOU ranges between 0 and 1 where 0 indicates no overlapping and 1 indicates both the anchor box and ground-truth BBs are equal. IOU is taken as an evaluation metric to gauge the accuracy of the YOLO detector (Fig. 6.8).

$$IOU = \frac{\text{Area of overlap}}{\text{Area of union}} \qquad (6.2)$$

6.6.5 Defining network layers

Once the best maximum number of anchors and the size of the anchor boxes are estimated, the deep NN training process commences. YOLOV2 uses ResNet-50 pretrained network as the base network. Activation layer 49 of the Resnet-50 network is used as the feature extraction layer. The consequent layers are frozen and a sequence of convolution, ReLu, batch normalization, YOLOv2 layers is added with the feature layer of the base network by passing the anchor boxes. Other network training options such as optimizer, mini batch size, learning rate, epochs count are fixed based on tuning and investigation.

6.7 Experimental outcomes—training

Table 6.1 presents the outcome of the training done on a single CPU for detecting the object class-"Head" using the manually crafted ground-truth BBs. This is a snap shot of training for an epoch of 20. The Root Mean Square Error (RMSE) is 1.12 and the loss is 1.3.

6.7.1 Testing

The trained YOLOV2 detector for fetal head localizations was applied to the test images to detect the BBs and their corresponding scores. The confidence score shows the likelihood of the presence of fetal head in the box and the accurate fitness of the object in the BB. The conditional class probability represents the likelihood of the object that is detected fit into a particular class (one probability per category for each cell). To make final prediction, we only have the images with high box confidence scores and suppress the rest (Fig. 6.9).

IOU and NMS can be used to improve the outcome of predicted box. The area of the IOU is calculated for the actual and predicted boxes as defined by Eq. (6.2). If the calculated IOU is >0.5, it is considered to be a good prediction (Fig. 6.10).

In Fig. 6.10, NMS was applied and the BB with high confidence is taken for account by discarding the boxes having probabilities less than the predefined threshold. We sift through all the BBs using NMS concept, keeping only the near-perfect fit boxes, and getting rid of the overlapping ones. Anchor boxes for each grid are not applied here as the ultrasound image does not have more than one instance of "head" in the grids and the region of interest in only one object appearing only one time in the image. Hence, anchor box concepts are not considered.

Table 6.1 Training YOLOv2 object detector for head localization.

Checking the training data
Training is done on single CPU
Initializing input data normalization

Epoch	Iteration	Time elapsed (hh:mm:ss)	Mini batch RMSE	Mini batch loss	Base learning rate
1	1	00:00:26	8.32	69.3	0.0010
6	50	00:26:33	2.02	4.1	0.0010
12	100	00:46:34	2.32	5.4	0.0010
17	150	01:02:54	2.37	5.6	0.0010
20	180	01:12:39	1.12	1.3	0.0010

Detector training complete.

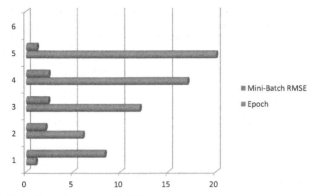

Figure 6.9 Plot of epochs versus RMSE.

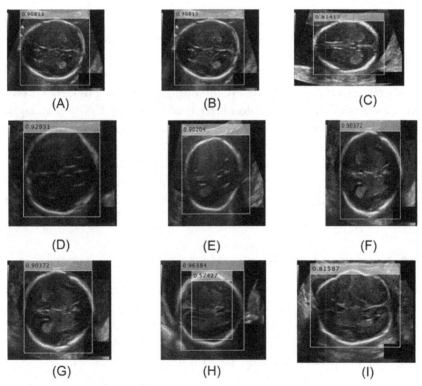

Figure 6.10 Figure (A) to (I) shows US images with the bounding boxes and their confidence scores. *YOLO*, You Only Look Once. Fro.

6.7.2 Hyperparameters setting—YOLOV2 object detector

The configurable parameters for optimizing the network are experimented and evaluated. These include CNN architecture (layers and activations), anchor boxes, number of classes, learning rate, probability score threshold, batch size, etc.

- Optimize and learning rate: sgdm with learning rate of 0.001
- Loss function: RMSE
- Mini batch size: 64
- Max epochs: 50
- Shuffling: Yes
- Validation data: 70−30 Split

6.7.3 Training the convolutional neural network for regression

ConvNets are vital tools in deep learning for analyzing image data. To predict the HC as continuous numerical values, a regression layer at the end of the network is incorporated (Fig. 6.11).

These localized ROIs (head) are cropped and given as input to a CNN regressor to estimate the HC by fitting a regression model. To predict continuous data, a regression layer is added to the CNN. The layers of the CNN architecture are defined: an input image layer, the middle layers; convolution layer, normalization, and ReLU layers with average pooling layers that performs the computation are added. Finally, a FC layer, subsequently followed by a regression layer, is created. Fig. 6.12 depicts the training progress plot to monitor network accuracy during the training.

6.8 Results and discussion

The model created for estimating the fetal HC was tested against the test data of 335 images to predict the HC value and the performance is evaluated. RMSE is the standard deviation of the errors and is used to measure the differences in observed and predicted value. The RMSE obtained was 0.43. The lower the RMSE, the better is the predicted model.

$$RMSE = \sqrt{(f-o)^2} \qquad (6.3)$$

where f forecast values and o observed values. Since most of the grids are empty with no objects, this makes the probabilities of those grid cells to

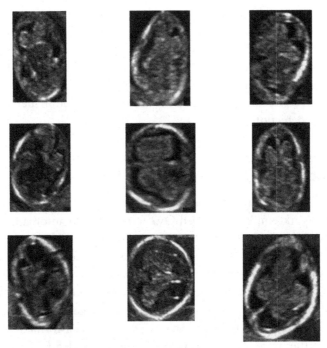

Figure 6.11 Localized and cropped ROI.

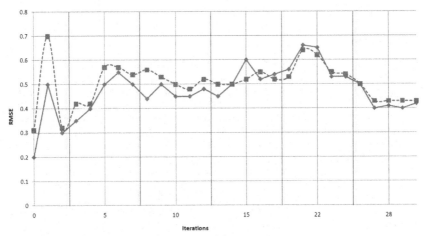

Figure 6.12 CNN Training Progress for estimation of HC.

zero that perfectly match with the ground-truth confidence. This guides to divergence and so we pick a threshold value of 0.5 to decrease the loss involved. Moreover, as Pretrained YOLOV2 weights were used with

Darknet-19 network, the advantages are numerous. Improved BB prediction and more accurate class predictions further lead to improved mean average precision. With the substantial experimentation of tuning of hyperparameters, the network yielded a good performance of 88%. The Challenges involved are inflated storage, hardware costs, and time consumption. Despite the challenges, the futuristic algorithm performs real-time object detection with ease.

6.9 Conclusion

Fetal HC is one of the vital parameters that decide the well-being of the fetal. This work employs the YOLOv2 framework to automatically localize the head of the fetal on a 2D Ultrasound scan image using CNN and produce a BB. This ROI is segmented further and fed to a CNN regressor for estimation of fetal head size. The work can be further improved by tuning and investigating the hyperparameters and optimizing the performance of the CNN. Data augmentation strategies could also be integrated to have more training data to form a well-formed prediction model. Furthermore, it has provided a smooth tradeoff between speed and accuracy.

References

Aji, C. P., Fatoni, M. H., & Sardjono, T. A. (2019). *Automatic measurement of fetal head circumference from 2-dimensional ultrasound,"* 2019 *International Conference on Computer Engineering, Network, and Intelligent Multimedia (CENIM)* (pp. 1–5). IEEE. doi:10.1109/CENIM48368.2019.8973258.

Li, J., Wang, Y., Lei, B., Cheng, J.-Z., Qin, J., Wang, T., Li, S., & Ni, D. (2018). Automatic fetal head circumference measurement in ultrasound using random forest and fast ellipse fitting". *IEEE Journal of Biomedical and Health Informatics, 22*(1), 215–223. Available from https://doi.org/10.1109/JBHI.2017.2703890.

Nadiyah, P., Rofiqah, N., Firdaus, Q., Sigit, R., & Yuniarti, H. (2019). *Automatic detection of fetal head using haar cascade and fit ellipse.* 2019 *International Seminar on Intelligent Technology and Its Applications (ISITIA)* (pp. 320–324). IEEE.

Ni, D., Yang, Y., Li, S., Qin, J., Ouyang, S., Wang, T., & Heng, P. A. (2013). Learning based automatic head detection and measurement from fetal ultrasound images via prior knowledge and imaging parameters. In: *2013 IEEE 10th international symposium on biomedical imaging* (pp. 772–775). Available from https://doi.org/10.1109/ISBI.2013.6556589.

Rajinikanth, V., Dey, N., Kumar, R., Panneerselvam, J., & Sri Madhava Raja, N. (2019). Fetal head periphery extraction from ultrasound image using JAYA algorithm and Chan-Vese segmentation. *Procedia Computer Science, 152,* 66–73. Available from https://doi.org/10.1016/j.procs.2019.05.028.

Redmon, J., & Farhadi, A. (2018). Yolov3: An incremental improvement. arXiv preprint arXiv:1804.02767.

Sahli, H., Mouelhi, A., Hadada, F., Rachdi, R., Sayadi, M., & Fnaiech, F. (2018). Statistical analysis based on biometrie measures for fetal head anomaly characterization. In: *2018 IEEE 4th Middle East conference on biomedical engineering (MECBME)* (pp. 237−242). Available from https://doi.org/10.1109/MECBME.2018.8402440.

Satwika, I. P., Habibie, I., Ma'Sum, M. A., Febrian, A., & Budianto, E. (2014). Particle swarm optimation based 2-dimensional randomized hough transform for fetal head biometry detection and approximation in ultrasound imaging. In: *2014 International conference on advanced computer science and information system* (pp. 468−473). Available from https://doi.org/10.1109/ICACSIS.2014.7065898.

Sinclair, M., Baumgartner, C. F., Matthew, J., Bai, W., Martinez, J. C., Li, Y., Smith, S., Knight, C. L., Kainz, B., Hajnal, J., King, A. P., & Rueckert, D. (2018). Human-level performance on automatic head biometrics in fetal ultrasound using fully convolutional neural networks. In: *2018 40th annual international conference of the IEEE engineering in medicine and biology society (EMBC)* (pp. 714−717). Available from https://doi.org/10.1109/EMBC.2018.8512278.

Sobhaninia, Z., Rafiei, S., Emami, A., Karimi, N., Najarian, K., Samavi, S., & Reza Soroushmehr, S. M. (2019). Fetal ultrasound image segmentation for measuring biometric parameters using multi-task deep learning. In: *2019 41st annual international conference of the IEEE Engineering in Medicine and Biology Society (EMBC)* (pp. 6545−6548). Available from https://doi.org/10.1109/EMBC.2019.8856981.

Sobhaninia, Z., Emami, A., Karimi, N., & Samavi, S. (2019). Localization of fetal head in ultrasound images by multiscale view and deep neural networks. *arXiv*, 1−5.

Van Den Heuvel, T. L. A., Dagmar De, B., De Korte, C. L., & Bram Van, G. (2018a). Automated measurement of fetal head circumference using 2D ultrasound images. *PLoS One, 13*(8), e0200412.

Van Den Heuvel, T. L. A., Dagmar De, B., De Korte, C. L., & Bram Van, G. (2018b). *Automated measurement of fetal head circumference [Data set].* Zenodo. Available from https://doi.org/10.5281/zenodo.1322001.

Zhang, J., Petitjean, C., & Lopez, P. (2020). Direct estimation of fetal head circumference from ultrasound images based on regression CNN. In: *Proceedings of machine learning research* (pp. 1−9).

Further reading

Bell, S., Zitnick, C. L., Bala, K., & Girshick, R. (2015). Insideoutside net: Detecting objects in context with skip pooling and recurrent neural networks. *arXiv preprint arXiv:1512.04143.*

Bourdev, L., & Malik, J. (2009). Poselets: Body part detectors trained using 3d human pose annotations. In: *International conference on computer vision* (ICCV).

Ginosar, S., Haas, D., Brown, T., & Malik, J. (2014). *Detecting people in cubist art. Computer vision—ECCV 2014 workshops* (pp. 101−116). Springer.

Girshick, R., Donahue, J., Darrell, T., & Malik, J. (2014). Rich feature hierarchies for accurate object detection and semantic segmentation. In: *Computer vision and pattern recognition (CVPR), 2014 IEEE conference on* (pp. 580−587). IEEE.

Hoiem, D., Chodpathumwan, Y., & Dai, Q. (2012). *Diagnosing error in object detectors. Computer vision—ECCV 2012* (pp. 340−353). Springer.

Krizhevsky, A., Sutskever, I., & Hinton, G. E. (2012). ImageNet classification with deep convolutional neural networks. In: *Advances in neural information processing systems* (pp. 1097−1105).

Lin, T.-Y., Maire, M., Belongie, S., Hays, J., Perona, P., Ramanan, D., Dollar, P., & Zitnick, C. L. (2014). *Microsoft coco: Common objects in context. European conference on computer vision* (pp. 740–755). Springer.

Redmon, J. (2013–2016). *Darknet: Open source neural networks in C.* http://pjreddie.com/darknet/.

Schmidt, U., Temerinac, D., Bildstein, K., Tuschy, B., Mayer, J., Sütterlin, M., Siemer, J., & Kehl, S. (2014). Finding the most accurate method to measure head circumference for fetal weight estimation. *European Journal of Obstetrics & Gynecology and Reproductive Biology, 178*, 153–156. Available from https://doi.org/10.1016/j.ejogrb.2014.03.047.

CHAPTER 7

FunNet: a deep learning network for the detection of age-related macular degeneration

Anju Thomas, P.M. Harikrishnan and Varun P. Gopi
Department of Electronic and Communication Engineering, National Institute of Technology
Tiruchirappalli, Tiruchirappalli, India

7.1 Introduction

Vision helps us to provide knowledge about our surroundings. We learn almost 80% of things through our sight. Macula is a thin oval-shaped structure found in the retina's central region, as shown in Fig. 7.1, which provides the central and high-resolution color vision. The damage to the macula leads to the impairment of vision, and it is called macular degeneration; this is commonly occurring in aged people. According to new studies on age-related macular degeneration (AMD), total of 8.7% of the global population has undergone AMD disease, and about 196 million in the year 2020, and that will rise to 288 million in the year 2040 (Wong et al., 2014).

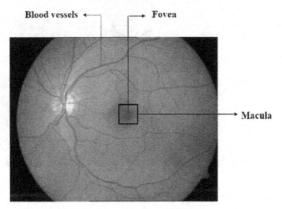

Figure 7.1 Eye anatomy (ARIA Dataset, Walter, 2002).

Edge-of-Things in Personalized Healthcare Support Systems
DOI: https://doi.org/10.1016/B978-0-323-90585-5.00006-0

AMD are of two types: wet AMD (WAMD) and dry AMD (DAMD), as shown in Fig. 7.2. The primary stage of AMD is called DAMD, or it is less severe than WAMD. Among that, 85%—90% of AMD belongs to the dry category. The main causes of DAMD are thinning of macular tissue, the deposit of pigments in the macula, or the combination of both. The diagnosis of DAMD is to find a yellowish spot (which is known as drusen) accumulate in and around the macula. Abnormal blood vessel growth beneath the retinal part leads to blood leakage, which is the main reason for WAMD. This leakage of blood results in permanent damage to the retina cell's light-sensitive part and generates a central blind spot. AMD's permanent solution is not present, and we can control it by using a healthy diet and exercise. So the early detection of AMD is very important, and the severity is detected by analyzing the size and number of drusen present in the retina. Dr is widely classified into two types, nonproliferative Dr and proliferative Dr (PDR) (IDF Diabetes Atlas, 2006), or six stages by the occurrence of various symptoms in the retina. Severe damages due to Dr can be reduced or excluded if appropriately identified and frequently at the primary step. Fig. 7.1 shows a normal and a diseased eye.

Manually the AMD detection is possible by identifying the presence of drusen in the fundus image. This process can be changed into an automatic detection method, but the main problem is that the machine cannot identify the drusen easily like a human. The machine needs some more

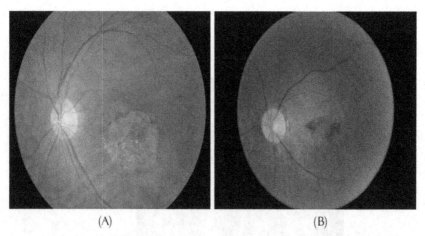

(A)   (B)

Figure 7.2 Types of AMD: (A) dry and (B) wet (Hoover & Goldbaum, 2003; Hoover et al., 2000).

parameters to distinguish the AMD images from normal, so the relevant features like size, shape, color, textures, etc. help classify images more precisely (Pead et al., 2019). There is no distinct shape for drusen, so the conventional method, like edge detection and morphological operations, is not an effective method to detect AMD from fundus photographs.

One of the most trending research fields is artificial intelligence (AI) and deep learning (DL) techniques (Tan et al., 2018). AI mainly uses human intelligence, which is applied to the computer program for a specific task to achieve a particular goal. Machine learning (ML) works under AI, and DL is a subset of ML. Usually, the medical field prefers DL techniques because many data are used to detect a particular disease, which provides an accurate result. The human brain inspires the working of DL; messages are passed through neurons in the neural system. The DL has three sections: input layers, hidden layers, and the output layers [fully connected layers (FCLs)]. Normally each layer performs convolutional operations on the image to be known as a convolutional neural network (CNN). Each layer produces different features, and machines can discover a particular pattern corresponding to each category. If an unknown data is given to that network, it creates a pattern of the input image and tries to find out which category pattern is similar to that and map it into that category.

7.2 Background

There are a lot of works carried out in the area of AMD detection. Those methods are listed in Table 7.1.

From Table 7.1, it is clear that there are a lot of works carried out in fundus images for the detection of AMD using different type of CNN network. Hijazi et al. (2014), Mookiah et al. (2015a), Mookiah et al. (2015b), Tan et al. (2018), and Zheng et al. (2012) are got noticeable accuracy which is greater than 95%, but highly sensitive to noise. Acharya UR (Acharya et al., 2016) proposed an algorithm using random transform and discrete wavelet transform for feature extraction. The extracted features were subjected to the locality-sensitive discriminant analysis method to reduce the dimension by utilizing a t-test. Finally, different classifiers were used for the input image classification. The technique produced 100% accuracy in the STARE dataset. The limitation of the work may be to identify the early stage of AMD.

Table 7.1 Summary of existing methods in AMD detection.

Author	Methods	Accuracy	Dataset	Comments
Kose et al. (2008)	Region growing	90%	(Private)	Sensitive to threshold value
Hijazi et al. (2010)	Dynamic time warping (DTW)	75%	ARIA	Optic disk and blood vessel removal is required before DTW
Hijazi et al. (2012)	Spatial histogram	74%	ARIA	Removal of blood vessels and optic
Zheng et al. (2012)	Hierarchical decomposition	99.6%	ARIA + STARE	Blood vessel removal is required
Hijazi et al. (2014)	Wavelet and GLCM	99.9%	ARIA + STARE	Tree-based approach
Mookiah et al. (2014a) and Mookiah et al. (2014b)	Wavelet transform	93.70%	Private	Tested only on private dataset
Mookiah et al. (2014a) and Mookiah et al. (2014b)	Gabor wavelet, bispectrum entropy, fractal dimension	90.19% 95.07% 95%	Private ARIA STARE	This method is more robust to noise
Mookiah et al. (2015a) and Mookiah et al. (2015b)	Linear configuration model	93.52% 91.36% 97.78%	Private ARIA STARE	22 features were used
Acharya UR (Acharya et al., 2016)	Radom transform and discrete wavelet transform	99.49% 96.89% 100%	Private ARIA STARE	Limited to defining the early phase of AMD

(*Continued*)

Table 7.1 (Continued)

Author	Methods	Accuracy	Dataset	Comments
Tan et al. (2018)	14-layer deep CNN	95.45%	Private	More learnable parameters due to 14-layer CNN
Garcia-Floriano et al. (2019)	Morphological operations	92.15%	Private	Sensitive to threshold value
Peng et al. (2019)	1. Drusen-Net (D-Net) 2. Pigment-Net (P-Net) 3. Late AMD Net (LA-Net)	85.45%	NIH-AREDS	High-quality image required.
Grassmann et al. (2018)	Preprocessing done on input images, six different networks are used for feature extraction 1. AlexNet 2. GooLeNet 3. VGG-16 (11 CONV. only) 4. Inception v-3 5. ResNet (101 layers) 6. Inception ResNet v-2, random forest (RF) classifier	89.25%	NIH-AREDS	13-class classification with quadratic weighted K of 92%
Yoo et al. (2019)	Transfer learning using VGG-16 and RF classifier	83.5%	Project macula	1. Input images are cropped manually for the requirement 2. Worked on small dataset

(Continued)

Table 7.1 (Continued)

Author	Methods	Accuracy	Dataset	Comments
Burlina et al. (2017)	Pretrained overfeat features and linear SVM classifier (LSVM)	85%	NIH-AREDS	Less accuracy
Xu et al. (2020)	ResNet-50 and RF classifier	75%	Private	1. Only worked with a limited collection of data 2. Data imbalance in dry and wet AMD

The literature demonstrated the usefulness of the conventional CNN in detecting AMD from fundus images due to its effective high-level feature extraction. In the proposed work utilizes the CNN network as the feature extraction method and different types of classifier named as support vector machine (SVM), random forest, and J48.

7.3 Proposed method

Generally a CNN network consists of different layers like convolution, pooling, and FCLs for the deep study of the features in the image. Each layer has its own purpose; mainly detailed features are extracted using convolution layer (Tan et al., 2018). Pooling will help to reduce the dimension and finally relevant features are selected in the fully connected section. There has been several work consist of the detection of diseases using medical imaging. FunNet is proposed in this work for the feature extraction purpose and different classification method has performed on this. FunNet has a multipath (MP) structure for the extraction of global features from the fundus image. The concatenation function is performed to concatenate the features from the primary path and MP. Different classifiers like J48, SVM, and random forest are used for the classification of images into AMD or normal.

7.3.1 Details of FunNet

The FunNet network architecture is shown in Fig. 7.3. The input size of FunNet is chosen as 112×112 by using trial and error method. The FunNet consist of two feed-forward paths. The primary path performs normal convolution-based feature extraction. The primary path performs normal convolution-based feature extraction. The hierarchical neural network-based primary path contains three successive single-scale convolutional layers (SSCL) trailed by two FCL. The other path is meant for the extraction of the MP feature. This alternative path is formed by connecting the first SSCL in the primary path to the first FCL in the primary path. The 3×3 filter size is performed on the first SSCL and generates feature maps of 12. Likewise, to create 12 function maps, the second SSCL uses the same filter size. Then, to generate 24 function maps, one more SSCL of the same filter size of padding follows this.

The two paths are concatenated and connected to the first FCL. After two FCL, the final output features are given to classifiers to classify the fundus images into AMD and normal. The first and second FCL contains 128 and 64 hidden neurons. Due to the full connectivity structure, the denser FCL leads to over-fitting. To avoid over-fitting, in the proposed CNN architecture, there is a dropout factor of 0.5 to the first and second FCL (Srivastava et al., 2014). The activation function used to activate the CL is the rectified linear unit (ReLU). The ReLU keeps positive values and suppresses negative values to 0. Compared to other activation functions, ReLU shows better gradient changes (Krizhevsky et al., 2012). Besides, it is easy to implement, and it improves the speed performance of training. After every CL, a max-pooling layer (MPL) is employed to diminish the network's subsequent layers' spatial resolution to keep more significant local structures. Furthermore, the pooling process upturns the receptive field size. Because of this, from the input, the network can learn complex local structures.

7.3.2 Extraction of multipath feature

A method of extracting MP functions is achieved by concatenating the feature maps from various layers. The MP feature extraction permits CNN to merge more robust features regarding the sparse local and fine global structures. The conventional CNN has a direct path to the FCLs to pass extracted features. In a conventional CNN, the initial layers' extraction of features is about the fine global structure. The extracted features

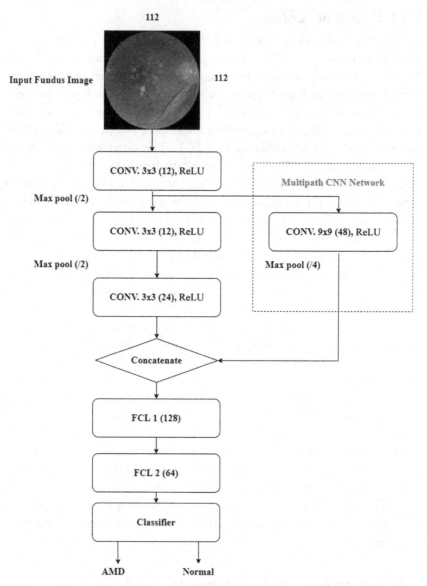

Figure 7.3 Proposed multipath CNN architecture.

become localized and sparser as CNN progresses to the deeper layer, and the global architectures are diminished. The conventional CNN is suitable for extracting local structure features; this may be insufficient to do AMD analysis. The proposed MP method compensates for the losses of global structures through an alternate path. As shown in Fig. 7.3, the alternative route ties the first SSCL and binds the first FCL to a bounce node. MPL is inserted in the alternative direction to concatenate features from two subdivisions (branches). Passing the early layer features into the first FCL, the final merged features can define local and global structures. Thus we get more features, like unique local textures and global features of the AMD. In an alternate path, sometimes passing the early layer's feature may misrepresent the CNN to analyze unlinked global structures. An extra CL with a 9×9 filter size with 48 feature map is inserted into the alternate path to polish the early layer's features for improved depictions of the global structure to avoid this issue. For classification, the 64 relevant features from the FunNet second FCL are fed into various classifiers.

7.3.3 Classifiers
7.3.3.1 J48 classifier
It is a decision tree classifier that makes a binary tree (Sharma et al., 2013). Once the tree constructs, it is given to each sequence in the dataset and results in the classification for that sequence. When building a tree, the classifier disregards the missing values. The fundamental idea is to break the data into ranges depending on the characteristic values contained in training sets for that object (Gayathri et al., 2020a; Gayathri et al., 2020b; Gayathri, Krishna, et al., 2020).

7.3.3.2 Random forest
Random forest is a combination of learning procedure for classification, which consists of a large number of decision trees at training period and outputs the class that is the mean prediction (regression) or the type of the classes (classification) by the distinct trees (Roychowdhury & Banerjee, 2018). The forest picks the classification having the majority votes over all the trees in the forest.

7.3.3.3 Support vector machine
SVM (Kandhasamy et al., 2019) is a supervised learning scheme used for classification purpose. It deals with more input data efficiently and contains a nonlinear transformation. The classification is performed by finding the hyper plane that differentiates the two classes very well (Akbar et al., 2015; Daqi & Tao, 2007).

7.3.4 Feature extraction

The performance of the classification is evaluated based on 10-fold evaluation (Gayathri et al., 2020a; Gayathri et al., 2020b; Gayathri, Krishna, et al., 2020). The proposed work's findings are assessed using the confusion matrix (Gayathri et al., 2020a; Gayathri et al., 2020b; Gayathri, Krishna, et al., 2020; Visa et al., 2011). In Table 7.2, the uncertainty matrix structure representing the characteristics of a binary classifier is seen. In that matrix, the number of True Negatives (TN), True Positives (TP), False Positives (FP), and False Negatives (FN) is P, M, 0, and N, respectively. TP and TN give properly classified data results, while FN and FP give the wrongly classified information. We can measure the precision, F1-score, recall, and accuracy using these values to analyze machine performance. TP and TN generate correctly classified data, while FN and FP produce incorrectly classified data. These values can be used to assess system efficiency by measuring precision, F1-score, recall, and accuracy.

$$\text{Accuracy} = \frac{P+M}{M+N+O+P}, \tag{7.1}$$

$$\text{False-positive rate (FPR)} = \frac{P}{P+O}, \tag{7.2}$$

$$\text{Precision} = \frac{M}{M+N}, \tag{7.3}$$

$$\text{Recall} = \frac{M}{M+O}, \tag{7.4}$$

$$F1 - \text{score} = \frac{2M}{2M+O+N}. \tag{7.5}$$

Table 7.2 The confusion matrix of binary classification.

| | | Predicted class | |
		Normal	Disease
Targeted Class	Normal	M	N
	Disease	O	P

7.3.5 Results

The proposed CNN is implemented using python DL framework. CNN training and validation are performed on a Windows 10 system with Intel i7 @ 3.40 GHz, 64 GB RAM, and Nvidia Geforce RTX 2080 11 GB GPU. The proposed method is tested using standard databases like ARIA, STARE, and a PRIVATE dataset obtained from Joseph eye hospital Tiruchirappalli. The details of the dataset are given in Table 7.3. The backpropagation is used for the FunNet training purpose (Walter, 2002) with 10 as batch size. The FunNet is trained using stochastic gradient descent (SGD) (Bottou, 2012). Learning rate alpha = 0.001, exponential decay rates beta1 = 0.9, beta2 = 0.999, and the numerical value epsilon = 10^{-8} are the default settings for model. Because of its consistency and strong convergence, SGD with momentum is commonly used, despite the fact that it converges slower than the other optimization algorithms. The cumulative number of learnable parameters is 62,87,072 for the proposed MP CNN.

The confusion matrices of the ARIA, STARE, and the PRIVATE dataset is listed in Tables 7.4—7.6, respectively. Table 7.7 lists the weighted average values of output parameters. It demonstrates that the SVM and random forest classifier's output is insufficient for the classification of extraction features. It is clear that the J48 classifier produces fewer misclassifications across all datasets. The average FPR for the J48 classifier is lower, while the other output metrics (precision, recall, and F1-score) are all around 1. The J48 classifier's greatest performance with the suggested CNN feature extraction is narrated by the system's consistency. The performance assessment shows that the suggested technique works well regardless of the amount of training tests, implying that the suggested CNN with J48 classifier is likely to be the best choice for AMD classification across all datasets.

Table 7.3 Distribution of dataset.

Dataset	Category	Number of images
ARIA	Normal	60
	AMD	101
STARE	Normal	37
	AMD	64
PRIVATE	Normal	800
	AMD	461

Table 7.4 Confusion matrix of ARIA dataset with different classifiers.

Classifier	Predicted class	Targeted class Normal	AMD
J48	Normal	60	0
	AMD	1	100
SVM	Normal	57	3
	AMD	2	99
Random forest	Normal	58	2
	AMD	0	101

Table 7.5 Confusion matrix of STARE dataset with different classifiers.

Classifier	Predicted class	Targeted class Normal	AMD
J48	Normal	37	0
	AMD	1	63
SVM	Normal	31	6
	AMD	2	62
Random forest	Normal	31	6
	AMD	2	62

Table 7.6 Confusion matrix of PRIVATE dataset with different classifiers.

Classifier	Predicted class	Targeted class Normal	AMD
J48	Normal	800	0
	AMD	1	460
SVM	Normal	786	14
	AMD	11	450
Random forest	Normal	800	0
	AMD	3	458

7.4 Discussion

FunNet: a DL technique for the detection of AMD from the fundus images is proposed. The FunNet has MP structure for feature extraction, which creates a separate path to concatenate the features obtained from the early and later layers to maintain the later layers' global structures losses. Table 7.8 shows a comparison of the proposed approach to recent

Table 7.7 Confusion matrix of PRIVATE dataset with different classifiers.

Classifier	Dataset	FP rate	Precision	Recall	F1-score	Accuracy (%)
J48	ARIA	0.004	0.994	0.994	0.994	99.38
	STARE	0.006	0.990	0.990	0.990	99.00
	PRIVATE	0.001	0.999	0.999	0.999	99.92
SVM	ARIA	0.039	0.969	0.969	0.969	96.89
	STARE	0.114	0.922	0.921	0.920	92.08
	PRIVATE	0.022	0.980	0.980	0.980	98.02
Random forest	ARIA	0.002	0.996	0.996	0.996	99.61
	STARE	0.114	0.922	0.921	0.920	92.08
	PRIVATE	0.004	0.998	0.998	0.998	99.76

Table 7.8 Comparison with existing methods.

Author	Accuracy(%)	Dataset
Hijazi et al. (2010)	75.00	
Hijazi et al. (2012)	74.00	
Mookiah et al. (2014a); Mookiah et al. (2014b)	95.07	
Mookiah et al. (2014a); Mookiah et al. (2014b)	91.36	ARIA
Mookiah et al. (2015a); Mookiah et al. (2015b)	85.09	
Acharya et al. (2016)	96.89	
Proposed method	99.38	
Mookiah et al. (2014a); Mookiah et al. (2014b)	95.00	
Mookiah et al. (2014a); Mookiah et al. (2014b)	97.78	
Mookiah et al. (2015a); Mookiah et al. (2015b)	100	STARE
Acharya et al. (2016)	100	
Proposed method	99.00	
Proposed method	99.92	Private

methods. Table 7.8 reveals that the proposed system has an accuracy of 99.38% for the ARIA data collection, which is higher than other approaches. The suggested approach obtained an accuracy of 99.00% for the STARE data collection, which is lower than approaches in Mookiah et al. (2015a) and Mookiah et al. (2015b). Also, we tested the proposed method on a Private dataset and achieved an accuracy of 99.92%. From Table 7.8, it is clear that the proposed method is providing better accuracy compared with the existing methods. Since 64 important features of the proposed classification time is reduced when FunNet is supplied to the classifier—as a result, adopting CNN as a feature extractor decreases the time and complexity of computing. By creating high precision and recall ratings, the suggested technique compensates for the issues encountered in DL classification.

7.5 Conclusion

The FunNet framework is designed to simplify the early detection of AMD. The proposed CNN is an MP CNN, containing a feed-forward primary path and an alternate path. The performance of the proposed architecture is evaluated through 10-fold cross-validation methods using different classifiers. The proposed method is tested on ARIA, STARE, and a PRIVATE dataset and obtained an accuracy of 99.38%, 99.00%, and 99.92%, respectively, with the J48 classifier. The efficiency of the FunNet for the detection of AMD was demonstrated by comparisons with alternative approaches. The FunNet with J48 classifier can be used for the eye screening method in AMD detection task.

References

Acharya, U. R., Mookiah, M. R. K., Koh, J. E., Tan, J. H., Noronha, K., Bhandary, S. V., Rao, A. K., Hagiwara, Y., Chua, C. K., & Laude, A. (2016). Novel risk index for the identification of age-related macular degeneration using radon transform and DWT features. *Computers in Biology and Medicine, 73*, 131–140.

Akbar, B., Gopi, V. P., & Babu, V. S. (2015). Colon cancer detection based on structural and statistical pattern recognition. In: *Proceedings of the 2nd international conference on electronics and communication systems* (pp. 1735–1739).

ARIA Dataset. <http://www.eyecharity.com/aria>.

Bottou, L. (2012). *Stochastic gradient descent tricks. Neural networks: Tricks of the trade* (pp. 421–436). *Springer.*

Burlina, P., Pacheco, K. D., Joshi, N., Freund, D. E., & Bressler, N. M. (2017). Comparing humans and deep learning performance for grading AMD: A study, in using universal deep features and transfer learning for automated AMD analysis. *Computers in Biology and Medicine, 82*, 80–86.

Daqi, G. & Tao, Z. (2007). Support vector machine classifiers using RBF kernels with clustering-based centers and widths. In: *Proceedings of the IEEE international joint conference on neural network* (pp. 2971–2976).

East, M. & Africa, N. (2017). Idf diabetes atlas, *Diabetes, 20*, 79.

Garcia-Floriano, A., Ferreira-Santiago, A., Camacho-Nieto, O., & Yanez-Marquez, C. (2019). A machine learning approach to medical image classification: Detecting age-related macular degeneration in fundus images. *Computers & Electrical Engineering, 75*, 218–229.

Gayathri, S., Gopi, V. P., & Palanisamy, P. (2020a). A lightweight CNN for diabetic retinopathy classification from fundus images. *Biomedical Signal Processing and Control, 62*, 102–115.

Gayathri, S., Gopi, V. P., & Palanisamy, P. (2020b). Automated classification of diabetic retinopathy through reliable feature selection. *Physical and Engineering Sciences in Medicine*, 1−19.

Gayathri, S., Krishna, A. K., Gopi, V. P., & Palanisamy, P. (2020). Automated binary and multiclass classification of diabetic retinopathy using Haralick and multiresolution features. *IEEE Access*, *8*, 57497−57504.

Grassmann, F., Mengelkamp, J., Brandl, C., Harsch, S., Zimmermann, M. E., Linkohr, B., Peters, A., Heid, I. M., Palm, C., & Weber, B. H. (2018). A deep learning algorithm for prediction of age-related eye disease study severity scale for age-related macular degeneration from color fundus photography. *Ophthalmology*, *125* (9), 1410−1420.

Hijazi, M. H. A., Coenen, F., & Zheng, Y. (2010). Retinal image classification using a histogram based approach. In: *Proceedings of the* IEEE *international joint conference on neural networks* (pp. 1−7).

Hijazi, M. H. A., Coenen, F., & Zheng, Y. (2012). Data mining techniques for the screening of age-related macular degeneration. *Knowledge Based System*, *29*, 83−92.

Hijazi, M. H. A., Coenen, F., & Zheng, Y. (2014). Data mining for amd screening: A classification based approach. *International Journal of Simulation, Systems, Science and Technology*, *15*(2), 57−69.

Hoover, A., & Goldbaum, M. (2003). Locating the optic nerve in a retinal image using the fuzzy convergence of the blood vessels. *IEEE Transactions on Medical Imaging*, *22* (8), 951−958.

Hoover, A., Kouznetsova, V., & Goldbaum, M. (2000). Locating blood vessels in retinal images by piecewise threshold probing of a matched filter response. *IEEE Transactions on Medical Imaging*, *19*(3), 203−210.

Kandhasamy, J. P., Balamurali, S., Kadry, S., & Ramasamy, L. K. (2019). Diagnosis of diabetic retinopathy using multi-level set segmentation algorithm with feature extraction using SVM with selective features. *Multimedia Tools and Applications*, 1−16.

Kose, C., Sevik, U., & Gencalioglu, O. (2008). Automatic segmentation of age related macular degeneration in retinal fundus images. *Computers in Biology and Medicine*, *38* (5), 611−619.

Krizhevsky, A., Sutskever, I., & Hinton, G. E. (2012). Imagenet classification with deep convolutional neural networks. *NeurIPS*, 1097−1105.

Mookiah, M. R. K., Acharya, U. R., Fujita, H., Koh, J. E., Tan, J. H., Chua, C. K., Bhandary, S. V., Noronha, K., Laude, A., & Tong, L. (2015a). Automated detection of age-related macular degeneration using empirical mode decomposition. *Knowledge-Based System*, *89*, 654−668.

Mookiah, M. R. K., Acharya, U. R., Fujita, H., Koh, J. E., Tan, J. H., Noronha, K., Bhandary, S. V., Chua, C. K., Lim, C. M., & Laude, A. (2015b). Local configuration pattern features for age-related macular degeneration characterization and classification. *Computers in Biology and Medicine*, *63*, 208−218.

Mookiah, M. R. K., Acharya, U. R., Koh, J. E., Chandran, V., Chua, C. K., Tan, J. H., Lim, C. M., Ng, E., Noronha, K., Tong, L., & Laude, A. (2014a). Automated diagnosis of age-related macular degeneration using greyscale features from digital fundus images. *Computers in Biology and Medicine*, *53*, 55−64.

Mookiah, M. R. K., Acharya, U. R., Koh, J. E., Chua, C. K., Tan, J. H., Chandran, V., Lim, C. M., Noronha, K., Laude, A., & Tong, L. (2014b). Decision support system for age-related macular degeneration using discrete wavelet transform. *Medical & Biological Engineering & Computing*, *52*(9), 781−796.

Pead, E., Megaw, R., Cameron, J., Fleming, A., Dhillon, B., Trucco, E., & MacGillivray, T. (2019). Automated detection of age-related macular degeneration in color fundus photography: A systematic review. *Survey of Ophthalmology*, *64*(4), 498−511.

Peng, Y., Dharssi, S., Chen, Q., Keenan, T. D., Agrón, E., Wong, W. T., Chew, E. Y., & Lu, Z. (2019). Deepseenet: A deep learning model for automated classification of patient-based age-related macular degeneration severity from color fundus photographs. *Ophthalmology, 126*(4), 565–575.

Roychowdhury, A. & Banerjee, S. (2018). Random forests in the classification of diabetic retinopathy retinal images. In: *Proceedings of the international conference in advanced computational and communication paradigms* (pp. 168–176).

Sharma, S., Agrawal, J., & Sharma, S. (2013). Classification through machine learning technique: C4. 5 algorithm based on various entropies. *International Journal of Computing Applications, 82*(16).

Srivastava, N., Hinton, G., Krizhevsky, A., Sutskever, I., & Salakhutdinov, R. (2014). Dropout: A simple way to prevent neural networks from overfitting. *Journal of Machine Learning Research, 15*(1), 1929–1958.

Tan, J. H., Bhandary, S. V., Sivaprasad, S., Hagiwara, Y., Bagchi, A., Raghavendra, U., Rao, A. K., Raju, B., Shetty, N. S., Gertych, A., et al. (2018). Age-related macular degeneration detection using deep convolutional neural network. *Future Generation Computing System, 87*, 127–135.

Visa, S., Ramsay, B., Ralescu, A. L., & Van Der Knaap, E. (2011). Confusion matrix-based feature selection. In: *Proceedings of the midwest artificial intelligence and cognitive science* (Vol. 710, pp. 120–127).

Walter, T., Klein, J. C., Massin, P., & Erginay, A. (2002). A contribution of image processing to the diagnosis of diabetic retinopathy-detection of exudates in color fundus images of the human retina. *IEEE Transactions on Medical Imaging, 21*(10), 1236–1243.

Wong, W. L., Su, X., Li, X., Cheung, C. M. G., Klein, R., Cheng, C. Y., & Wong, T. Y. (2014). Global prevalence of age-related macular degeneration and disease burden projection for 2020 and 2040: A systematic review and meta analysis. *The Lancet Global Health, 2*(2), 106–116.

Xu, Z., Wang, W., Yang, J., Zhao, J., Ding, D., He, F., Chen, D., Yang, Z., Li, X., Yu, W., et al. (2020). Automated diagnoses of age-related macular degeneration and polypoidal choroidal vasculopathy using bi-modal deep convolutional neural networks. *The British Journal of Ophthalmology, 105*(4), 561–566.

Yoo, T. K., Choi, J. Y., Seo, J. G., Ramasubramanian, B., Selvaperumal, S., & Kim, D. W. (2019). The possibility of the combination of OCT and fundus images for improving the diagnostic accuracy of deep learning for age-related macular degeneration: A preliminary experiment. *Medical & Biological Engineering & Computing, 57*(3), 677–687.

Zheng, Y., Hijazi, M. H. A., & Coenen, F. (2012). Automated "disease/no disease" grading of age-related macular degeneration by an image mining approach. *Investigative Ophthalmology & Visual Science, 53*(13), 8310–8318.

CHAPTER 8

An improved method for automated detection of microaneurysm in retinal fundus images

Avinash A.[1], Biju P.[1], Prapu Premanath[2], Anju Thomas[3] and Varun P. Gopi[3]
[1]Department of ECE, College of Engineering Thalassery, Thalassery, India
[2]Department of ECE, College of Engineering Vadakara, Vatakara, India
[3]Department of Electronic and Communication Engineering, National Institute of Technology Tiruchirappalli, Tiruchirappalli, India

8.1 Introduction

Diabetic retinopathy (Dr) is one of the most common complications associated with diabetes. Dr has been the leading cause of new instances of impaired vision in persons aged 20−74 years (Walter et al., 2007). Structural transformation in vascular branches in the retina increases with the duration and severity of hyperglycemia. Capillary microaneurysms (MAs) grow, and later there may be blood vessel proliferation. Fibrosis, second retinal detachment, and hemorrhages may occur, leading to retinal relapse and vision loss. Timely diagnosis over consistent screening is suggested to diabetic patients, helping them preclude blindness and visual loss. However, many diabetic patients needed to be screened yearly and pose a hefty load for ophthalmologists. Thus screening of patients using manual sorting is cautious and resource-demanding. Therefore considerable efforts have been put to frame consistent computer-assisted screening schemes using retinal images (Chris DB).

Dr is widely classified into two types: nonproliferative Dr (NPDR) and proliferative Dr (IDF Diabetes Atlas, 2006), or six stages by the occurrence of various symptoms in the retina. If severe Dr damages are detected correctly and frequently at the original stages, they can be mitigated or eliminated. Fig. 8.1 shows a normal and a diseased eye.

Edge-of-Things in Personalized Healthcare Support Systems
DOI: https://doi.org/10.1016/B978-0-323-90585-5.00007-2

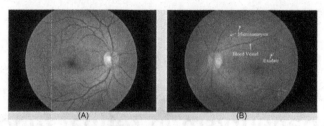

Figure 8.1 (A) Normal eye. (B) Diseased eye.

MA is a critical marker and one of the earliest signs of Dr in the retina. MAs are seen in the early stages of NPDR, which is physical flagging within capillary walls. They are dark red, small, and round spots resulting due to swellings in retinal vessels. The MA diameter may vary from 10 to 125 μm. Since MA is one of the first lesions to develop in Dr, identifying it will assist in early detection. One of the most common clinical procedures in the diagnosis of Dr is the analysis of eye fundus images.

In contrast to fluorescein angiography (FA), fundus images' acquisition is fast, inexpensive, and painless (Hipwell et al., 2000; Sinthanayothin et al., 2002). Besides, the FA is not applicable for everyone, such as pregnant women (Frame et al., 1998; Mendonca et al., 1999). Thus the adaption of the eye fundus images is the better choice for screening purposes.

Preprocessing, candidate extraction, feature extraction, and classification are the four key phases in MA diagnosis. Preprocessing is used to eliminate noise from retinal fundus images and to adjust for nonuniform illumination. Several substeps will be in the preprocessing stage. MA candidate extraction is trying to find an initial set of regions where MA candidates are existing. Efficient candidate extraction will detect most of the MA candidates and a few false-positive areas. For classification purposes, the feature extraction stage is utilized to locate features linked to MA. Candidate classification uses different classifiers to achieve better accuracy, specificity, and sensitivity of the system by reducing the number of false positives from the candidate extraction stage. By choosing suitable features, the classier accurately isolates the real candidates and false predictions.

The purpose of this study is to develop a detection approach based on segmentation and morphological operations that is effective. In this method mainly five stage processing is carried out on fundus images. Different statistical parameters from images are measured, such as mean, standard deviation, energy, etc. Several classifiers are used for classification of non-MA and MA images, and the best classifier is chosen according to the performance. Baudoin et al. (1984) proposed a computerized MA detection approach in fluorescein angiograms that depended on morphological operations. MAs were recognized from FA images using various top-hat transforms. However, as discussed above, FA is not suitable for screening systems.

Further computation time was very high. Purwita et al. (2011) introduced a method for detecting MA. The mathematical morphology is chosen because the MA appears in specific shapes in the fundus images. This technique primarily has three stages: preprocessing, candidate extraction followed by postprocessing. Canny edge detection and area filling algorithms extract the MA candidate by eliminating vessels and minor objects. Unusable objects were eliminated using a postprocessing step. In this procedure precise identification of MAs was contingent on proper segmentation of the optic disk and vascular structures.

8.1.1 Background

Pallawala et al. (2005) introduced a scheme for detecting MA based on generalized eigenvectors. They used generalized eigenvector-based segmentation. The output can provide a sign of the position of MAs. These localities are evaluated based on the definite features of the MA to recognize the exact MAs. Tests on 70 retinal subimages are carried out. This method can achieve 93% accuracy in the detection. Quellec et al. (2008) suggested an MA detection method based on optimal wavelet transform. They proposed a template matching procedure to identify MAs. The input images have illumination variations along with the presence of noise. In the wavelet domain a template matching technique is proposed, which decreases the issues caused by illumination fluctuation and noise. This method achieved 89.62% accuracy.

Spencer et al. (1992) introduced an MA detection methodology using adaptive filters. On the top-hat converted image, Gaussian matched filtering was used. The performed region on the binary image was filtered using this filter. Different features like shape, size, and intensity of MA candidates were calculated. With this method, they achieved 82% sensitivity at 86% specificity. B. Antal et al. (Antal & Hajdu, 2012) proposed an ensemble-based MA detection system. A set of preprocessing techniques and candidate extraction procedures was proposed to detect MA. Using this method, they achieved an area under the ROC curve (AUC) value of 0.90, modest with the early reported outcomes. However, because the system requires a huge amount of training and testing data, it takes a long time to complete.

Lazar et al. (Lazar & Hajdu, 2013) proposed a technique for MA detection in retinal images, using cross-section proles analysis on candidate pixels. The local maximum pixels are the candidates under consideration. Thirty proles were generated using cross-sectional scanning on each candidate. Then each prole's peak is detected, and its statistical measures are calculated to feed it into a Bayesian classier. The feature values that were used in a classification step are to abolish false candidates. This method showed substantial performance compared to previous methods. Wu et al. (2017) proposed an automated process for detecting MA. This paper is an extension of the technique based on cross-section prole's analysis. Two additional steps are added: (1) Region growing method to develop initial candidate pixels back to the original pathologies. The extra postprocessing step is used to eliminate false preliminary candidates. (2) Except for three new prole features, 17 local features are extracted to describe the MA. This method achieved 97% accuracy, and while comparing with the Lazar method (Lazar & Hajdu, 2013), it gave better performance results.

In Veiga's approach, MA identification is based on the Laws texture features-based methodology (Veiga et al., 2018). They presented a two-stage method to make the method less burdensome, and they speculated that a high-order statistical method, such as Laws masks, may fit well to this case. Three retinal image recordings were utilized to test the proposed method: LaTIM, ROC, and e-ophtha. The textural features of the laws were shown to be capable of recognizing MAs and obtaining sensitivities close to 100% using this technique. While comparing the existing method, we can see that these methods cannot achieve better performance with minimum features and low time consumption. So we propose a novel method for automated detection of MA from retinal fundus images with a minimum number of features and low time consumption. Studies on the detection of MA from fundus images have become one of the most relevant research topics in recent years. After detecting whether the image is diseased or not, it is essential to evaluate the image's features to classify the loaded input images, whether it is a diseased or not.

Eftekhari et al. (2019) utilizes two-stage CNN network for the detection of MA. The network parameters were chosen using trial and error methods. The sensitivity value of 0.8 was obtained by this method. In Melo et al. (2020) MA candidates are extracted using a sliding band filter, and promising candidates are identified by integrating the filter responses with shape and intensity information. Thirty-one MA candidate features are then retrieved, and an RUSBoost classifier is used to distinguish MA from non-MA objects and achieved a detection accuracy of 83.1%. Derwin et al. (2020) proposed a method for MA detection where to improve the image quality, preprocessing was used. The texture-based characteristics of MA were then extracted using a rotational invariant local binary pattern. Finally, SVM was used for the classification. The method obtained an AUC value of 0.906. The eigen values of Hessian matrix were used to enhance the blood vessel portion in an image (Long et al., 2020). Then the highlighted blood vessels were removed from the image by employing the shape characteristics and connected components analysis. MA candidate detection was done by using directional local contrast method and the extracted features were fed to three different classifiers namely SVM, KNN, and Naive Bayesian, and obtained highest AUC value of 0.87. A unique one-stage encoder−decoder network for MA detection is presented in Liao et al. (2021). The suggested approach uses skip subtraction to enhance MA information that could otherwise be lost in the encoder process. The true MAs were extracted from the MA candidates using a multiscale residual network as discussed in Xia et al. (2021). A multiscale efficient network that serves as a classifier for distinguishing fake MAs from MA candidates found by the preceding network and obtained an accuracy of 97%.

The proposed method contains five different stages: preprocessing, segmentation, candidate selection, extraction of feature, and classification. Preprocessing is employed to generate suitable images for the selection of candidates and feature extraction. Gabor filter is used as image enhancement method to make MA more visible. The edge detection process is carried in the segmentation stage. Multilayer perceptron (MLP) classifier is used for image classification.

8.1.2 Materials and methods

This paper introduces a novel automated method for the accurate detection of MA from fundus images. The flow diagram of the proposed method is shown in Fig. 8.2. The inverted green channel of retinal images is used as input in this approach because it provides higher background contrast for MA identification than the red and green channels.

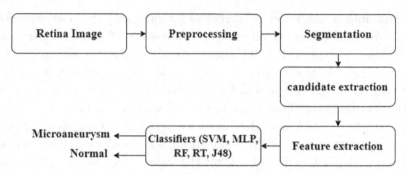

Figure 8.2 Flow diagram of proposed system.

8.1.2.1 Preprocessing
In preprocessing two subsections include adaptive histogram equalization and Gabor filtering; commonly, green channel images are preferred in preprocessing.

Contrast limited adaptive histogram equalization (CLAHE)

The significance of this stage in preprocessing is that many MAs in the green channel images do not have enough contrast with the nearby surroundings, which affects the detection step that follows. To solve this problem, CLAHE is used (Walter et al., 2007). Due to its effectiveness in image enhancement, this is widely used in medical imaging techniques.

Gabor filtering

Gabor filter is a linear filter used for texture analysis. The motive for selecting this filtering technique instead of others is that it is very beneficial in optical character recognition (Kamarainen, 2012).

8.1.2.2 Segmentation
Image segmentation mentions the decomposition of a scene into its components. It is a significant step in image analysis. Segmentation is normally used to categorize boundaries and objects in the images. This stage contained segmentation as explained below.

Edge detection

It is mainly used to detect the boundaries of the objects in images. Detection works by finding the discontinuities in brightness. Edge detection is mainly used for image segmentation. Several edge detection algorithms are available; among these canny edges, detection method is used in this work. Canny is an edge detection method used to extract useful physical information from images. It has been commonly used in numerous computer vision systems. Canny has found that the requirements for the claim of edge detection on various vision systems are comparatively akin. The general benchmark for edge detection includes the following:

1. low error rate while detecting edges that are accurately detecting among many edges.
2. the edge point spotted should exactly localize on the midpoint of the edge.
3. the Canny edge detection procedure is one of the supreme stringently defined approaches that offers fair and consistent detection (Gonzalez & Richard, 2006).

Optic disk removal

For more accurate analysis of filtered images, optic disk removal is an essential step. The top-hat transform is employed to extract minute bright and dark structures in an image; this is useful when variations in the background are much higher.

8.1.2.3 Candidate extraction

Candidate extraction is an important role here. In this stage mainly two steps are carried out:

- morphological erosion on the binary image
- morphological opening on the binary.

Morphological operations and connected component analysis are useful to abolish noises and obtain complete areas of MA candidates. Morphological processes modify the intensities of definite regions of an image. Morphological operations separate specific segments of an image and then enlarge or contract just those areas to realize the preferred effect. Morphological functions change an object's shape by performing dilation, erosion, maximum, and minimum processes. Dilation expands objects. Erosion contracts them. Erosion and dilation are the mathematical morphological operators in image processing. It is mainly applied to binary images. The erosion operator erodes the boundaries of regions of foreground pixels (i.e., white pixels) in a binary image. Thus foreground pixel area shrinks in size, and holes within those areas become more extensive. The opening is a significant operator in mathematical morphology. It is a dilation of the erosion of a set by a structuring element. Here opening is used to remove the small object (MA) from the foreground. Subtract the morphologically closed image's output from the output of the morphologically opened image after the morphological opening process, and we get an image that only contains MA.

8.1.2.4 Feature extraction

The local features, shape, and intensity features are considered in this work. Therefore we can extract the MA, its surroundings, and several overall features.

Hessian matrix-based features

A Hessian matrix is a set of second-order partial derivatives of a scalar-valued function. It denotes the local curvature of a multivariable function. The Hessian matrix of a two-dimensional image is obtained by taking the second-order partial derivative of the image pixel, which is given as

$$H(x, y) = \left\{ \begin{matrix} D_{xx} & D_{xy} \\ D_{yx} & D_{yy} \end{matrix} \right\}, \tag{8.1}$$

$$D_{xx} = G_{xx}(x, y; \sigma) * I(x, y), \tag{8.2}$$

$$D_{xy} = G_{xy}(x, y; \sigma) * I(x, y), \tag{8.3}$$

$$D_{yy} = G_{yy}(x, y; \sigma) * I(x, y). \tag{8.4}$$

Here G_{xx}, G_{xy}, and G_{yy} are second-order Gaussian derivatives in each direction, $*$ is the convolution operator, and $I(x,y)$ is the preprocessed image.

Energy: Energy of MA candidate region is

$$E_{\text{cand}} = \sum I(x, y)^2 \tag{8.5}$$

where $I(x,y)$ is extracted image.

Mean: The mean intensity of MA candidate region is

$$M_{\text{cand}} = \frac{\sum_{j \in \Omega} g_j}{N}. \tag{8.6}$$

Standard deviation: The MA candidate region's standard deviation of intensity is

$$\sigma_{\text{cand}} = \frac{\sum_{j \in \Omega} \sqrt{(g_j - M_{\text{cand}})^2}}{N}. \tag{8.7}$$

8.1.2.5 Classification

Classification is performed to classify whether the candidate is diseased or not based on the extracted features. Mainly five different types of classifiers are used in this work. Brief descriptions of classifiers used for the proposed method are given below. Among these classifier's output, we can finalize which one is better for detection.

Support vector machine (SVM).

SVM is a supervised learning scheme used for classification purposes. It deals with more input data efficiently and contains a nonlinear transformation. The classification is performed by finding the hyperplane that differentiate the two classes very well.

MLP

The MLP classifier is a feed-forward artificial neural network. It used a single hidden layer for simplicity and went for two hidden layers for higher classification performance. The hidden units were picked differently for every dataset. The MLP training utilizes the back-propagation learning procedures that come close to the Bayesian theory's optimal discriminant function. The outputs approximate posterior probability functions of the classes being trained.

Random tree and random forest classifier

Random trees are a group of tree predictors called a forest. In this classifier there is no need for any accuracy estimation algorithm or a different test set for evaluating the training error. The error is internally assessed during the training phase. Random trees are trees that are constructed randomly. The popular machine learning framework Weka, on the other hand, uses the word to refer to a decision tree constructed from a random selection of columns. Random forest is a combination of learning process that consists of numerous decision trees during the training phase and outputs the class: mean prediction (regression) or type of class (classification) by the distinct trees. The forest picks the classification having the majority votes over all the trees in the forest.

J48 classifier

It is a decision tree classifier that makes a binary tree. Once the tree constructs, it is given to each sequence in the dataset and results in the classification for that sequence. When building a tree, the classifier disregards the missing values. The basic idea is to divide the data into ranges based on the characteristic values obtained in training sets for that item.

8.1.2.6 Performance analysis
Proposed methodology uses accuracy, sensitivity, and specificity as the performance inspectors. These are calculated by the following equations:

$$\text{Accuracy} = \frac{\text{TP} + \text{TN}}{\text{TP} + \text{FP} + \text{TN} + \text{FN}}, \tag{8.8}$$

$$\text{Sensitivity} = \frac{TP}{TP + FN}, \tag{8.9}$$

$$\text{Specificity} = \frac{TN}{FP + TN}. \tag{8.10}$$

True positive denotes TP, true negative denotes TN, false positive signifies FP, and false negative signifies FN.

8.2 Results and discussion

This section deals with the results of the proposed method and a comparison with the existing methods. All experiment except classification was implemented using MATLAB® (R 2014a) and the classification using WEKA 3.7.4 (WEKA, 2011). A total of 130 samples are used with 10-fold cross-validation. Mainly two steps are performed in the preprocessing stage: CLAHE and Gabor filtering of image. The preprocessed images are shown in Fig. 8.3.

Canny edge detection procedure is performed in the preprocessed image in the segmentation stage. The segmented image is shown in Fig. 8.4. The MA detected and outlined images are shown in Figs. 8.5 and 8.6, respectively. The classification process is carried in five different classifiers. The performance of the different classifiers is tabulated in

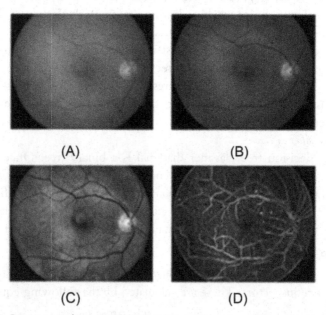

(A)   (B)

(C)   (D)

Figure 8.3 Preprocessed images: (A) input image, (B) green channel image, (C) histogram equalized image, and (D) Gabor filtered image.

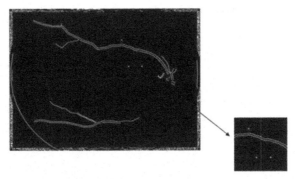

Figure 8.4 Canny edge detected image.

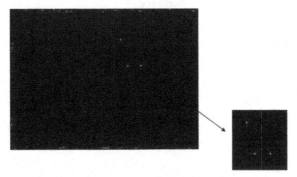

Figure 8.5 MA detected image.

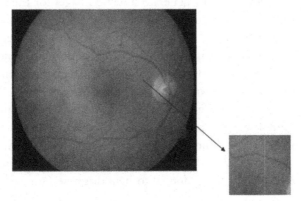

Figure 8.6 Outlinedi.

Table 8.1. From these, we can conclude that multilayer preparation classifier gives better classification with 99.02% accuracy. Comparison of the proposed work with existing methods is listed in Table 8.2. From these results, it is clear that the proposed method is better in MA classification.

Table 8.1 Assessment on different classifiers.

Classifier	Accuracy	Sensitivity	Specificity
MLP	99.23%	100%	98.03%
Random tree	93.84%	71.4%	97.15%
Random forest	97.69%	82.14%	92.15%
J48	90%	82.14%	93.13%
SVM	96.15%	85.71%	84.31%

Table 8.2 Performance analysis.

Method	Accuracy	Dataset
Pallawala et al. (2005)	ROC	93%
Lazar and Hajdu (2013)	diaretdb1_v_1_1	95.3%
Wu et al. (2017)	e-ophtha and ROC	97%
Veiga et al. (2018)	LaTIM e-ophtha and ROC	97%
Proposed method	e-ophtha and diaretdb1_v_1_1	99.23%

8.3 Conclusion

This paper proposes a novel method for MA detection in retinal images. Preprocessing is employed to generate suitable images for the selection of candidates and feature extraction. An advantage of the proposed method is that image enhancement steps in the preprocessing and segmentation process. The enhancement method makes MA more visible and quashing noise. Gabor filtering is more helpful for enhancement. Eventually, the enhancement makes the profile analysis eliminate more noise. The edge detection process is carried in the segmentation stage. For accurate detection of the candidate, morphological operations are performed in the candidate extraction step. Among the five classifiers, MLP classifier gives better accuracy of 99.23%. The results display that the proposed method is very effective and has the potential to help diagnose Dr.

8.4 Future scope

The future scope includes the detection of other abnormalities like exudates, hemorrhage, and neovascularization to grade the Dr images.

References

Antal, B., & Hajdu, A. (2012). An ensemble-based system for microaneurysm detection and diabetic retinopathy grading. *IEEE Transactions on biomedical Engineering, 59*(6), 1720.

Baudoin, C., Lay, B., & Klein, J. (1984). Automatic detection of microaneurysms in diabetic fluorescein angiography. *Revue d'Epidemiologie et de Sante Publique, 32*(3/4), 254–261.

Chris D. B., 10 facts about diabetes. WorldHealthOrganization [homepage on the internet]. <http://www.who.int/features/factfiles/diabetes/facts/zh/>.

Derwin, J. D., Tami Selvi, S., & Jeba Singh, O. (2020). Secondary observer system for detection of microaneurysms in fundus images using texture descriptors. *Journal of Digital Imaging Springer, 33*(1), 159–167.

Eftekhari, N., Pourreza, H.-R., Masoudi, M., Ghiasi-Shirazi, K., & Saeedi, E. (2019). Microaneurysm detection in fundus images using a two-step convolutional neural network. *Biomedical Engineering Online, 18*(1), 1–16.

Frame, A. J., Undrill, P. E., Cree, M. J., Olson, J. A., McHardy, K. C., Sharp, P. F., & Forrester, J. V. (1998). A comparison of computer based classification methods applied to the detection of microaneurysms in ophthalmic fluorescein angiograms. *Computers in Biology and Medicine, 28*(3), 225–238.

Gonzalez R. C. & Richard E. W. (2006). *Digital image processing* (3rd ed.). New Jersey: Prentice-Hall.

Hipwell, J., Strachan, F., Olson, J., McHardy, K., Sharp, P., & Forrester, J. (2000). Automated detection of microaneurysms in digital red-free photographs: A diabetic retinopathy screening tool. *Diabetic Medicine, Wiley Online Library, 17*(8), 588–594.

IDF Diabetes Atlas. (2006). *International diabetes federation*. Cape Town, South Africa: Press Release.

Kamarainen J. (2012). Gabor features in image analysis. In: *Proceedings of the international conference on image processing theory, tools and applications* (pp. 13–14).

Lazar, I., & Hajdu, A. (2013). Retinal microaneurysm detection through local rotating cross-section profile analysis. *IEEE Transactions on Medical Imaging, 32*(2), 400–407.

Liao, Y., Xia, H., Song, S., & Haisheng, L. (2021). Microaneurysm detection in fundus images based on a novel end-to-end convolutional neural network. *Biocybernetics and Biomedical Engineering, 41*(2), 589–604.

Long, S., Chen, J., Hu, A., Liu, H., Chen, Z., & Zhen, D. (2020). Microaneurysms detection in color fundus images using machine learning based on directional local contrast. *Biomedical Engineering Online, 19*, 1–23.

Melo, T., Mendonça, A. M., & Campilho, A. (2020). Microaneurysm detection in color eye fundus images for diabetic retinopathy screening. *Computers in Biology and Medicine, 126*, 103–115.

Mendonca A. M., Campilho A., & Nunes J. (1999). Automatic segmentation of microaneurysms in retinal angiograms of diabetic patients. In *Proceedings of the international conference on image analysis and processing* (pp. 728–733).

Pallawala P., Hsu W., Lee M. L., & Goh S. (2005). Automated microaneurysm segmentation and detection using generalized eigenvectors null. In *Proceedings of the IEEE workshop on applications of computer vision* (Vol. 1, pp. 322–327).

Purwita A. A., Adityowibowo K., Dameitry A., & Atman M. W. S. (2011). Automated microaneurysm detection using mathematical morphology. In *Proceedings of the international conference on instrumentation, communications, information technology, and biomedical engineering* (pp. 117–120).

Quellec, G., Lamard, M., Josselin, P. M., Cazuguel, G., Cochener, B., & Roux, C. (2008). Optimal wavelet transform for the detection of microaneurysms in retina photographs. *IEEE Transactions on Medical Imaging, 27*(9), 1230–1241.

Sinthanayothin, C., Boyce, F., Williamson, T. H., Cook, H. L., Mensah, E., Lal, S., & Usher, D. (2002). Automated detection of diabetic retinopathy on digital fundus images. *Diabetic Medicine, 19*(2), 105−112.

Spencer, T., Phillips, R. P., Sharp, P. F., & Forrester, J. V. (1992). Automated detection and quantification of microaneurysms in fluorescein angiograms. *Graefe's Archive for Clinical and Experimental Ophthalmology = Albrecht von Graefes Archiv fur Klinische und Experimentelle Ophthalmologie, 230*(1), 36−41.

Veiga, D., Martins, N., Ferreira, M., & Monteiro, J. (2018). Automatic microaneurysm detection using laws texture masks and support vector machines. *Computer Methods in Biomechanics and Biomedical Engineering: Imaging and Visualization, 6*(4), 405−416.

Walter, T., Massin, P., Erginay, A., Ordonez, R., Jeulin, C., & Klein, J. C. (2007). Automatic detection of microaneurysms in color fundus images. *Medical Image Analysis, 11*(6), 555−566.

WEKA [homepage on the internet]. Weka-3−7-4 [updated (2011) June 30]. Available from: <https://sourceforge.net/projects/weka/files/weka-3-7/3.7.4/>.

Wu, B., Zhu, W., F. Shi, F., Zhu, S., & Chen, X. (2017). Automatic detection of microaneurysms in retinal fundus images. *Computerized Medical Imaging and Graphics: The Official Journal of the Computerized Medical Imaging Society, 55*(2017), 106−112.

Xia, H., Lan, Y., Song, S., & Haisheng, L. (2021). *A multi-scale segmentation-to-classification network for tiny microaneurysm detection in fundus images. Knowledge Based System* (pp. 107−140).

CHAPTER 9

Integration and study of map matching algorithms in healthcare services for cognitive impaired person

Ajay Kr. Gupta and Udai Shanker
Department of Computer Science and Engineering, M.M.M. University of Technology, Gorakhpur, India

9.1 Introduction

The use of different intelligent Internet of Things (IoT) future innovations (Gupta & Shanker, 2020a) for machine learning has become fairly important in most practical applications these days owing to their high performance and low cost for various devices. Wearable cameras, actuators, the IoTs, and automation technologies are now developing the foundations for modern healthcare technologies such as pattern analysis and prognostic modeling. In the near future, smart assistive devices can allow cognitive impaired person, that is, elderly and disabled participants to interact in everyday activities, participate in friendly events, entertain, and socialize themselves while preserving their health and well-being. The involvement of health-related practices (Kefi & Asan, 2021), which necessitates the use of communication and collaboration technology (Eikey et al., 2015), complements those smart environments. Sophisticated networks and IoT systems are often used in these approaches to improve health and well-being. Intellectual disabilities are a category of conditions marked by reduced cognitive and adaptive functioning. They are also known as mental retardation or general intellectual disability. Mild (approximately 80%), moderate (14%), severe (approximately 4%), and extreme (approximately 2%) developmental disorders are the most prominent. The bulk of cases are associated with mild to moderate form of disorder. For individuals with mild to severe conditions in the early stages with intellectual disabilities, speculative computation cognitive assistant (SCCA) module is a form of orientation facility (American Psychiatric Association, 2013; Carmien

Edge-of-Things in Personalized Healthcare Support Systems
DOI: https://doi.org/10.1016/B978-0-323-90585-5.00008-4

187

et al., 2005; Dawe, 2006; Liu et al., 2009). In an advanced, serious, or drastic phase of life, the person is unable to use a mobile screen making orientation inevitable. The main characteristics of this type of location based services (LBS) (Gupta & Shanker, 2018a; Gupta & Shanker, 2020c; Gupta & Shanker, 2021a) are that it is a continuous process (Ramos et al., 2013). In LBS, augmented reality user interfaces were used to guide cognitively challenged individuals. It diminishes the amount of mental activity needed to comprehend the way to approach. A screen may appear to alert the user to wait as the system decides the best travel direction. The moving path must be fed into the speculative computational module to predict user errors and ensure proper travel direction. The trajectory data mining approach was applied and tested with the goal of completing the travel path as per the user preferences. The trajectory (Lou et al., 2009) of a person with cognitive impairments is denoted as T from their log L. It is a set of GPS points with a fixed distance between them that does not exceed a certain threshold T_h. The trajectory is written as $T: p_1 \rightarrow p_2 \rightarrow \ldots \rightarrow p_m$, where, with $p_i \in P \subset L$, $(m > i \geq 1)$, and $T_h > p_{i+t} \cdot t > 0$. The term $(|T| = m)$, and t are known as the number of samplings and the interval of the sampling point, respectively. The point structure $P = p_1, p_2 \cdot \cdot \cdot, p_m$ are known as a GPS log, in which, each point $p_i \in P$ includes p_i lng, p_i lat, p_i t as longitude, latitude, and time-stamp, respectively. Path W [W: $e_1 \rightarrow e_2 \rightarrow \ldots \rightarrow e_n$] is a connected street segment arrangement that begins at the vertex V_i and ends at the vertex V_j, where, $1 \leq k \leq n$, e_1 start $= V_i$, e_n end $= V_j$, and e_k end $= e_{k+1}$ start. Fig. 9.1 provides a description of the GPS client movement log and trajectory.

In 1999 ESRI published ArcGIS, which is a geospatial tool written in the C++ programming language. It effectively aids us in capturing, archiving, analyzing, sorting, and handling vast quantities of mapped data on the LBS framework. Assume a person is moving down a finite street path, N, which we are unfamiliar with. We also had an ArcGIS map system with road network geographic information system (GIS) instead

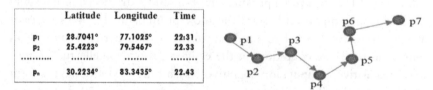

	Latitude	Longitude	Time
p₁	28.7041°	77.1025°	22:31
p₂	25.4223°	79.5467°	22.33
....	....	....	....
pₙ	30.2234°	83.3435°	22.43

Figure 9.1 Mobile user log and trajectory.

(*V*, *E*). For the sake of convenience, arcs from *G* and roads from *N* are believed to have a one-to-one correspondence. As seen in Fig. 9.2, $G(V, E)$ is made up of dimensional arcs that represent paths. The finite sequence of values for labeled points $(A_0, A_1, \ldots, A_n)$ will define increasing arc, $A \in E$. These are the junctions or dead ends of an arc used for the switching to another arc. In the GPS, P_t denotes the actual position of the individual at time t, and P_{et} is the approximate location at this moment. The increasing arc, $A \in E$, will be defined by a finite sequence of values for labeled points $(A_0, A_1, \ldots, A_n)$. These are all the arc junctions or dead ends that are used to turn from one arc to another. In GPS, P_{et} signifies the individual's approximate location at time t and P_t denotes corresponding position.

One may obtain route to the destination by regularly gathering information from the GPS over a server linked with a cell phone. Recalculations of frequent visit positions will result in greater accuracy at the cost of a substantial amount of time and effort (Gupta, 2020). To correct the user's position, identify a path for the user, and effectively use resources, it is indeed essential to implement a complex location management system with continuous monitoring of GPS requests. Aside from the main users, the framework may be required by the caretaker of person. As a consequence, the device has a localization feature that tells the caretaker of the cognitive impairment person's current location in real time. This role is particularly helpful because understanding a person's position helps the caretaker to build a suitable practice. More functions may be applied to the proposed SCCA from external systems such as ambient intelligence (AmI). AmI attracts research support in fields related to an IoT system's ability to interpret surrounding environments and respond to people's

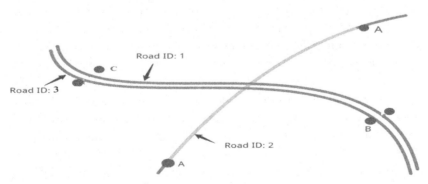

Figure 9.2 Road segments illustration.

presence including human-centered computation, context perception, artificial intelligence, pervasive computing, and ubiquitous computing (Gupta & Shanker, 2018b; Gupta & Prakash, 2018; Gupta & Shanker, 2020e).

The proposed LBS provides a predictive computational functionality that directs the user through error detection and alerting capabilities. The aim of this work is to propose a module for incorporating software as a tool for disabled or elderly people with diminished cognitive ability by having access to basic and intuitive services that are customized as per individual's needs. The suggested approach assists in the user's freedom and well-being in the home or outside, as well as reducing his/her loneliness by encouraging him/her to comfortably interact with others. Previously identified options lacked orientation and also lacked a module for monitoring cognitively challenged individuals as they went outside. Therefore it is important to consider the model features to improve its accessibility and usefulness while designing apps for people with cognitive disabilities. Owing to the LBS application's minimization of mental control, the user's physical and/or cognitive disabilities could be minimized.

The SCCA's map matching (MM) module's objective is to find the path that leads to T from Graph G by comparing the actual position P_t to the estimated location P_e with its true direction. By creating a spatial map of roads from the traces, the chain of user positions is connected on a virtual map. There is a preprocessing step for any LBS MM. The need for an effective MM algorithm is crucial for two reasons. The main reason is GPS device failures which result in a person's incorrect location. The second reason is that the route recommending protocol is hindered by the low sampling rate, bandwidth restriction, and storage expense (Gupta & Shanker, 2021b). Monitoring of data collection for traffic monitoring and suggesting routes for a traveling person on the map is often a crucial task. If the LBS MM imprecision is greater due to a computer malfunction or insufficient bandwidth, the machine should be able to reliably identify the moving human on the map. As a result, LBS with the capability of constant tracking of GPS location for moving persons and corresponding route creation needs is to be introduced. The sensor data is translated via specific MM algorithms, such as particle swarm optimization, Kalman filter, Bayesian inference, Dempster method, fuzzy logic, and Shafer's principle of increasing the precision to minimize the MM error. The existing policies simply neglect the error factors involved with the road map scheme and even the form of orientation. This chapter suggests a multicriteria fuzzy-logic-based MM framework and sequential pattern mining and clustering (SPMC) based path recommendation to address identified limitations (Gupta & Shanker, 2020d). The road network and trajectory data can be obtained from stakeholders such as road builders, road users, and traffic cops. These data could be used to combine the MM structure and road network components to create a qualitative modeled method. This system could potentially reduce the number of road accidents. With given remarks, the chapter's key contributions are highlighted as follows:

- an exhaustive survey of MM solutions
- incorporation of the SPMC process to mine frequent mobility patterns (FMPs) and frame mobility rules (R) to be used in route recommendation
- feeding of the user movement direction into the self-trained virtual reality system to predict user movement errors
- study and comparison of proposed module with already existing compatible algorithms for addressing the MM problem
- future research recommendations

The remainder of paper are written as follows. The related works linked to this research are given in Section 9.2. In Section 9.3 the fuzzy-based MM algorithm and SPMC for route guidance are presented. Also, we explain simulation results with a comparative study in Section 9.4. Section 9.5 gives the conclusion and discusses the future scope of the work in this domain.

9.2 Literature review

MM methods use map digital information combined with GPS receiver information to locate a moving vehicle on a route. Several investigators researched and worked on MM approaches in depth. These approaches are supported by MM modules, including travel time estimation (Rahmani et al., 2017; Tang et al., 2018), path-selection-based evaluation (Oyama & Hato, 2018), gridlock network analysis (Oyama & Hato, 2017), and motion anomaly identification (Qin et al., 2019). In India, we can say that our cities are now being converted into smart cities; as a result, a broad GIS network is needed to turn a simple city into a smart city. Further than that, technical awareness of QGIS, R, PostGIS, and SQL, etc. are also needed. These feature-based MM frameworks support excellent functionality in GIS for GPS trajectories with low sampling frequency.

Probabilistic-approach-based MM strategies (Yin et al., 2016) are designed to work with complicated path or intersection segments. They use extensive calculations to improve matching accuracy. Hidden Markov Model (HMM)-dependent MM and multiple hypotheses (Yuan et al., 2010) are examples of probabilistic-approach-based MM for GPS positioning (Ren & Karimi, 2009). Instead of estimating individual locations and neighboring candidate road links, these methods depend on an aggregate view of the situation for all candidate road links and all positioning specifics (Qinglin et al., 2015). The accuracy of positioning data is often influenced by the sample size of positioning data (Hu et al., 2009). In case of low sampling rate, GPS mapping for mobile phone positioning needs procedure for incremental improvements in terms of accuracy and reliability. To solve complex road mapping, several sophisticated MM methodologies have been proposed including HMM (Mohamed et al., 2017; Yang & Gidófalvi, 2018), fuzzy logic model (Quddus et al., 2006), and Kalman filtering (Cho & Choi, 2014). HMM (Gupta & Shanker, 2021c, 2021d) is commonly used to describe the best matched route by taking into account the noise and sparse location data.

Inconsistencies in acceleration, traveling speed, expected distance between points, shifts in direction, and other factors are all present in the trajectories. As a result, a filtering phase is required in priory in the map generation process, in which traces are examined for any abnormality. All of these methods usually follow a three-step method as follows:

- The first step is used to find the direct relation with constrained optimization problems that are close to the positioning point's distance. This ensures the collection of true trajectories with a high degree of duplicity.
- The next step is to add a practicality value to each relation. The consistency is based on topological MM practicability properties such as topology connectivity, angle intersections, directional similarity, and projective deviation.
- The third one entails calculating all of the ties that have accumulated at certain period of interval.

In the early stages of this area of study, optimization is one of the most popular methods. MM is an important and challenging topic of target recognition in the field of robot mapping. By calculating the maps correspondence, Georgiou et al. (2017) reported that the existing device design needs improvement in the effectiveness of robot—human cooperation. Path matching is a related topic in a topological map and it can be used to coordinate maps (Schwertfeger & Yu, 2016). Using multiple hypotheses methodology, Schuessler and Axhausen, (2009) presented a computational algorithm. To deal with issue of low sampling, Lou et al. (2009) implement ST-MM with speed (i.e., temporal) and geometric (i.e., spatial) constraints. The ST-MM algorithm can be broken down into two stages: candidate filtering and ST analysis. For two consecutive GPS sampling positions, ST-MM calculates the weight function and then candidate points (CPs). The ST-MM recognizes only the spatial—temporal constraints while calculating the weight function. CPs count and error circle radius (ECR) are the internal variables used to identify the most likely segments of candidate roads within the road network. ST-MM generates a real course by following the largest number of weight points. Doing the observational assessment, the weight function in the ST-MM methodology uses the shortest path queries (SPQs) among two consecutive GPS points. This is the most expensive part of the MM system; however, it decreases the overall matching process' response time and improves QoS. Due to a fixed count of CPs and ECR, the MM technique performs a large number of SPQs. The GridST aims to improve the management of the first step namely, the candidate step filtering (Chandio et al., 2015). The ECR as well as the applicant points count selected are dynamically adjusted based on the road system geography. As a result, the number of shortest path estimates is reduced by lowering the overall run-time complexity. GridST divides the road system diagram into grids to determine the road system's locality. The information of all grids is calculated and organized before the actual MM operation to ensure that the method's run-time complexity is reduced. Where a grid appears to have a greater number of road segments, candidate extraction has a greater chance of picking a larger number of CPs in a reduced error area.

Present MM strategies can be divided into two groups based on the variety of trajectories used: local/incremental MM (Liang et al., 2016) and global MM (Liu et al., 2017). The first group is relied on a greedy method that chooses the initial assertion in MM depending on the latest trajectory point, which is then proceeded by the subsequent points. It uses turn-by-turn navigation in real-time application scenarios. With a higher rate of sampling, the local/incremental approach has greater precision. As the rate of sampling decreases, the problem of arc-skipping (Lou et al., 2009) arises to occur. In this case, the car moves from one highway segment to the next when it is still far from the intersection being absurd and resulting in a significant loss of precision. Because of the higher rate of sampling, the local/incremental approach will have a reasonably high energy requirement.

Weight-based MM (Sharath et al., 2019), spatiotemporal feature-based mapping (Hsueh & Chen, 2018), and topology mapping are some of the latest methods that have been introduced to improve matching accuracy (Wang et al., 2017). A global mapping tool is used to find a curve that is as close to the vehicle direction as possible within the network of roads. It entails a large number of track points before determining the most common trajectory direction on the road system. In global matching methods, the similarities between multiple line segments are determined using Frechet separation (Wei et al., 2013), long popular subsequence (Zhu et al., 2017), or likelihood function (Knapen et al., 2018; Millard-Ball et al., 2019; Rappos et al., 2018).

In Gong et al. (2017), AntMapper-matching utilizing ant colony synthesis approach had been considered for local and global geometric/topological details to achieve accurate mapping results in a relatively short period of time. The accuracy of global algorithms improved as more input became available; however, they experienced long computing and processing times. By the observation of the above works and their reported results, the goal of this chapter is basically to pinpoint in reducing the computation and processing time in SCCA using fuzzy inference method (FIS)-MM. In a metropolitan area with a higher rate of sampling, the SCCA's MM methodology improves the matching accuracy of high-frequency data and employs the intersection fragment matching method to improve the matching accuracy for trajectory data.

9.3 Speculative computation cognitive assistant module with trajectory mining methodologies

If an individual is using this service for the first time, the device may not have any user information. When information about the customer is available ahead of time, the software uses GraphHopper to find the shortest path. Predicting object movement from observational locations necessitates the use of a model, for which a large number of spatiotemporal (Gupta & Shanker, 2020b) options are available as detailed in the literature review section. The linear regression is one of the simplest and most widely used techniques. To adjust the path to be used as raw data for the extraction process, a data preprocessing step is necessary. The outliers in the preprocessed trajectory data are viewed as random motion that do not follow other patterns and do not add to the processing effect. Any trajectories see an error higher than the usual GPS error (where if the indicated GPS measurement is more than 50 m away from the median point between prior and consecutive GPS observations), which is viewed as noise. Spatial trajectories are rarely perfect due to sensor noise and other factors such as receiving poor positioning signals in a reduced bandwidth environment. In certain situations, as seen in Fig. 9.3, a noise point errors like $p4$ is too high to obtain useful information such as travel speed.

As a result, we should consider in removing these noise points from trajectories prior to actually beginning the mining process. This stage is known as trajectory preprocessing and is an important part from many tasks in trajectory mining. Removing the noise of the data, deletion of points that do not improve trajectory detail, and MM are critical. Noise filtering is used to exclude all noise points from such a trajectory that may be caused by a faulty GPS signal. Though this problem of noise reduction has not been fully resolved, the particle filters, Kalman, outlier detection based on heuristics, and mean (or median) filter are among the most popular methods for removing trajectory noise.

9.3.1 Speculative computation cognitive assistant module

Through SCCA module the system is able to predict and anticipate possible user mistakes via the use of trajectory pattern mining and shortest route estimation process. The SCCA module depends on the accuracy of next-location estimation and MM method. By creating a spatial map of roads from the traces, the MM process creates the chain of user positions on a virtual map. The next-location/place estimation through pattern mining helps in finding the solution of best path selection. With the help

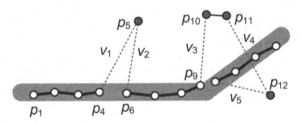

Figure 9.3 Noise points in trajectory.

of SCCA module, the system is capable of predicting the client's error and alerting them continuously.

Fig. 9.4 depicts the procedure for trajectories mining. LBS uses as an open-source tool called GraphHopper Directions API (Karich & Schroder, 2017) for the first stage of preprocessing, which includes a map feature. The MM protocol is the next step, which involves gathering the remaining data and comparing geographical position to the database of Open Street Map. After that, the information is retrieved using the SPMC method. SPMC method is a prevalent trajectories mining method that includes SPMC algorithm for grouping of cluster locations. The SPMC method uses the SPMF (Fournier-Viger et al., 2016) open-source API. The regular trajectories will be derived from the clients' past moving trajectories using SPMC technique. The SPMC technique employs the input called support threshold of a sequential pattern to determine the optimal number of sequences needed to show as a pattern. For customer information that would be driven by the framework, the system employs a data mining strategy. As a result, the user's preferences are applied to the route based on the dataset's patterns threshold, where a route created by utilizing the mined patterns is longer than the preceding shortest path. Here, the system will take the shortest path. The data is clustered using the SPMC cluster-based framework, which takes into consideration the input parameters. To acquire a path that leads the individual to the intended destination from his present location, the mining method must be used to establish connections between the clusters.

This section covers the MM methodology as well as the trajectory data mining techniques that will be used in SCCA. MM is the method of matching observed GPS coordinates to real road network functional maps

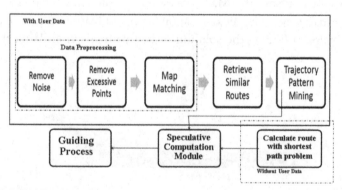

Figure 9.4 Trajectories mining processes.

to make urban computations simpler such as traffic applications. As seen in Fig. 9.5, the MM issue can be built on. The traveling path is fed into a speculative computation model in the MM algorithm for forecasting the user errors and preserving the correct travel path.

The suggested system employs the FIS-MM methodology. This MM is based on the ST–MM technique, which is a global MM methodology (Lou et al., 2009). The SPQs are the most expensive component of MM service between two consecutive GPS sites and are used by the weight feature in the ST–MM strategy. Overall response time and QoS for matching are reduced by enhancing the SPQ computation. Because of the MM approach's fixed count restrictions of CPs and ECR, a significant number of SPQs are conducted. Overall response time and QoS for matching are reduced by enhancing SPQ computation. Because of the MM approach's fixed count restrictions of CPs and ECR, a significant number of SPQs is conducted.

Integration of modern map information in MM GPS odometer reading (Δd), gyro rate reading ($\Delta \theta$), and associated errors (ε_θ) are the crucial variables to affect the navigation performance. For this mixture of GPS and Dr, Zhao et al. (2014) apply EKF algorithm. Finding the exact location of any moving person on a route is a crucial prerequisite of any MM method. The error variances, northing (N), easting (E), heading (H), and person's velocity (v) are few of the GPS/Dr outputs (i.e., σ_v, σ_e, σ_n, and σ_h). The performance from MM is the correct road connection with a trust value greater than the given threshold. If there is not a connection in the candidate position field, the software assumes that the car is off the

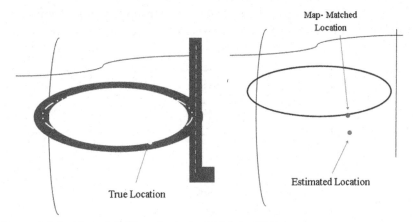

Figure 9.5 Problem definition of map matching in SCCA.

road and away from established avenue connections. In a case where the candidate sector only has one segment, the selection technique will be simple. When there are multiple connections with fixed locations, an approximation tool like FIS might be helpful to determine the true link in between candidate ties. There are two sections of each model for matching the maps. The first section contains the steps for appropriate link identification, while the second section contains the steps for determining the location of the chosen link. The MM system is the position fix function to the perpendicular distance (PD) of the direction, the path bearing, and the individual's orientation, as well as the GPS receiver's first fixed time. The search space is derived from the error variance. A few fixes of the first link are used to identify the subsequent link identification.

9.3.2 Fuzzy inference system enrooted map matching

An FIS can be used in Matlab 6.5 to optimize functions of fuzzy memberships. The collection of membership functions is accomplished by utilizing at least squares. The if-then rules require the precedent (input) for the corresponding relationship (output) within the FIS. The weights in FIS are set based on the criticality. The concrete fuzzy logic requires several significant steps, which are shown in Fig. 9.6.

FIS may be based on Mamdani, that is, the output of weighted expression or Sugeno, that is, the output of the fuzzy set. To apply the fuzzy logic, we used Mamdani-based FIS. It fuzzifies the value of the parameter within [0, 1] range. Various types of membership functionality may be used; however, we prefer triangular membership function (TMF)—the more straightforward type functions to describe membership for the

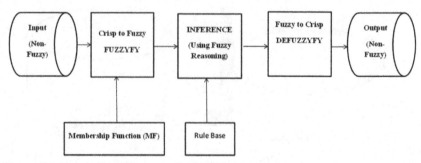

Figure 9.6 Fuzzy inference system for SCCA implementation.

convenience of the user. TMF is widely used to describe membership and can be mathematically defined as follows:

$$\text{Membership Function}(\mu f) = \begin{cases} 0, & b \leq x \leq a \\ \dfrac{b-x}{c-b}, & a \leq x \leq c \text{ or } c \leq 0 \leq b \\ 0, & b \leq x. \end{cases}$$

FIS fuzzifies the corresponding value vi in the range $[0, 1]$ from the value set v between $[a, b]$ for each parameter (Pt_i) from the parameter set (PS). The selection of boundaries should be achieved with the consideration of tradeoff between running time complexity and accuracy. The TMF depiction is given in Fig. 9.7. The range of mapping value of membership function, that is, $v_i \in v$ lies in between 0 and 1 from value set $[a, b]$. This mapping technique is called variable fuzzification. If a function has a condition of n boundaries, then the parameters of the $2n$ boundary must be optimized. More formally, if $Pt_i \in PS$ has bi boundaries, then the following equation gives the required number of optimization variables to produce FIS production in a fuzzy form in a maximum limit and minimum values. The weighted average procedure is used for crisp output computation of associated likelihood for a link.

$$\text{Number of optimization variables} = 2 * \sum_{i=1}^{n} b_i + 2.$$

The weighted average procedure is used for crisp output computation of associated likelihood for a link. For every fuzzy rule, min() function obtains the degree of applicability computation. For the calculation of

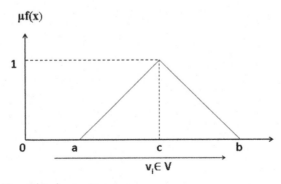

Figure 9.7 TMF graphical overview.

probability value, the following terms are extracted from a to b, given a certain set of membership functions. The Pt_{fn} reflects parameter Pt_n fuzzified value. The Pt_{min} is the minimum value for all parameters after fuzzification:

$$Pt_{f1} = \frac{v_1 - a}{b - a} \quad Pt_{f2} = \frac{v_2 - a}{b - a} \quad Pt_{fn} = \frac{v_n - a}{b - a}.$$

- Range $(Pt_{f1}, Pt_{f2}, Pt_{f3}) \rightarrow [0, 1]$
- $Pt_{min} = \min (Pt_{f1}, Pt_{f2}, Pt_{f3})$

$$\text{Likelihood } T\text{-value} = 10 \times \left(1 - \frac{\int_a^b [Pt_{min} \times (Pt_{min} \times (b - a) + a)] dx}{\int_a^b [Pt_{min} \, dx]} \right).$$

The fuzzified function of a parameter Pt_n is represented by Pt_{fn}. For all parameters, Pt_{min} is the minimum value after fuzzification. The epochs, GPS fix time (procured from GPS), heading error in degree, and heading change (obtained from the gyro) are all input variables in the fuzzy logic MM algorithm, PD (m), horizontal dilution of precision, velocity direction with respect to links, movement direction with respect to links, and velocity of moving person v (m/s). Three types of fuzzy rule sets are implemented by the fuzzy logic MM process. First set of fuzzy rules can be used to locate the source connection, the second set of fuzzy rules is used to pinpoint the position of the individual on this link, and the third one is used to identify the link junctions to locate the next link.

Through simulation of moving persons, a dependency of likelihood-value on v_i is assigned during creation of the rule base. For the number assignment in restricted boundary such as high, mild, and moderate, fuzzy value of the parameter are used as shown in Fig. 9.8. For the first

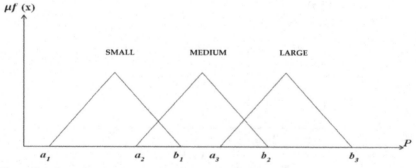

Figure 9.8 Boundary representation for parameter (P).

parameter vector, ith element of a fuzzy set is represented as $S_1[i]$. In the same way, for nth parameter vector, ith element of a fuzzy set is represented as $S_n[i]$ for $i \leftarrow 1, 2, 3 \ldots \ldots$ Let S_1, S_2, and S_3 represent boundary set for the three parameters, then rules in the rule base for the general category can be defined by rules as given in Table 9.1. The FIS-MM output is computed probability for fixing of the location on the link. Constants that fit are taken as 10 as SMALL, 50 as MEDIUM, and 100 as LARGE on the Suzeno fuzzy map.

P stands for the initial successful fix as seen in Fig. 9.9. For position fix P, FIS measures edge 1 as the true link. In the corresponding step, the FIS

Table 9.1 Rule set for SCCA MM.

Perpendicular distance	Horizontal dilution of precision	Gyro processed heading change (Hi)	L1 (rule-set out)
SMALL	SMALL	SMALL	SMALL
SMALL	SMALL	MEDIUM	SMALL
SMALL	SMALL	LARGE	MEDIUM
SMALL	MEDIUM	SMALL	SMALL
SMALL	MEDIUM	MEDIUM	MEDIUM
SMALL	MEDIUM	LARGE	MEDIUM
SMALL	LARGE	SMALL	SMALL
SMALL	LARGE	MEDIUM	SMALL
SMALL	LARGE	LARGE	LARGE
MEDIUM	SMALL	SMALL	SMALL
MEDIUM	SMALL	MEDIUM	MEDIUM
MEDIUM	SMALL	LARGE	MEDIUM
MEDIUM	MEDIUM	LARGE	MEDIUM
MEDIUM	MEDIUM	MEDIUM	MEDIUM
MEDIUM	MEDIUM	LARGE	LARGE
MEDIUM	LARGE	SMALL	SMALL
MEDIUM	LARGE	MEDIUM	LARGE
MEDIUM	LARGE	LARGE	LARGE
LARGE	SMALL	SMALL	SMALL
LARGE	SMALL	MEDIUM	SMALL
LARGE	SMALL	LARGE	MEDIUM
LARGE	MEDIUM	SMALL	MEDIUM
LARGE	MEDIUM	MEDIUM	LARGE
LARGE	MEDIUM	LARGE	LARGE
LARGE	LARGE	SMALL	LARGE
LARGE	LARGE	MEDIUM	LARGE
LARGE	LARGE	LARGE	LARGE

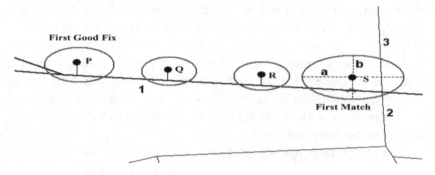

Figure 9.9 An initial MM condition in SCCA.

treats edge 1 as a true relation for position fixes Q and R. As a result, it can be shown that the position fix S exists on the edge 1.

9.3.3 Pattern mining with clustering-based next-location estimation

The proposed policy consists of a procedure to mine frequent mobility. Let $P_{k-1} = <(l_1, t_1), (l_2, t_2) (l_{k-1}, t_{k-1})>$ is the frequent $(k-1)$-pattern, and $V(l_{k-1})$ is the coverage region cell l_{k-1} neighbor set of cells. The pattern mining involves attaching $p = (v, t_k)$ at the end to generate a candidate k-pattern. Here, t_k is the timestamp of v, and t_{k-1} is the timestamp of l_{k-1}. The value of $supp_{min}$ is used for candidate k-patterns selection. The mobility patterns rules can be defined as follows:

R: $X \rightarrow Y$, where X and Y are two FMPs, and $X \cap Y = \emptyset$.

Let the all FMP set is represented by S. Then, mobility rules can be defined as follows:

$$Z \rightarrow (S - Z) \text{ for all } (Z \subset S) \text{ having } (Z \neq \emptyset).$$

ALGORITHM A1: FMPs (L) Extraction

Input: Mobility Log (D) with 1- patterns

$Supp_{min}$

$L_1 \leftarrow$ Legth-1 *large* pattern set   // Let $L_k =$ Legth-K *large patterns set*

$C_1 \leftarrow$ Legth-*1* candidate patterns set // Let $Ck =$ Legth-K candidate patterns set

$G =$ Directed graph for neighboring information,

Output: $L =$ User FMP set.

(Continued)

ALGORITHM A1: (Continued)

Begin

    k = 0

Repeat

    k = k + 1

    For All *Legth-k-large patterns set* $F_{k=} < (p_1, t_1), (p_2, t_2), \ldots, (p_k, t_k) > \in$ L_k do

        $N(p_k) \leftarrow \{n|$ n is the adjacent cell of $p_{k-1}\}$

        For All node $n \in N(l_{k-1})$ do

            $U(p_k) \leftarrow \{s = (n, t_k)| t_{k+1} > t_k$ and $< (n, t_k) > \in N(p_{k-1})\}$

            For each $s = (n, t_k) \in U(p_k)$ do

                $C = < (p_1, t_1), (p_2, t_2), \ldots, (p_k, t_k), (n, t_{k+1}) >$

                $C_{k+1} = C \cup C_{k+1}$

            End For_All

        End For_All

    End For_All

    For All mobility pattern $B \in D$ do

        $F \leftarrow \{q \mid q \in C_{k+1}$ and q is a subsequence of B$\}$

        For All $c \in C$ do

            q.count = q.count + 1

        End For_All

    End For_All

    $L_{k+1} \leftarrow \{q \mid c.support \geq supp_{min}$ and $q \in C_{k+1}\}$

    $L = L_{k+1} \cup L$

While $Lk \neq \varnothing$

Return L

End

ALGORITHM A2: Mobility Rules Formulation

Input: *L:* User FMPs set.

    conf_min: Minimum confidence threshold.

Output: *Rules*: frequent mobility rules set.

Begin

   Rules $\leftarrow \phi$

   For All User FMPs $M_{k=} < (p_1, t_1), (p_2, t_2), \ldots, (p_k, t_k) > \in L_k, k > = 2$

        $M_l \leftarrow M_k$

        y = k

(Continued)

ALGORITHM A2: (Continued)
 Repeat
  //A $\leftarrow$ (y-1) sub pattern of M$_l$.
  A $\leftarrow$ $< (p_1, t_1), (p_2, t_2), \ldots, (p_{y-1}, t_{y-1}) >$.
  conf = support.(P_k)/support.(A).
  If conf $\geq$ *conf_min* **then**.
   R $\leftarrow$ $\{ < (p_1, t_1), (p_2, t_2), \ldots, (p_{y-1}, t_{y-1}) > \rightarrow < (c_l, t_l), \ldots,$
   $(c_k, t_k) > \}$ //R $\leftarrow$ (A $\rightarrow$ P$_k$-A).
   R.w = (RuleTime-MinTime) / (MaxTime-MinTime) $\times$ 100.
   Rules = Rules $\cup$ R.
  Else break.
  End if.
  y = y − 1.
 While y > 1.
End For_All
Return *Rules*.
End

The respective value of confidence (R) is associated with every rule (R):

$$\textbf{Confidence } (R) = \frac{\textit{\textbf{support}} (A \cup B)}{support(A)} \times 100.$$

Confidence (R) is the confidence threshold for rule (R). Frequent mobility rules filtering is achieved by Confidence (R). The value w_i is temporal weighted and associated with every rule, which is estimated by the equation as follows:

$$\text{weight}(R) = \frac{\text{RuleTime} - \text{MinTime}}{\text{MaxTime} - \text{MinTime}} * 100$$

where $\text{Rule}_{\text{Time}}$ is the time of rule's tail the last point. In log file, the recorded minimum and maximum time is represented by Max_{Time} and Min_{Time}, respectively.

ALGORITHM A3: Prediction_Next_Location(P, R)
Input: Rules Set = R
 Prediction Count Threshold = m
 Moving client's trajectory P = $< (l_1, t_1), (l_2, t_2), (l_3, t_3), \ldots,$
 $(l_{n-1}, t_{n-1}) >$.
Output: Next_Cell_ID L$_p$

(Continued)

ALGORITHM A3: (Continued)
Begin
   $L_p = \phi$
   $j = 1$
   For All rule r in R: $< (r_1, t_1'), (r_2, t_2'),...,(r_k, t_k') > \rightarrow <(r_{k+1}, t_{k+1}'),...,(r_v, t_t') >$.
      If $< (r_1, t_1'), (r_2, t_2'),...,(r_k, t_k') >$ is element of P = $< (l_1, t_1), (l_2, t_2),$
        $(l_3, t_3),...,(l_{n-1}, t_{n-1}) >$ and $r_k = l_{n-1}$.
      $T_{diff} = 1/(|t_{n-1} - t_k'| + 1)$.
      r.matching_score = r.w + T_{diff}.
      r.total_score = r.matchingscore + r.support + r.confidence.
      Matching_Rules = Matching_Rules $\cup$ r.
      Next_Cells[j] = (r.total_score, a_{k+1})
      j = j + 1.
      End if
   End For_All
   Descending order Sorting of NextCells array.
   i = 0
   Select the starting x TuplesArray elements
   While (i < x and i < TuppleArray.length) do
   $L_p \leftarrow L_p \cup$ TuppleArray[i]
   Cell_ID = L_p
   Return Cell_ID
End

Table 9.2 Example trajectory.

Sequence ID	Mobility patterns
1.	$< (L_5, t_3), (L_1, t_5), (L_{15}, t_7) >$
2.	$< (L_5, t_3), (L_1, t_5), (L_{13}, t_9) >$
3.	$< (L_5, t_3), (L_2, t_4), (L_1, t_5), (L_{15}, t_7) >$
4.	$< (L_5, t_3), (L_2, t_4), (L_1, t_5), (L_{13}, t_9) >$
5.	$< (L_2, t_4), (L_1, t_5), (L_{15}, t_7) >$
6.	$< (L_2, t_4), (L_1, t_5), (L_{13}, t_9) >$
7.	$< (L_3, t_2), (L_2, t_4), (L_{12}, t_6), (L_{13}, t_9) >$
8.	$< (L_3, t_2), (L_2, t_4), (L_{12}, t_6), (L_{14}, t_8) >$
9.	$< (L_4, t_1), (L_3, t_2), (L_2, t_4), (L_{12}, t_6), (L_{13}, t_9) >$
10.	$< (L_4, t_1), (L_3, t_2), (L_2, t_4), (L_{12}, t_6), (L_{14}, t_8) >$

Table 9.3 For $k = 1$, frequent mobility1-patterns.

C_1		L_1	
Candidates 1-patterns	**Support**	**Frequent 1-patterns**	**Support**
$< (L_5, t_3) >$	4	$< (L_5, t_3) >$	4
$< (L_2, t_4) >$	8	$< (L_2, t_4) >$	8
$< (L_3, t_2) >$	4	$< (L_3, t_2) >$	4
$< (L_4, t_1) >$	2	$< (L_1, t_5) >$	6
$< (L_1, t_5) >$	6	$< (L_{15}, t_7) >$	3
$< (L_{15}, t_7) >$	3	$< (L_{13}, t_9) >$	3
$< (L_{13}, t_9) >$	3	$< (L_{12}, t_6) >$	3
$< (L_{12}, t_6) >$	3		
$< (L_{14}, t_8) >$	2		

The support value of $< (L_4, t_2) >$, $< (L_{14}, t_8) >$ is less than $supp_{min} = 3$. So they are discarded and not to be a part of the next step.

Table 9.4 For $k = 2$, patterns of mobility and FMPs.

C_2		L_2	
Candidates 2-patterns	**Support**	**Frequent 2-patterns**	**Support**
$< (L_5, t_3), (L_2, t_4) >$	2	$< (L_2, t_4), (L_{13}, t_9) >$	4
$< (L_5, t_3), (L_1, t_5) >$	2	$< (L_2, t_4), (L_{12}, t_6) >$	4
$< (L_5, t_3), (L_{13}, t_9) >$	2	$< (L_1, t_5), (L_{15}, t_7) >$	3
$< (L_2, t_4), (L_1, t_5) >$	2	$< (L_1, t_5), (L_{13}, t_9) >$	3
$< (L_2, t_4), (L_{13}, t_9) >$	4		
$< (L_2, t_4), (L_{12}, t_6) >$	4		
$< (L_3, t_2), (L_5, t_3) >$	0		
$< (L_3, t_2), (L_2, t_4) >$	2		
$< (L_1, t_5), (L_{15}, t_7) >$	3		
$< (L_1, t_5), (L_{13}, t_9) >$	3		
$< (L_{15}, t_7), (L_{13}, t_9) >$	0		
$< (L_{12}, t_6), (L_{13}, t_9) >$	2		

We have used a mobility log file dataset of users to model client movement behavior and explain process in accordance with the methods presented here. Table 9.2 shows a reference trajectory log sheet. Frequent mobility 1-patterns and frequent mobility 2-patterns for the considered trajectory example are shown in Tables 9.3 and 9.4. Algorithm 1 describes the procedure for acquisition of FMPs and then achieves candidate k-patterns from a series of FMPs (L). The next section gives a detailed analysis of the experimentation results, as well as insights into the benefits and drawbacks of the current MM approaches, which encourages participants to choose the best MM algorithm for LBS.

9.4 Experimental evaluations

On a computer with 32 GB RAM, an i7−4770 K CPU, Intel Core, and 3.50 GHz, Microsoft Visual Studio 2005 (C++) with ArcGIS setup is used to apply algorithms. R-tree spatial indexing is used to filter the data in the table (Wolfson et al., 1999). The coverage area in SPMC is deployed using the Microsoft Research Asia public GPS Geolife dataset. This dataset includes 182 individuals, 18,670 trajectories, a total distance of nearly 1,200,000 km, and a total time of 48,000 h and more. The GPS points in the path are organized into a triplet with three parts: longitude, latitude, and timestamp. We used 3465 trajectory dataset streams in the simulator with a random collection of start-end position pairs and a given individual speed range. We considered the trajectory dataset to construct a cumulative length of 15 visited places to be used in simulation. Shorter segments are more likely to provide inaccurate findings because GPS traces are noisier right after a capturing gap. As a result, they are often overlooked. Using clustered sequential pattern mining, the CPU with the appropriate configuration took nearly 18 min to a peak of 254 min to frame the mobility rules in our experiment. The FIS-based MM techniques explored the shortest route using the A* algorithm (Hart et al., 1968). In this research, two analyses are carried out. First, the MM (precision and effectiveness) estimations are calculated using actual driving experience for a predetermined path. The second analysis examined at how much the accuracy had improved since the parameters had been learned in the experiment. A navigating device, which includes a 16-bit 80C196 KB microcontroller, user interface device, positioning system, and path guidance system, is needed for execution of algorithm. The shifting human's speed ranges from 1 to 5 km per hour (Carmien et al., 2005; Liu et al., 2009; Ramos et al., 2013). Table 9.5 shows the amount of data and

Table 9.5 Nodes count and road map data.

Area	Application	Road count	Count of nodes
Delhi	Source data	72	60
	Target data	64	54

Table 9.6 Confusion matrix in MM.

		Actual	
		Match	Mismatch
Computed	Matches	T_1	F_1
	Mismatch	F_2	T_2

the count of nodes surrounding the path map. For measuring precision and recall, the confusion matrices for proposed FIS–MM is depicted in Table 9.6.

T_1 comprises the proportion that a given approach correctly accepts with in true match and T_2 defines the amounts that the given approach correctly rejects the true mismatches, in this matrix representation. F_1 and F_2 are the count that arise inaccurately as a true mismatch in an accurate match and a true match to accurate mismatch, respectively, from the defined strategy. Scientists use some of its most popular metric systems to measure the algorithm's performance such as consistency, recall, and precision. For evaluating the corresponding findings in this article, the formulas are provided by the equations mentioned below:

$$\text{Accuracy} = \frac{T_1 + T_2}{T_1 + T_2 + F_1 + F_2}$$

$$\text{Positive Recall} = \frac{T_1}{T_1 + F_2}$$

$$\text{Negative Recall} = \frac{T_2}{T_2 + F_1}$$

$$\text{Positive Precision} = \frac{T_1}{F_1 + F_2}$$

$$\text{Negative Precision} = \frac{T_2}{T_2 + F_2}.$$

The navigation system's efficiency is measured in terms of recall and precision. The recall refers to the proportion of results that were correctly observed by the algorithm, while the precision corresponds to the proportion

Table 9.7 Fuzzy dependent and traditional MM algorithm comparative study in SCCA.

Method	Recall	Precision	F-Measure	Accuracy
Fuzzy-based MM	94	90	92	91
Traditional MM	60	51	55	56

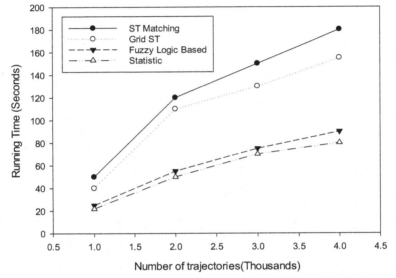

Figure 9.10 Run-time mesurement of different MM approaches used in SCCA.

of outcomes that were truly estimated, that is, true fit and true mismatch irrespective of whether or not these estimates are correct. Table 9.7 shows the outcomes of the experiments. The reciprocal precision of the fuzzy-logic-based MM algorithm is increased from 51% to 92% as a result of algorithm parallelism. The proposed system's recall score, on the other hand, has increased significantly from 60% to 94%. F-measure, which is a measure of Recall and Precision, can be calculated using the equation given below:

$$F_{Measure} = 2 * \frac{Precision \times Recall}{Precision + Recall}.$$

The key evaluation criteria for online MM strategies in this study are running time and accuracy. The horizontal axis depicts the number of trajectories used in the study, while the vertical axis depicts the running time. The effects of trajectories count on running time of conventional and fuzzy-logic-based MM algorithm in SCCA are depicted in Fig. 9.10.

Table 9.8 Effects of trajectories count in SCCA with traditional and fuzzy-logic-based MM algorithms.

Number of trajectory (in thousands)	Running time comparison of MM algorithms (in seconds)			
	Statistic	Grid ST	ST-MM	Fuzzy logic MM
1	22	40	50	25
2	50	110	120	55
3	70	130	150	75
4	80	155	180	90

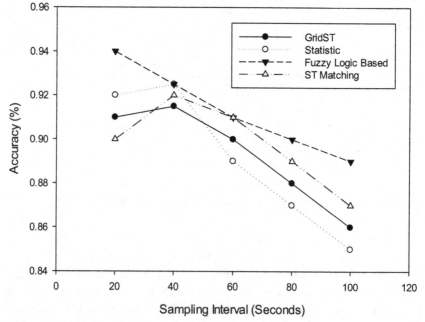

Figure 9.11 Evaluation of influence of sampling frequency on accuracy for various MM algorithms used in SCCA.

Table 9.8 shows a comparison of the four algorithms based on run-time efficiency.

We contrasted the proposed MM algorithm with the three algorithms GridST (Chandio et al., 2015), statistics (Schuessler & Axhausen, 2009), and ST-matching (Lou et al., 2009). According to statistics, the run-time performance of GridST, ST-matching, and fuzzy-logic-based MM is

Table 9.9 Effect of sampling interval variance on precision of past MM and SCCA's FIS-MM.

Sampling interval (in seconds)	Accuracy of MM methods (in %)			
	ST-MM	FIS-MM	Statistics	Grid ST
20	0.90	0.94	0.92	0.91
40	0.92	0.925	0.925	0.915
60	0.91	0.91	0.89	0.90
80	0.89	0.90	0.87	0.88
100	0.87	0.89	0.85	0.86

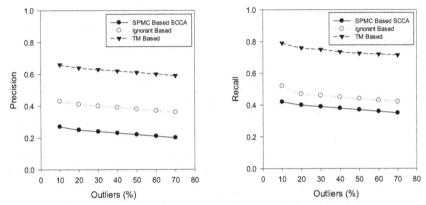

Figure 9.12 Precision and recall analysis of SPMC-based SCCA against various next-location assertion approaches.

proportional to the amount of navigation unit trajectories. The ST-matching and GridST algorithms necessitate a time-consuming matrix estimation, and they lag the FuzzyLogic technique's running time complexity including its statistics.

Fig. 9.11 depicts the accuracy measurement of different MM methods on varying sampling interval. The vertical axis shows the accuracy variations, while the sample time variation is shown in the horizontal axis. Table 9.9 compares the four algorithms based on their accuracy performance.

The simulation results demonstrate that the four methods under consideration have similar patterns in matching precision. The FIS-MM method is more reliable than the previous methods. Due to related simple logic, the GridST and ST-matching methods have equal matching accuracy. The lowest matching precision is shown by statistics MM.

Fuzzy-logic-based SCCA has a higher level of accuracy and is less vulnerable to sampling interval variation. Fig. 9.12 depicts the impact of an outlier on results in terms of recall and precision for SPMC-based SCCA compared to previous next-place prediction based route recommendation policies such as TM-based and ignorant-based assistant services.

9.5 Conclusion

In any route recommendation-based system, the notable difficulties exist are MM procedure complexity and precision. It includes first-location selection of candidate segments, next-place selection of candidate segments, identification of crossing intersections, determination of the best section of initial GPS point, and locating a given part from GPS site. As compared to past practices like GridST, statistics, and LBS ST-matching, the experimental findings demonstrate that the MM algorithm built into SCCA based on fuzzy logic achieves significant performance improvements in the terms of run time and precision. To recover the individual's actual path with vast trajectory sets of specimens, the fuzzy MM algorithm based reasoning of SCCA module isolates every sampling point to the proper position across the real street. The SPMC, on the other hand, has been implemented and evaluated with the aim of adapting the path to customer choices. Assessments of state-of-the-art real-time MM approach in proposed SCCA show certain open questions that require further study.

- The mandatory condition of the next section relative to the previous one has been taken care in proposed module. However, for a moving vehicle from the last location already traveled more than a segment or longer distance between adjacent successive GPS points, this mandatory requirement cannot be met. This deserves to be solved in future MM approaches.
- The MM algorithms discussed in prior study have a mandatory requirement of vehicle turning. It is required to be turned only on valid section of road. However, this requirement may not be fulfilled if the driver unintentionally moved to a wrong part of the road. This dilemma must be solved by the designing of the potential algorithms.
- For better precision, large datasets is needed to be incorporated in real-world fuzzy log based MM into SCCA, which would necessitate a significant amount of data cleaning. Thus a possible area of research is to simplify data cleaning procedures particularly for heterogeneous sources of data.

References

American Psychiatric Association. (2013). *Diagnostic and statistical manual of mental disorders (DSM-5)* (5th ed., p. 991)Washington, DC: American Psychiatric Publishing.

Carmien, S., Dawe, M., Fischer, G., Gorman, A., Kintsch, A., & Sullivan, J. F. (2005). Socio-technical environments supporting people with cognitive disabilities using public transportation. *ACM Transactions on Computer-Human Interaction, 2005*(12), 233−262.

Chandio, A. A., Tziritas, N., Zhang, F., & Xu, C.-Z. (2015). An approach for map-matching strategy of GPS-trajectories based on the locality of road networks. In *Proceedings of the international of vehicles.* (Vol. 9502, pp. 234−246). Cham: Springer.

Cho, Y., & Choi, H. (2014). Accuracy enhancement of position estimation using adaptive Kalman filter and MM. *International Journal of Control Automation, 7*(7), 167−178.

Dawe, M. (2006). Desperately seeking simplicity: How young adults with cognitive disabilities and their families adopt assistive technologies. In: *Proceedings of the SIGCHI conference onhuman factors in computing systems* (pp. 1143−1152), April 22−27, 2006, Montréal, QC, Canada. New York: ACM.

Eikey, E. V., Reddy, M. C., & Kuziemsky, C. E. (2015). Examining the role of collaboration in studies of health information technologies in biomedical informatics: A systematic review of 25 years of research. *Journal of Biomedical Informatics, 57,* 263−277. Available from https://doi.org/10.1016/j.jbi.2015.08.006.

Fournier-Viger, P., Lin, C. W., Gomariz, A., Soltani, A., Deng, Z., & Lam, H. T. (2016). The SPMF open-source data mining library version 2. *In: Proceedings 19th European conference on principles of data mining and knowledge discovery (PKDD 2016)* (pp. 36−40). Springer.

Georgiou, C., Anderson, S., & Dodd, T. (2017). Constructing informative Bayesian map priors: A multi-objective optimisation approach applied to indoor occupancy grid mapping. *The International Journal of Robotics Research, 36*(3), 274−291.

Gong, Y.-J., Chen, E., Zhang, X., Ni, L. M., & Zhang, J. (2017). AntMapper: An ant colony-based MM approach for trajectory-based applications,. *IEEE Transactions on Intelligent Transportation Systems, 19*(2), 390−401.

Gupta, A. K. & Shanker, U. (2018a). Location dependent information system's queries for mobile environment. In: *23rd international conference on database systems for advanced applications (DASFAA) international workshops* (pp. 1−9), May 21−24, 2018, Gold Coast, QLD, Australia.

Gupta, A. K., & Shanker, U. (2018b). CELPB: A cache invalidation policy for location dependent data in mobile environment. *ACM International Conference Proceeding Series.* Available from https://doi.org/10.1145/3216122.3216147.

Gupta, A. K., & Prakash, S. (2018). Secure communication in cluster-based ad hoc networks: a review. In D. Lobiyal, V. Mansotra, & U. Singh (Eds.), Next- Generation Networks. In: *Advances in Intelligent Systems and Computing.* (Vol. 638). Springer.

Gupta, A. K. (2020). Spam mail filtering using data mining approach: A comparative performance analysis. In S. Shanker, U., Pandey (Ed.), Handling Priority Inversion in Time-Constrained Distributed Databases (pp. 253−282). Hershey, PA: IGI Global, 2020. Available from https://doi.org/10.4018/978-1-7998-2491-6.ch015

Gupta, A. K., & Shanker, U. (2020a). A literature review of location-aware computing policies: Taxonomy and empirical analysis in mobile environment. *International Journal of Mobile Human Computer Interaction, 12*(3), 21−45.

Gupta, A. K., & Shanker, U. (2020b). MAD-RAPPEL: Mobility aware data replacement & prefetching policy enrooted LBS. *Journal of King Saud University—Computer and Information Sciences.*

Gupta, A. K., & Shanker, U. (2020c). Study of fuzzy logic and particle swarm methods in map matching algorithm. *SN Applied Sciences* 2, 608 (2020). Available from https://doi.org/10.1007/s42452-020-2431-y.

Gupta, A. K., & Shanker, U. (2020d). Some issues for location dependent information system query in mobile environment. 29th ACM International Conference on Information and Knowledge Management (CIKM '20), 4. Available from https://doi.org/10.1145/3340531.3418504.

Gupta, A., & Shanker, U. (2020e). OMCPR: Optimal mobility aware cache data prefetching and replacement policy using spatial K-Anonymity for LBS. *Wireless Personal Communications*, *114*(2), 949−973. Available from https://doi.org/10.1007/s11277-020-07402-2.

Gupta, A. K., & Shanker, U. (2021a). Prediction and anticipation features-based intellectual assistant in location-based services. *International Journal of System Dynamics Applications (IJSDA)*, *10*(4), 1−25. Available from http://doi.org/10.4018/IJSDA.20211001.oa4.

Gupta, A. K., & Shanker, U. (2021b). Mobility-aware prefetching and replacement Scheme for location-based services: MOPAR. In P. Saravanan, & S. Balasundaram (Eds.), *Privacy and Security Challenges in Location Aware Computing* (pp. 26−51). IGI Global. Available from http://doi:10.4018/978-1-7998-7756-1.ch002.

Gupta, A. K., & Shanker, U. (2021c). An efficient Markov Chain model development based prefetching in location-based services. In P. Saravanan, & S. Balasundaram (Eds.), *Privacy and Security Challenges in Location Aware Computing* (pp. 109−125). IGI Global. Available from http://doi:10.4018/978-1-7998-7756-1.ch005.

Gupta, A. K., & Shanker, U. (2021d). CEMP-IR: a novel location aware cache invalidation and replacement policy. *International Journal Computational Science and Engineering*, *24*(5), 450−462.

Hart, P. E., Nilsson, N. J., & Raphael, B. (1968). A formal basis for the heuristic determination of minimum cost paths. *IEEE Transactions on Systems Science and Cybernetics*, *4*(2), 100−107.

Hsueh, Y. L., & Chen, H. C. (2018). MM for low-sampling-rate GPS trajectories by exploring real-time moving directions. *Information Sciences*, *433*, 55−69.

Hu, J., Cao, W., Luo, J., & Yu, X. (2009). Dynamic modeling of urban population travel behavior based on data fusion of mobile phone positioning data and FCD. In: *Proceedings of the 2009 17th international conference on geoinformatics* (pp. 1−5), Fairfax, VA, USA, August 12−14, 2009.

Kefi, S. E., & Asan, O. (2021). How technology impacts communication between cancer patients and their health care providers: A systematic literature review. *International Journal of Medical Informatics,* 104430. Available from https://doi.org/10.1016/j.ijmedinf.2021.104430.

Knapen, L., Bellemans, T., Janssens, D., & Wets, G. (2018). Likelihood-based offline MM of GPS recordings using global trace information. *Transportation Research Part C Emerging Technologies*, *93*, 13−35.

Karich, P. & Schroder, S. (2017). *GraphHopper directions API with route optimization.* http://www.graphhopper.com, accessed March 2021.

Liang, B., Wang, T., Chen, W., Li, H., Lei, K., & Li, S. (2016). Online learning for accurate real-time MM. In: *Proceedings of the 20th Pacific−Asia conference on advanced knowledge discovery and data mining* (Vol. 9652, pp. 67−78). New York: Springer-Verlag.

Liu, A. L., Hile, H., Borriello, G., Kautz, H., Brown, P. A., Harniss, M., & Johnson, K. (2009). Informing the design of an automated wayfinding system for individuals with cognitive impairments. In: *Proceedings of the 3rd international conferenceon pervasive computing technologies for healthcare* (Vol. 9, p. 8), April 1−3, 2009, London, UK.

Liu, X., Liu, K., Li, M., & Lu, F. (2017). A ST-CRF map-matching method for low-frequency floating car data. *IEEE Transactions Intelligent Transportation Systems, 18*(5), 1241−1254.

Lou, Y., Zhang, C., Zheng, Y., Xie, X., Wang, W., & Huang, Y. (2009). Map-matching for low-sampling-rate GPS trajectories. In: *Proceedings of 17th ACM SIGSPATIAL international symposium on advances in geographic information systems, ACM GIS 2009* (pp. 352−361), November 4−6, 2009, Seattle, Washington, USA.

Millard-Ball, A., Hampshire, R. C., & Weinberger, R. R. (2019). Map-matching poor-quality GPS data in urban environments: The pgMapMatch package. *Transportation Planning and Technology, 42*, 539−553.

Mohamed, R., Aly, H., & Youssef, M. (2017). Accurate real-time MM for challenging environments. *IEEE Transactions on Intelligent Transportation Systems, 18*(4), 847−857.

Oyama, Y., & Hato, E. (2017). A discounted recursive logit model for dynamic gridlock network analysis. *Transportation Research Part C Emerging Technologies, 85*, 509−527. Available from https://doi.org/10.1016/j.trc.2017.10.001.

Oyama, Y., & Hato, E. (2018). Link-based measurement model to estimate route choice parameters in urban pedestrian networks. *Transportation Research Part C Emerging Technologies, 93*, 62−78. Available from https://doi.org/10.1016/j.trc.2018.05.013.

Quddus, M. A., Noland, R. B., & Ochieng, W. Y. (2006). A high accuracy fuzzy logic based MM algorithm for road transport. *Journal of Intelligent Transportation Systems, 10* (3), 103−115.

Qin, G., Huang, Z., Xiang, Y., & Sun, J. (2019). ProbDetect: A choice probability-based taxi trip anomaly detection model considering traffic variability. *Transportation Research Part C Emerging Technologies, 98*, 221−238. Available from https://doi.org/10.1016/j. trc.2018.11.016.

Qinglin, T., Zoran, S., & Kevin, I. W. (2015). A hybrid indoor localization and navigation system with MM for pedestrians using smartphones. *Sensors, 15*, 30759−30783.

Rahmani, M., Koutsopoulos, H. N., & Jenelius, E. (2017). Travel time estimation from sparse floating car data with consistent path inference: A fixed point approach. *Transportation Research Part C Emerging Technologies, 85*, 628−643. Available from https://doi.org/10.1016/j.trc.2017.10.012.

Ramos, J., Anacleto, R., Novais, P., Figueiredo, L., Almeida, A., & Neves, J. (2013). Geo-localization system for people with cognitive disabilities. In: *Proceedings of the trends in practical applications of agents and multiagent systems* (Vol. 221, pp. 59−66). Cham: Springer.

Rappos, E., Robert, S., & Cudré-Mauroux, P. (2018). A force-directed approach for off-line GPS trajectory MM. In: *Proceedings of the 26th ACM SIGSPATIAL international conference on advances in geographic information systems* (pp. 319−328), November 6−9, 2018, Seattle, WA, USA.

Ren, M., & Karimi, H. A. (2009). A hidden Markov model-based map-matching algorithm for wheelchair navigation. *Journal of Navigation, 62*, 383−395.

Schuessler, N., & Axhausen, K. W. (2009). Map-matching of GPS traces on high resolution navigation networks using the Multiple Hypothesis Technique (MHT). Working Paper: Transport and Spatial Planning. Zurich. Retrieved from http://www.baug. ethz.ch/ivt/ivt/vpl/publications/reports/ab568.pdf.

Schwertfeger, S. & Yu, T. (2016). Matching paths in topological maps. In: *Proceedings of the 9th symposium on intelligent autonomous vehicles (IAV)*. IFAC.

Sharath, M. N., Velaga, N. R., & Quddus, M. A. (2019). A dynamic two-dimensional (D2D) weight-based map-matching algorithm. *Transportation Research Part C Emerging Technologies, 98*, 409−432.

Tang, K., Chen, S., Liu, Z., & Khattak, A. J. (2018). A tensor-based Bayesian probabilistic model for citywide personalized travel time estimation. *Transportation Research Part C Emerging Technologies, 90*, 260−280. Available from https://doi.org/10.1016/j.trc.2018.03.004.

Wang, H., Li, J., Hou, Z., Fang, R., Mei, W., & Huang, J. (2017). Research on parallelized real-time MM algorithm for massive GPS data. *Cluster Computing, 20*, 1123–1134.

Wei, H., Wang, Y., Forman, G., & Zhu, Y. (2013). *MM by Fréchet distance and global weight optimization. Technical paper* (p. 19). Department of Computer Science and Engineering.

Wolfson, O., Sistla, A. P., Chamberlain, S., & Yesha, Y. (1999). Updating and querying databases that track mobile units. *Distributed and Parallel Databases, 7*(3), 257–387.

Yang, C., & Gidófalvi, G. (2018). Fast MM, an algorithm integrating hidden Markov model with precomputation. *International Journal of Geographical Information Science, 32* (3), 547–570.

Yin, Y., Shah, R. R., & Zimmermann, R. (2016). A general feature-based MM framework with trajectory simplification. In: *Proceedings of the ACM SIGSPATIAL international workshop on geostreaming* (p. 7).

Yuan, J., Zheng, Y., Zhang, C., Xie, X., & Sun, G. Z. (2010). An interactive-voting based MM algorithm. In: *Proceedings of the 2010 eleventh international conference on mobile data management* (pp. 43–52), May 23–26, 2010, Kansas City, MO, USA.

Zhao, S, Hrbek, S, Lu, M, & Akos, D. (2014). Deep integration of GPSINS based on a software defined receiver—Implementation and test results. In: *Proceedings of the 27th international technical meeting of the satellite division of the Institute of Navigation.* ION GNSS.

Zhu, L., Holden, J. R., & Gonder, J. D. (2017). Trajectory segmentation map-matching approach for large-scale, high-resolution GPS data. *Transportation Research Record, 2645*, 67–75.

Further reading

Yin, Y., Shah, R. R., Wang, G., & Zimmermann, R. (2018). Feature-based MM for low-sampling-rate GPS trajectories. *Transactions on Spatial Algorithms and Systems, 4*, 1–24.

Algizawy, E., Ogawa, T., & El-Mahdy, A. (2017). Real-time large-scale MM using mobile phone data. *ACM Transactions on Knowledge Discovery Data, 11*, 52.

CHAPTER 10

Emotion-recognition-based music therapy system using electroencephalography signals

Swatthi Vijay Sanker, Nivetha B. Ramya Sri Bilakanti, Anju Thomas, Varun P. Gopi and Palanisamy P.
Department of Electronic and Communication Engineering, National Institute of Technology Tiruchirappalli, Tiruchirappalli, India

10.1 Introduction

Music therapy/recommendation systems are a popular means of helping patients suffering from mental health issues such as depression and anxiety. These systems play music according to the patient's current emotional state which helps calm them down. Therefore, to perform music therapy, knowing the patient's current emotional state is necessary. The emotional state of a person can be identified using various types of inputs such as their facial expressions, electrooculography (EOG) signals, and electroencephalography (EEG) signals. Out of these, EOG and EEG are physiological signals obtained from the peripheral and central nervous systems (Tang et al., 2017). Studies show that EEG is more informative than EOG and facial expressions for high-level brain activities and hence more effective in capturing emotions. Placing electrodes on the scalp of the brain helps capture EEG signals. It records the electrical activity of the brain through voltage fluctuations occurring within the neurons. EEG signals can be decomposed into five frequency subbands, namely, delta (1−4 Hz), theta (4−8 Hz), alpha (8−13 Hz), beta (13−30 Hz), and gamma (30−80 Hz). The brain information is contained in these subbands.

Knowledge-based techniques using domain knowledge, statistical methods using algorithms, and hybrid approaches are used for detecting the emotion. Machine learning classification algorithms are popular among these. However, training machine learning algorithms require large amounts of data. There are many emotion analysis data sets available such as HUMAINE, Belfast, SEMAINE, IEMOCAP, ENTERFACE, DEAP,

Edge-of-Things in Personalized Healthcare Support Systems
DOI: https://doi.org/10.1016/B978-0-323-90585-5.00009-6

217

DREAMER, and SEED which help train machine learning algorithms. Out of these, we use the DEAP data set, which is one of the most commonly used data sets because it has both EEG signals and also the facial expressions of selected participants (Tang et al., 2017). Rather than directly feeding the EEG signal data to the machine learning classifier, features that may be either statistical such as mean and standard deviation or nonstatistical such as entropy, power spectral density (PSD) are extracted from these data points and given to the classifier. The accuracy levels of emotion detection using this technique are hardly ideal due to the interference caused by multiple neuronal activities in the brain. Another factor that contributes to the nonideal accuracy levels is the drastic variation of EEG signal changes within short periods. To improve emotion recognition accuracy levels, our work uses a combined set of statistical and nonstatistical features which are Hjorth Mobility Parameter (HMP), Kurtosis Parameter (KP), Spectral Entropy (SE), and PSD. This set of features has provided a very satisfactory accuracy value.

People experience many types of emotions such as angry, sad, happy, and relaxed. Various emotional models exist in literature which may be used to represent emotions quantitatively. We have used one such model, namely, James Russell's valence-arousal model (Russell, 1980). This model classifies emotions on a valence-arousal scale. The extracted set of features is fed to a machine learning classifier which classifies the emotion on the valence-arousal scale into four quadrants. Once the emotion is identified, music/songs are played accordingly to help change the mood of the user. Music/songs are classified into various genres which create different impacts on people. Each song stimulates the brain differently and changes the mood of the person accordingly. In this paper, after identifying the current emotional state, songs are played which help enhance the mood of the user. This helps relieve stress and cure diseases.

Various frequency bands obtained from the EEG signals are detected during various activities of the brain. For instance, delta waves are detected when a person is in sleep or coma. Theta waves are detected when creatively thinking or focusing on something. Alpha waves are detected during the state of relaxation and calm. Beta waves are associated with anxious thinking and active concentration. Gamma waves are associated with multitasking and conscious waking state.

This shows that only one of these frequency bands has the highest energy in accordance with the task performed. When a participant is watching a video, he/she may be in any of these states—calm if the video is a peaceful one, anxious and active when watching an action movie, etc. So, it is not possible to select one of the frequency bands randomly since the energy might be concentrated in any of these bands.

Since the two nonstatistical parameters, namely, SE and PSD are directly related to the energy of the signal, we have considered the frequency band with maximum energy. This ensures that we get proper data.

We can explain this in two scenarios:

- Selecting all the frequency bands for every participant

    In this case the computational intensity increases fivefold and hence the time taken to compute the emotion is longer. The other frequency bands except the desired one have very less energy and do not have proper values, since they are not in orientation with the desired task.

- Selecting one frequency band by trial and error

If we select one of the frequency bands randomly, say we select the beta band for a participant—if suppose the video was an action-packed one, then there would be no problem. But if the video was a very subtle and nature-oriented one, the participant would have been calm and the energy or information contained in the beta band would be minimal. All the information would have lied in the alpha band.

Many other research works have considered only gamma waves as it is associated with a conscious waking state which happens when the participant is watching a video, however, to improve the accuracy of detection, this work has gone in for the maximum energy frequency band.

10.2 Background
10.2.1 Electroencephalography and emotion recognition

Emotion recognition is a developing application under human–computer interaction and machine learning (Fig. 10.1). There are many ways by which human emotion can be gauged, including texting behavior, facial cues, body temperature, and heart rate (Koelstra et al., 2011). Since brain activity is one such parameter that is closely linked to emotion and the neurological working of living beings are yet to be decoded fully, EEG-based emotion recognition is a hot topic among researchers today. EEG signals are used to measure the electrical activity of the brain through voltage fluctuations from brain neurons by placing electrodes on the brain scalp (Tang et al., 2017). Since it possesses a significant correlation to the state of mind of the individual, EEG signals can be used to understand the working of the brain in a noninvasive manner (Bird et al., 2018).

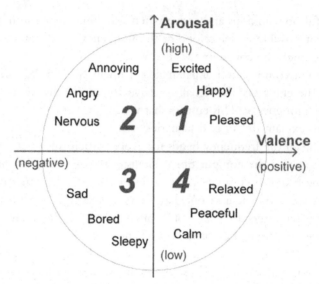

Figure 10.1 The Russell valence-arousal model. *From Tang, C., Wang, D., Tan, A.-H., & Miao, C. (2017). EEG-based emotion recognition via fast and robust feature smoothing. Brain Informatics: International Conference, 11, 83–92 as adapted Yang, Y.-H., & Chen, H. H. (2012). Machine recognition of music emotion: A review. ACM Transactions on Intelligent Systems and Technology (TIST), 3(3), 1–30.*

One of the most common data sources for EEG-related emotion recognition studies is the DEAP data set (Koelstra et al., 2011), followed by others such as SEED (Qing et al., 2019). DEAP is preferred for its partially preprocessed data, volume, and labeled entries. Moreover, the DEAP data set used the prevalent Russell's valence-arousal model which increases the ease of classification. While some studies used the EEG data directly as an input for feature extraction, those which split them into subbands (theta, alpha, beta, gamma, and delta) before feature extraction had shown better results (Abo-Zahhad et al., 2015; Al-Nafjan et al., 2017). This was because only certain bands show significant features in correlation to the subject's emotion, which was often the subband that possesses the maximum energy (Koelstra et al., 2011). Also, data from not all EEG electrodes were necessary to predict their emotion content, as proved by papers that used data from only a limited number of electrodes (the ones that play the greatest roles in detecting emotion) but got results comparable to others which use all of them. This saved processing power and time, especially for researchers without ready access to them. Some electrode positions that contributed the most to human emotion include FP1, FP2, F3, F4, F7, F8, Fz, C3, C4, T3, T4, and Pz (Hou & Chen, 2019).

Since emotions are not objectively quantitative by nature, a framework for classifying them scientifically is essential. James Russell's valence-arousal model is one of the most widely used, which plots different emotions against a two-axis graph. It was first described in an article by Russell in 1980 (Russell, 1980). The X and Y axes measure valence (positive/negative emotion) and arousal (agitated/calm), respectively. The entire gradient of emotions is thus divided into four quadrants (as shown in Fig. 10.1) and named (Koelstra et al., 2011; Qing et al., 2019):

- High valence, high arousal: Quadrant I
- Low valence, high arousal: Quadrant II
- Low valence, low arousal: Quadrant III
- High valence, low arousal: Quadrant IV

Some major emotion labels belonging to each of the above quadrants are included in Fig. 10.1.

Hence, using the valence-arousal model, emotions can be classified with ease.

10.2.2 Feature extraction and classification

Features of EEG signals need to be elucidated to correlate them with emotions. Signal features can be either statistical or nonstatistical. The former is a popular choice due to simplicity in extraction, but nonstatistical features including energy-based ones hold useful insights about EEG data, particularly when emotion recognition is considered (Tang et al., 2017). Key features that are often used include mean, standard deviation, PSD, and first and second difference means (Tang et al., 2017). Previous literature is, for the greater part, concerned largely with statistical features, but it is generally noticed that combining nonstatistical features gives better results. For example, Yang et al. (2019) use just statistical features whereas Koelstra et al. (2011) use PSD. It can be noticed that the classification accuracy of the former is less than that of the latter. Moreover, the number of features extracted and used only loosely correlates with the final accuracy of the model (Chao et al., 2018; Yulita et al., 2019). Hence the features are to be chosen with care, not just relying on sheer volume to get better results. Thus an optimized approach involves using a carefully selected set of both statistical and nonstatistical (energy-based) features.

Classification is often performed using machine learning or deep learning methods. Often different types of classifiers are tried out on the extracted features and the ones giving the best results are applied for the end purpose. Some classifiers used in prominent recent literature include support vector machine (SVM), random forest, J48, and random tree (Qing et al., 2019). Neural net frameworks such as convolutional neural networks, deep neural networks, and deep belief networks (Mousavinasr et al., 2019), and their variations are also used. Cross-validation during training ensures that over-fitting is reduced. A combination of features for each of the subjects in DEAP while they watch each video are extracted and fed into a machine learning classifier such as random forest, SVM, K-nearest neighbor, or J48. The individual's emotional state can then be predicted after training the model suitably (optimally with 5- or 10-fold cross-validation) (Chao et al., 2018).

10.2.3 Music therapy and recommendation systems

Music therapy is a recently developing popular technique used to help patients suffering from various diseases including mental hazards and cancers. It makes use of music to improve patients' quality of life and healing capability. In most music recommendation systems, a weighted sum calculation based on a cooccurrence matrix was performed to determine if a song is a suitable match for a particular user (Shakirova, 2017). While presenting the user with a relevant song matching his/her interest is of top priority, the system must also ensure that the individual does not get bored with the same type of songs and hence introduce some amount of diversity to keep them whetted (Shakirova, 2017).

Three major types of music recommendation systems (Shakirova, 2017; Yang & Chen, 2012) were described in the relevant literature, namely
- Content-based filtering: The categorization and recommendation of songs were based on metadata (tags or genres attached to each song) or the actual audio content (such as the rhythm or tempo).
- Collaborative filtering: Recommendation of songs was based on the user's previous watch history, what type of users listen to a particular piece of music, and the correlation between the two. Needless to say, this type requires a large amount of music and user data to function.
- Hybrid filtering: It combined both the above methods so that the simplicity of the former could complement the accuracy of the latter.

Any one of these is put to use by most recommendation systems (not just music-based). Most of them are hybrids, which contain a unique mix of content-based and collaborative methodologies as per need and necessity.

The proposed work primarily focused on improving the quality and accuracy of the features taken for classification and not on the classifiers. The reason being that, from past literature, it observed that the accuracy has been low not mainly because of the classification technique but rather because of the nature of the features and its values. The statistical parameters are computed in the time domain and nonstatistical features in the frequency domain. For instance, a signal with multiple spikes may have a high standard deviation but a low energy level in the frequency domain. This may mislead the classifier to wrong detection. However, statistical features cannot be completely neglected. Sometimes, information missing in the frequency domain may be found in the time domain. Hence, the proposed model has selected a mix of statistical and nonstatistical features and has tried to ensure that their values are accurate using maximum energy bands.

10.3 Proposed method

10.3.1 Emotion recognition

The first step toward building a music therapy system is to recognize the user's current emotion. This is done with the help of EEG signals acquired from the user (Fig. 10.2). The signals are preprocessed and given to a machine learning classifier to classify the emotion. A brief overview of our proposed method is shown in Fig. 10.2.

10.3.1.1 Data

Many data sets exist in academia with a focus on emotion recognition, including DEAP, SEED, HUMAINE, Belfast, and ENTERFACE (Qing et al., 2019). Foremost among these is DEAP (Koelstra et al., 2011) which consists of the physiological recordings of 32 participants as they watched

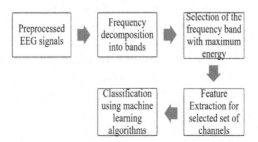

Figure 10.2 Proposed methodology.

40 video clips of 60 seconds each. These videos are meant to elicit emotion in the viewer, which is then reflected in their physiological data. The data consists of 40 channels: one each for the 32 EEG electrodes and the remaining 8 recording other physiological markers (e.g., body temperature) (Koelstra et al., 2011). Moreover, the data also contains valence, arousal, dominance, and liking labels for each entry (on a scale of 0 to 9), assigned by the participants themselves (Koelstra et al., 2011). For the sake of this emotion recognition model we need only valence and arousal levels, the other two features are discarded. Some amount of basic preprocessing has already been done on the DEAP data, by which the signals have been downsampled from 512 to 128 Hz. EOG artifacts have been removed and a bandpass filter of 4–45 Hz has been applied to provide only the required subbands. The data is further averaged to common reference and divided into 60-second segments, discarding the 3-second gap between trials (Koelstra et al., 2011). Classifications are performed on a valence-arousal scale. We utilized a formula from (Koelstra et al., 2011) applied to EEG data from DEAP as a label. According to Balasubramanian et al. (2018) and Liu et al. (2011), frontal EEG asymmetry plays a major role in determining the valence and arousal levels. In this paper, we make use of two formulae for calculating the valence and arousal values from the EEG signal data (Koelstra et al., 2011). For valence calculation, we make use of Hayfa et al.'s formula given in Blaiech et al. (2013).

$$\text{Valence} = \frac{\text{alpha(F4)}}{\text{beta(F4)}} - \frac{\text{alpha(F3)}}{\text{beta(F3)}} \qquad (10.1)$$

For arousal calculation, we make use of Vamvakousis's formula given in Vamvakousis and Ramirez (2012).

$$\text{Arousal} = \log_2\left(\frac{\text{beta(Fz)}}{\text{alpha(Fz)}}\right) \qquad (10.2)$$

where Fz electrode is used to signify frontal powers.

These values of valence and arousal calculated serve as ground truth for our computations. The correlation of the values calculated using the above equations with the DEAP self-assessment measurement was calculated using the mean square error (MSE) parameter. Valence calculated using the Hayfa equation resulted in an MSE of 8.490 and arousal calculated using the Vamvakousis equation resulted in an MSE of 7.513 (Koelstra et al., 2011). These two equations proved to be the best with the least MSE values (Koelstra et al., 2011). An average of all the participants' valence and arousal values were computed and are used for labeling. The classification of valence and arousal into high and low levels was performed based on whether the participant's rating was $>$ or $< =$ average, respectively. The contents of the DEAP data set as given in Koelstra et al. (2011) are summarized in Table 10.1.

10.3.1.2 Frequency decomposition of the electroencephalography signals

An EEG signal is a measure of the electrical activity of the brain, gauged using electrodes placed on the scalp of the subject (Tang et al., 2017; Fig. 10.3). The voltage fluctuations resulting from ionic current within brain neurons manifest as the electrical signal that is measured by EEG. Since it possesses a significant correlation to the state of mind of the individual, EEG signals can be used to understand the working of the brain in a noninvasive manner (Bird et al., 2018). EEG signals are usually divided into five major subbands, based on their frequency content (Al-Nafjan et al., 2017, Koelstra et al., 2011). These subbands are given in Table 10.2 along with their approximate frequency ranges.

Table 10.1 Contents of DEAP data set.

Array name	Array shape	Array contents
Data	40 × 40 × 8064	video/trial × channel × data
Label	40 × 4	video/trial × label (valence, arousal, dominance, liking)

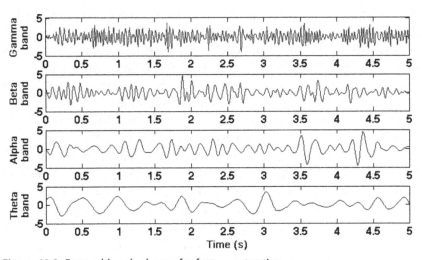

Figure 10.3 Four subbands chosen for feature extraction.

Table 10.2 Electroencephalography (EEG) subbands.

Subband of EEG signal	Frequency range (Hz)
Delta	1−4
Theta	4−8
Alpha	8−13
Beta	13−30
Gamma	30−80

The preprocessed EEG signal data is decomposed into four different frequency subbands—theta (4−8 Hz), alpha (8−13 Hz), beta (13−30 Hz), and gamma (30−80 Hz) using stationary wavelet transform (SWT) as shown in Fig. 10.3.

10.3.1.3 Selection of frequency band with maximum energy

EEG signals are nonstationary by nature and hence most of the energy of a signal is contained in one of the frequency bands. Therefore in this paper, we have selected the band with maximum energy to reduce the computation intensity and to provide better classification accuracy. The PSD and hence the average power was computed for each of the frequency bands. The frequency band with maximum energy was selected for every participant/user.

10.3.1.4 Feature extraction

Features are attributes of data that can be used to differentiate and classify data points. In the case of signals in general and a bulk amount of EEG data (such as for this problem statement) in particular, it is easier to extract such features and study them using a machine learning model than by simple visual examination (Chao et al., 2018; Yang & Chen, 2012). For efficient analysis, the EEG signals need to be split into their subbands before features are extracted (Al-Fahoum & Al-Fraihat, 2014; Koelstra et al., 2011). Filters can be applied to achieve this (Mousavinasr et al., 2019), but it can be performed more effectively by SWT (Tang et al., 2017). To save processing resources, usually, only the subbands containing high energy are considered for further analysis. The average power of each subband, required to determine the same, is calculated by PSD (Al-Nafjan et al., 2017; Koelstra et al., 2011). Moreover, data from only those EEG electrode channels that show the most significant impact on final emotion analysis can be used to yield good results with a lower processing time. Signal features can be broadly divided into two categories: statistical and nonstatistical. Some individual features that come under each of them are given in Table 10.3.

In this paper, we have used a combination of these features, namely, KP, HMP, SE, and PSD.

- KP

  The kurtosis is the fourth standardized moment, defined as

  $$K = \frac{1}{N} \sum_{i=1}^{N} \left(\frac{x_i - \mu}{\sigma} \right)^4 \qquad (10.3)$$

  where μ is the first central moment and σ is the standard deviation.

- HMP

  The Hjorth parameters are indicators of statistical properties used in signal processing in the time domain. These parameters can be of three types, namely, activity, mobility, and complexity. We have made use of the mobility parameter in this paper. This is defined as the square root of the variance of the first derivative of the signal $x(t)$ divided by variance of the signal $x(t)$.

  $$\text{Mobility} = \sqrt{\frac{\text{var}\left(\frac{dx(t)}{dt} \right)}{\text{var}(x(t))}} \qquad (10.4)$$

- SE

  SE indicates a measure of the spectral power distribution. The SE treats the signal's normalized power distribution in the frequency domain as a probability distribution and calculates the Shannon entropy of it.

  It is calculated using this formula:

  $$\text{Entropy} = \frac{\sum_{k=1}^{N} P_k \ln (P_k)}{\ln (N)} \qquad (10.5)$$

Table 10.3 Statistical and nonstatistical features.

Type	Features
Statistical	Mean, standard deviation, variance, first difference mean, second difference mean, Hjorth parameter, Kurtosis parameter, etc
Nonstatistical	Power spectral density, entropy, average energy of bands of signal, etc

$$P_k = \frac{Pw_k}{\sum_i Pw_i} \tag{10.6}$$

- PSD

PSD shows the distribution of the power of a signal in the frequency domain. It shows the energy of a function as a function of frequency. It is calculated from $S_{xx}(f)$ which is the power spectrum of a time series $x(t)$.

$$S_{xx}(w) = \lim_{T \to \infty} E\left[\left|x(w)\right|^2\right] \tag{10.7}$$

The following features were extracted from five specific electrodes which play a major role in human emotion—**Af3, F3, AF4, Fz, and F4** (Koelstra et al., 2011).

10.3.1.5 Classification

In our paper, various classifiers including SVM, J48, and random forest were trained out of which random forest proved to be the best. Hence a random forest classifier was trained using the above feature data set with 10-fold cross-validation.

10.3.2 Developing a music therapy system

Music therapy (Chao et al., 2018; Koelstra et al., 2011) is often used to treat patients in a state of clinical depression or anxiety and sometimes to soothe those in extended hospitalization. Music is known to affect one's emotional state, as attested by data sets such as DEAP wherein participants' EEG readings show variation from the normal mental state while they listen to different types of music (Hou & Chen, 2019). Normally music therapy is administered manually, according to the patient's mood. A significantly accurate automatic system to do the same makes the process easier and hands-free, which is the aim of the proposed system. In our proposed methodology, after the emotion of a user is identified, songs that have a happy mood are recommended.

10.3.2.1 Songs data set

Music recommendation involves playing appropriate music/songs which help change the mood of the user. For classifying songs into appropriate genres (Xue et al., 2018), we have made use of the NJU-MusicMood-v1.0 data set. This data set consists of 777 music clips classified into 4 mood categories—angry, happy, relaxed, and sad. Out of these, 400 clips were used as the training set and the remaining 377 clips were used for testing purposes. A text file consisting of a few lines from the song is provided for every song under the appropriate genres.

The data set NJU-MusicMood-v1.0 is an already classified one. It was obtained as a standard data set which contains the mood induced by various types of music. Once the emotion is predicted by using back-end matlab program, we suggest music from this data set accordingly.

In this paper, we have developed a website for music recommendation which first identifies the emotion of the user using the above-proposed methodology and recommends songs accordingly with the help of the above data set.

10.4 Results and discussion

10.4.1 Module1: emotion recognition

In this paper, we have achieved a significant improvement in accuracy levels for valence and arousal classifications. To perform emotion recognition, we first take in preprocessed EEG data from the DEAP data set and split it into various frequency subbands. We select only the frequency band which contains maximum energy out of these for further computations. Doing so ensures that the maximum amount of information is retained with a reduction in computational loads. After extracting the following features: HMP, KP, SE, and PSD, from Af3, F3, AF4, Fz, F4 electrodes, a random forest classifier is trained using the above feature data set with 10-fold cross-validation which provides results with an accuracy of 93.2% for valence classification and 95.3% for arousal classification. Our experiment provides a significant improvement in accuracy by using a small, but effective feature set. The confusion matrices and the weighted average values of the evaluation parameters of different classifiers for valence and arousal classification are given in Tables 10.4, 10.5, 10.6, and 10.7.

10.4.2 Music therapy website

After emotion detection, the next step is to recommend music/songs accordingly which help enlighten the user (Figs. 10.4 and 10.5). In this paper, we have developed a website that acts as a user interface for the same. First, a user has to register and log in using his account. Once he/she is signed in, the user's emotion is detected with the help of the emotion recognition module running in the back end (Fig. 10.4). After detecting the emotion, songs are suggested which tend to make the user

Table 10.4 Confusion matrix of valence classification for various classifiers.

Classifier	Predicted class	Targeted class	
		High	Low
J48	High	573	62
	Low	73	572
Random tree	High	539	96
	Low	87	558
Random forest	High	591	44
	Low	44	601

Table 10.5 Weighted average values of the evaluation parameters of different classifiers for valence classification.

Classifier	FP rate	Precision	Recall	F1-score	Accuracy (%)
J48	0.105	0.895	0.895	0.895	89.4
Random tree	0.143	0.857	0.857	0.857	85.7
Random forest	0.069	0.931	0.931	0.931	93.1

Table 10.6 Confusion matrix of arousal classification for various classifiers.

Classifier	Predicted class	Targeted class	
		High	Low
J48	High	579	38
	Low	50	613
Random tree	High	564	53
	Low	48	615
Random forest	High	592	25
	Low	35	628

Table 10.7 Weighted average values of the evaluation parameters of different classifiers for arousal classification.

Classifier	FP rate	Precision	Recall	F1-score	Accuracy (%)
J48	0.068	0.931	0.931	0.931	93.1
Random tree	0.079	0.921	0.921	0.921	92.1
Random forest	0.046	0.953	0.953	0.953	95.3

Figure 10.4 Emotion recognition web page.

Figure 10.5 Music therapy web page.

happy or relaxed accordingly (Fig. 10.5). Apart from this, the website provides other features such as Explore which, shows some of the best and popular songs, Feedback, and an About section. Currently, due to the impact of COVID-19, we were not able to acquire real-time EEG electrodes. Hence we have used a testing data set from DEAP which consists of a subset of 20 data points classified as follows: 1–5: Happy, 6–10: Relaxed, 11–15: Sad, and 16–20: Angry. To demonstrate our approach, the back-end code randomly selects one of these data points and computes the emotion as shown in Fig. 10.4.

Table 10.8 shows the comparison of classification with prior studies and it also shows that an increase in feature dimension does not necessarily result in improved accuracy (Chao et al., 2018). This table lists the most prominent works done in this field so far with the accuracy levels obtained. It shows the comparison of our work with some of the state-of-the-art works done in this field so far. The first study (Tang et al., 2017) has considered only statistical features and had obtained an accuracy of 82.3% using an SVM classifier. Yulita et al. (2018) have used principal component analysis technique to reduce the number of features and had

Table 10.8 Weighted average values of the evaluation parameters of different classifiers for valence and arousal classification.

Study	Features set	Classifiers used	Accuracy
Tang et al. (2017)	Mean, SD, mean of the absolute values of the first and second differences of raw and normalized signals	SVM with RBF/ polynomial kernel	82.3% (4 classes)
Yulita et al. (2018)	PCA and re-sampling (3 features)	K-star, KNN	81.2% (arousal) 81.6% (valence)
Al-Nafjan et al. (2017)	Power spectral density (FFT)	DNN	82% (valence, arousal)
Chao et al. (2018)	664 features (statistical features, power features, power differences, Hilbert−Huang spectrum (HHS) features)	Deep Belief Network with Glia Chains (DBN-GC) based ensemble, deep learning model	75.92% (arousal) 76.83% (valence)
Yang et al. (2019)	Downsampled, frequency decomposed signal points	CNN	69.37% (arousal) 71.67% (valence)
Proposed methodology	Hjorth mobility parameter, Kurtosis parameter, power spectral density, spectral entropy	Random forest	93.2% (valence) 95.3% (arousal)

CNN, Convolutional neural networks; *DBN*, deep belief networks; *DNN*, deep neural networks; *PCA*, principal component analysis; *SD*, standard deviation.

again obtained accuracy levels around 80%. The other studies (Chao et al., 2018; Koelstra et al., 2011; Yang et al., 2019) have considered a large number of features applied to deep learning-based classification techniques. Compared to all these techniques which have provided quite low accuracy levels, our proposed method has provided a significant rise in accuracy levels.

10.5 Conclusion

This paper describes an automated music therapy system that recommends songs based on the emotional state of the user, which is found out by analyzing his/her EEG activity. To quantify and classify emotions effectively, Russell's valence-arousal model is used. Input EEG data was taken from the DEAP data set, and subbands were split using an SWT. From the five electrode channels that play the most important part in human emotion, statistical and nonstatistical (spectral) features were extracted. The machine learning classifiers were tested with the data to identify the emotion of the subject, in which random forest with 10-fold cross-validation gave results with an accuracy of 93.2% for valence classification and 95.3% for arousal classification. To perform music therapy, a website that recommends songs based on the user's emotion is designed. Since only a limited number of key features are used, processing time and energy are reduced substantially. The system also has better accuracy than most other recent, prominent works done in the domain, thus making it effective and worthy of future research and development.

10.6 Future scope

The system developed and described here can be further enhanced by the usage of real-time live input from EEG electrodes. To increase accuracy and efficiency, more experiments with regards to the statistical and nonstatistical parameters (to determine which combination of features gives the best results) can be undertaken. Since the data set used employed video music clips to elicit emotion from the participants, they may also be reacting to the visual cues from the videos which will reflect in their brain activity. To restrict the response to that due to the music only, a new database where subjects listen to just audio can be formulated and constructed. This system of music therapy can be extended to other forms of recommendation systems to take into account movies, books, etc. where emotion tags can be attributed much more easily. The core idea can also be implemented in the form of a mobile application with a custom EEG electrode set, to make it more accessible.

References

Abo-Zahhad, M., Ahmed, S. M., & Abbas, S. N. (2015). A new EEG acquisition protocol for biometric identification using eye blinking signals. *International Journal of Intelligent Systems and Applications, 7*(6), 48.

Al-Fahoum, A., & Al-Fraihat, A. (2014). Methods of EEG signal features extraction using linear analysis in frequency and time-frequency domains. *ISRN Neuroscience.*

Al-Nafjan, A., Hosny, M., Al-Wabil, A., & Al-Ohali, Y. (2017). Classification of human emotions from electro-encephalogram (EEG) signal using deep neural network. *International Journal of Advanced Computer Science and Applications, 8*(9), 419−425.

Balasubramanian, G., Adalarasu, K., Mohan, J., & Seshadri, N. P. G. (2018). Music induced brain functional connectivity using EEG sensors: A study on Indian music. *IEEE Sensors Journal, 1−1,* 10.

Bird, J. J., Manso, L. J., Ribeiro, E. P., Ekart, A., & Faria, D. R. (2018). A study on mental state classification using EEG-based brain-machine interface. In: *2018 International conference on intelligent systems (IS)* (pp. 795−800).

Blaiech, H., Neji, M., Wali, A., & Alimi, A. (2013). Emotion recognition by analysis of EEG signals. In: *13th International conference on hybrid intelligent systems, HIS 2013, 12* (pp. 312−318).

Chao, H., Zhi, H., Dong, L., & Liu, Y. (2018). Recognition of emotions using multichannel EEG data and DBN-GC-based ensemble deep learning framework. *Computational Intelligence and Neuroscience, 2018,* 1−11.

Hou, Y., & Chen, S. (2019). Distinguishing different emotions evoked by music via electroencephalographic signals. *Computational Intelligence and Neuroscience, 2019,* 1−18.

Koelstra, S., Muhl, C., Soleymani, M., Lee, J.-S., Yazdani, A., Ebrahimi, T., Pun, T., Nijholt, A., & Patras, I. (2011). DEAP: A database for emotion analysis; using physiological signals. *IEEE Transactions on Affective Computing, 3*(1), 18−31.

Liu, Y., Sourina, O., & Nguyen, M. K. (2011). Real-time EEG-based emotion recognition and its applications. *Transactions on computational science XII, 6670,* 256−277.

Mousavinasr, S. M. R., Pourmohammad, A., & Saffari, M. S. M. (2019). Providing a four-layer method based on deep belief network to improve emotion recognition in electroencephalography in brain signals. *Journal of Medical Signals and Sensors, 9*(2), 77.

Qing, C., Qiao, R., Xu, X., & Cheng, Y. (2019). Interpretable emotion recognition using EEG signals. *IEEE Access, 7,* 94160−94170.

Russell, J. A. (1980). A circumplex model of affect. *Journal of Personality and Social Psychology, 39*(6), 1161.

Shakirova, E. (2017). Collaborative filtering for music recommender system. In: *2017 IEEE conference of Russian young researchers in electrical and electronic engineering (EIConRus), 01* (pp. 548−550).

Tang, C., Wang, D., Tan, A.-H., & Miao, C. (2017). EEG-based emotion recognition via fast and robust feature smoothing. *Brain Informatics: International Conference, 11,* 83−92.

Vamvakousis, Z., & Ramirez, R. (2012). A brain-gaze controlled musical interface. In: *Berlin BCI workshop 2012—Advances in neurotechnology, 4.*

Xue, H., Xue, L., & Su, F. (2018). Multimodal music mood classification by fusion of audio and lyrics. In: *International conference on multimedia modeling MMM 2015: MultiMedia modeling, 01 (2015)* (pp. 26−37).

Yang, H., Han, J., & Min, K. (2019). A multi-column CNN model for emotion recognition from EEG signals. *Sensors Multidisciplinary Digital Publishing Institute, 19*(21), 4736.

Yang, Y.-H., & Chen, H. H. (2012). Machine recognition of music emotion: A review. *ACM Transactions on Intelligent Systems and Technology (TIST), 3*(3), 1–30.

Yulita, I., Julviar, R., Triwahyuni, A., & Widiastuti, T. (2019). Multichannel electroencephalography-based emotion recognition using machine learning. *Journal of Physics: Conference Series, 1230*(1), 012008.

Yulita, N., Rosadi, R., Purwani, S., & Suryani, M. (2018). Multi-layer perceptron for sleep stage classification. *Journal of Physics: Conference Series, 1028*(1), 012212.

CHAPTER 11

Feedback context-aware pervasive systems in healthcare management: a Boolean Network approach

Fabio Alberto Schreiber[1] and Maria Elena Valcher[2]
[1]Department of Electronics, Information and Bioingeneering, Politecnico di Milano, Milan, Italy
[2]Department of Information Engineering, University of Padova, Padova, Italy

11.1 Introduction and related works

The growing complexity of modern software systems stimulated the use of component-based approaches and the enforcement of the separation of concerns (Djoudi et al., 2016). In context-aware computing the separation is made between the functions the system is built for, that can change in time, owing to different conditions, and the context in which the system must operate, which sets the current environmental situation (Cardozo & Dusparic, 2020; Li et al., 2020).

Context-aware databases have been used in different application domains, for example, to tailor application relevant data sets, to reduce information noise and increase the precision of information retrieval algorithms; to build smarter application environments; and to benefit from newly discovered web services. To better adapt to the specific problem, different models of the context have been proposed (see, e.g., Bolchini et al., 2007a, 2007b; Bolchini et al., 2009; Schreiber et al., 2012).

Among the most widely used definitions of context and of context-aware computing, those proposed in Dey (2001) state: "Context is any information that can be used to characterize the situation of an entity. An entity is a person, place, or object that is considered relevant to the interaction between a user and an application, including the user and applications themselves." and "A system is context-aware if it uses context to provide relevant information and/or services to the user, where relevancy depends on the user's task."

Edge-of-Things in Personalized Healthcare Support Systems
DOI: https://doi.org/10.1016/B978-0-323-90585-5.00010-2

237

New application domains such as self-adapting systems (Cardozo & Dusparic, 2020; Schreiber & Panigati, 2017), safety-critical applications, autonomous vehicle design and manufacturing (Tran et al., 2012), disaster prevention, or healthcare management, require a very high level of data quality and dependability that can only be achieved by formally determining their behaviors. Properties such as:

- the existence of *stable equilibrium points*;
- the absence of *undesired* oscillations (limit cycles);
- *observability*—the measure of how well internal states of a system can be inferred from the knowledge of its external outputs (and, possibly, of the corresponding data inputs);
- *controllability*—the capability of an external input data set (the vector of control variables) to drive the internal state of a system from any initial state to any other final state in a finite number of steps; and
- *reconstructability*—when the knowledge of the input and output vectors in a discrete time interval allows to uniquely determine the system final state

are only some of the features, together with the existence of *fault detection and identification* algorithms, which allow to guarantee the expected and safe operation of a system.

Several approaches to the formal definition/validation/verification of pervasive, context-aware, and self-adapting systems have been proposed; bigraphs and model-checking approaches using linear temporal logic have been proposed in Cardozo and Dusparic (2020), Cherfia et al. (2014), Djoudi et al. (2016), Li et al. (2020), Serral et al. (2010), Shehzad et al. (2004), Sindico and Grassi (2009), Tran et al. (2012), and Wang et al. (2011).

In Arcaini et al. (2015), formal properties such as validation, verification, and system correctness of self-adapting systems for systems specified by MAPE-K control loops are discussed using abstract state machines. In Padovitz et al. (2004) and Padovitz et al. (2005) a state-space approach is adopted to model the *situation* dimension and to determine the likelihood of transitions between *situation subspaces*, while keeping the other context dimensions constant. The probability of a transition is evaluated by resorting to concepts analogous to those of velocity and acceleration, typically adopted for mechanical systems. Table 11.1 summarizes the relevant features of some of the cited approaches.

Systems theorists are well acquainted with the techniques to prove the aforementioned properties, and in Diao et al. (2005) the authors have

Table 11.1 Comparison of approaches.

Reference	Main issues	Target architecture	Models and tools	Application domains
Arcaini et al. (2015)	Modeling, validation, verification, correctness	Distributed C–A self-adaptive systems	MAPE-K, simulation, ASMETA model checking (LTL)	Smart home
Cardozo and Dusparic (2020)	Development	Adaptive systems	COP	City transport management
Cherfia et al. (2014)	Modeling and verification	C–A systems	Bigraphs	Smart home
Diao et al. (2005)	Control theory	Self-managing systems	DTAC	IBM server control
Djoudi et al. (2016)	Specification and verification	C–A adaptive systems	Model based, CTXs-Maude	Cruise control
Filieri et al. (2011)	Control theory, Reliability	Self-adapting software	Discrete time Markov chains (DTMC)	SW reliability
Li et al. (2020)	Validation	Cyber-physical systems	Model checking	Motion control
Nzekwa et al. (2010)	Stability	C–A self-adaptive	Algorithm composition	Temperature control
Padovitz et al. (2004)	Stability	C–A pervasive	State space	Smart home
Serral et al. (2010)	Development	C–A pervasive	Model driven, OWL, UML, PervML	Smart home
Shehzad et al. (2004)	Need of formal model	C–A systems	CAMUS, Ontologies	Smart home
Sindico and Grassi (2009)	Development	C–A systems	CAMEL, UML	Personnel employment
Tran et al. (2012)	Modeling and Verification	C–A adaptive systems	ROAD4	Cruise control
Wang et al. (2011)	Formalization of structure and behavior	C–A systems	Bigraphs	Academic work

explored "... the extent to which control theory can provide an architectural and analytic foundation for building self-managing systems. ..."

Since context-aware systems are digital and mostly based on logics (Filieri et al., 2011; Nzekwa et al., 2010), Boolean Control Networks (BCN) seem to be the appropriate framework in which context-aware systems can be formalized. In recent times, through the introduction of the semitensor product of matrices, the representative equations of a logic system have been converted into an equivalent algebraic form (Cheng & Qi, 2010a; 2010b), and solutions to problems such as controllability, observability, stability, and reconstructability have been proposed (Cheng & Qi, 2009; Fornasini & Valcher, 2013a; Fornasini & Valcher, 2016; Zhang & Zhang, 2016).

In a previous paper (Schreiber & Valcher, 2019), we have proposed the use of BCNs to model open-loop context-aware systems, and we have illustrated our approach by focusing on the interesting example of an early-warning hydrogeological system, in which inputs are gathered as data provided by a set of physical sensors and data provided as messages by public web services. For this model, to which lots of open-loop context-aware systems can be reduced, we proved (1) the existence of equilibrium points corresponding to constant inputs; (2) the absence of limit cycles; (3) its reconstructability from the output measurements; and (4) the detectability of stuck-in-faults. In this paper, *we extend this technique to closed-loop feedback systems.*

The used approach is, by all means, general, but we found it beneficial to illustrate it by referring to a healthcare management example. Healthcare systems, which involve the interaction among several components, can benefit from formal design and verification methods to enhance their safety and efficacy properties. To make it understandable and focus on the methodology, we have oversimplified the example that therefore must be regarded as a *proof of concept*, rather than a realistic model.

The paper is organized as follows. In Section 11.2, we describe the fundamental steps of a BCN-based verification methodology for context-aware feedback systems and introduce the basics of BCN formalism. In Section 11.3, we describe the data structures of the case study; then we model the context as well as the functional system such as BCNs, as explained in Section 11.4. In Section 11.5 the mathematical formalization of real-life requirements is presented, while Section 11.6 brings some conclusive remarks.

11.2 A Boolean Control Network-based methodology

The fundamental structure of a feedback control system is shown in Fig. 11.1. The general idea is the following one: given a reference value (or set point), to make a selected internal variable achieve the desired value, one can feed back the controller with the output measurements taken on the system. The difference between the set point and the output measurement will be used by the controller to generate the appropriate control action to be applied to the target system.

The first step in our methodology is the identification of the functional blocks that can be mapped on the scheme of Fig. 11.1 and to use a formal model, such as sequential machines, to represent them.

In the second step the obtained model is translated into the BCN formalism, and in the third step the system properties are formally checked.

11.2.1 Boolean Control Networks

Before proceeding, we introduce a few elementary notions regarding the left semitensor product and the algebraic representations of Boolean Networks (BNs) and BCN. The interested reader is referred to Cheng et al. (2011) for a general introduction to this class of models and their basic properties. Additional references for the specific properties and results we will use in the paper will be introduced in the following.

We consider Boolean vectors and matrices, taking values in $\mathscr{B} = \{0, 1\}$, with the usual logical operations (And $\wedge$, Or $\vee$, and Negation $^-$). δ_k^i denotes the ith canonical vector of size k, namely, the ith column of the k-dimensional identity matrix I_k. $\mathscr{L}_k$ is the set of all k-dimensional canonical vectors, and $\mathscr{L}_{k \times n} \subset \mathscr{B}_{k \times n}$ the set of all $k \times n$ logical matrices, namely, $k \times n$ matrices whose n columns are canonical vectors of size k.

The (left) semitensor product $\ltimes$ between matrices (in particular, vectors) is defined as follows (Cheng et al., 2011): given $L_1 \in \mathscr{L}_{r_1 \times c_1}$

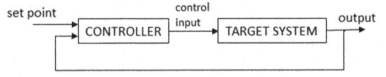

Figure 11.1 A feedback control system.

and $L_2 \in \mathscr{L}_{r_2 \times c_2}$, we set

$$L_1 \ltimes L_2 := (L_1 \otimes I_{T/c_1})(L_2 \otimes I_{T/r_2}), \text{ with } T := \text{l.c.m.} \{c_1, r_2\}.$$

The semitensor product generalizes the standard matrix product, meaning that when $c_1 = r_2$, then $L_1 \ltimes L_2 = L_1 L_2$. In particular, when $x_1 \in \mathscr{L}_{r_1}$ and $x_2 \in \mathscr{L}_{r_2}$, we have $x_1 \ltimes x_2 \in \mathscr{L}_{r_1 r_2}$.

A *BCN* is a logic state-space model taking the form:

$$X(t+1) = f(X(t), U(t)),$$

$$Y(t) = h(X(t), U(t)), t \in \mathbb{Z}_+, \tag{11.1}$$

where $X(t), U(t)$, and $Y(t)$ are the n-dimensional state variable, the m-dimensional input variable, and the p-dimensional output variable at time t, taking values in $\mathscr{B}^n$, $\mathscr{B}^m$, and $\mathscr{B}^p$, respectively. f and h are logic functions, that is, $f: \mathscr{B}^n \times \mathscr{B}^m \to \mathscr{B}^n$, while $h: \mathscr{B}^n \times \mathscr{B}^m \to \mathscr{B}^p$. By making use of the semitensor product $\ltimes$, the BCN (11.1) can be equivalently represented as (Cheng et al., 2011):

$$x(t+1) = L \ltimes u(t) \ltimes x(t),$$

$$y(t) = H \ltimes u(t) \ltimes x(t), \quad t \in \mathbb{Z}_+, \tag{11.2}$$

where $L \in \mathscr{L}_{N \times NM}$ and $H \in \mathscr{L}_{P \times NM}$, $N := 2^n, M := 2^m$ and $P := 2^p$. This is known as the *algebraic expression* of the BCN. The matrix L can be partitioned into M square blocks of size N, namely as

$$L = \begin{bmatrix} L_1 & L_2 & \cdots & L_M \end{bmatrix}.$$

For every $i \in \{1, 2, \cdots, M\}$ the matrix $L_i \in \mathscr{L}_{N \times N}$ represents the logic matrix that relates $x(t+1)$ to $x(t)$, when $u(t) = \delta_M^i$, namely

$$u(t) = \delta_M^i \Rightarrow x(t+1) = L_i x(t).$$

In the special case when the logic system has no input, its algebraic expression becomes

$$x(t+1) = Lx(t), \tag{11.3}$$

$$y(t) = Hx(t),$$

and it is called *BN*.

It is easy to realize that the previous algebraic expressions (11.2) and (11.3) can be adopted to represent any state-space model in which the state, input and output variables take values in finite sets, and hence the sizes of the state, input and output vectors N, M, and P need not be powers of 2. When so, oftentimes BCNs and BNs are called *multivalued (control) networks* (Cheng et al., 2011). With a slight abuse of terminology, in this work we will refer to them as BCNs and BNs. Also, in the following, capital letters will be used to denote the original vectors/variables, taking values in finite sets, and the same lowercase letters will be used to denote the corresponding canonical vectors.

To focus on the ideas and on the modeling techniques, rather than on the Boolean math, in Section 11.3, we have chosen to address the model structure and properties without assigning specific numerical values to the logic matrices involved in the system description. Thus we have derived general results that can be tailored to the specific needs and choices of the application. We believe that this is the power of the proposed modeling approach: its flexibility and generality.

Finally, we provide a deterministic model of the patient health evolution, which represents the evolution of the average case of a patient affected by a specific form of illness. Accordingly, interpretations to the patient symptoms, as captured by the values of his/her vital parameters, are given and, based on them, well-settled medical protocols, to prescribe therapies and locations where such therapies need to be administered, are applied. A probabilistic model of the patient reaction to therapies, that also keeps into account the probabilistic correlation between actual health status and the measured values of his/her vital parameters, requires the use of Probabilistic BNs and will be the subject of our future research. Thus our case study must be considered as a proof of concept and not necessarily as representative of a real system.

11.3 The case study

We consider a multiple feedback loops system as it naturally arises when modeling the evolution of a patient's health status, subjected to medical therapies, whose vital parameters are, in turn, used as inputs to update the therapies to be administered to the patient.

The model provides the mathematical formalization of a possible data tailoring algorithm, running on the mobile device of a nurse in a hospital, *aimed at providing him/her with all and only the information on the therapies the patients in his/her ward are to be given*. Fig. 11.2 shows the overall tailoring process.

A hospital keeps a database that stores all the data relevant both to the patients and to the administrative, medical, and assistance employees.

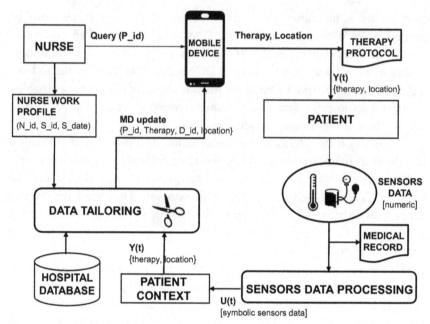

Figure 11.2 The tailoring process.

The work of a nurse is guided by an application on his/her mobile device. The app assists the nurse in his/her routine work following established medical protocols.

Each patient is provided with healthcare wearable sensors, measuring the variables that characterize his/her medical status, in our example: the body temperature (bt), the blood pressure (bp), and the heartbeat frequency (hf) (Baskar et al., 2020; Castillejo et al., 2013). For ease of representation, all of these variables are discretized and take values in the finite set $S = \{low, medium, high\}$. Therefore there are $3^3 = 27$ possible combinations (triples) of the sensors' symbolic data.

As detailed in Section 11.4, the patient context is constituted by a set of variables and it determines the therapy to be given (e.g., the drugs, their amount, and timing). The treatment should change the actual patient status—thus changing the sensors' output—and, possibly, could require a relocation of the patient in a different location, thus determining feedback loops. Moreover, the model is meant to be general and we intentionally neglect diagnose-prescription issues. We assume that *the therapies are effective and they will eventually lead the patient to be dismissed.* Fig. 11.3 shows the feedback schema of the case study.

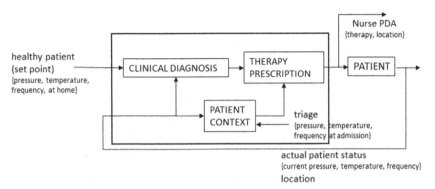

Figure 11.3 The case study general model.

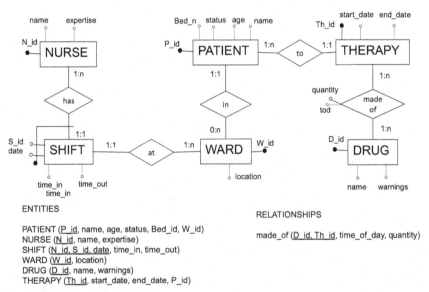

ENTITIES

PATIENT (P_id, name, age, status, Bed_id, W_id)
NURSE (N_id, name, expertise)
SHIFT (N_id, S_id, date, time_in, time_out)
WARD (W_id, location)
DRUG (D_id, name, warnings)
THERAPY (Th_id, start_date, end_date, P_id)

RELATIONSHIPS

made_of (D_id, Th_id, time_of_day, quantity)

Figure 11.4 The hospital database.

Fig. 11.4 shows a portion of the schema of the hospital database, which must be dynamically tailored to store, on the mobile device of each nurse in a shift, all and only the treatments each patient in his/her ward is to be given in that shift. Treatments are defined in the therapy protocol adopted for the diagnosed illness. The numerical values coming from the sensors—registered in the medical record—are converted into their symbolic aggregate counterparts {*low; medium; high*} in the sensors data processing block and affect the estimated patient status, which can take five values: healthy (*H*), convalescent (*C*), under observation (*UO*),

Ill (*I*), and life critical (*LC*). The estimated patient status determines the physician's decision on both the therapy and the patient location—at home (*h*), in the hospital ward (*hw*), in an intensive care unit (*ICU*). Of course, the prescribed therapies are also related to the current location and to the location recommended for the patient. For instance, some therapies can be given in a hospital *ICU* or in a *ward,* but cannot be given at *home.* On the other hand, the medical context can require a relocation of the patient. Thus data tailoring is made based on two different criteria:

- The work profile of the nurse, which is used to select all and only the patients he/she must attend; it is downloaded at the beginning of the shift and is not affected by external events (Listing 11.1).
- The medical status of the patient, which dynamically varies and hence requires different treatments.

Listing 11.1 Tailoring the nurse work profile.
```
select P_id, bed_n
    from nurse, shift, ward, patient
    where N_id = "A" AND S_id = "X" AND S_date = "yy/mm/dd"
```

The query on the nurse's mobile device is shown in Listing 11.2. In the rest of the paper, we focus only on medical and not on administrative issues. The schema of the tailored data, stored on the mobile device, is shown in Fig. 11.5.

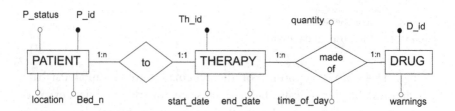

DATABASE ON MOBILE

PATIENT (P_id, Bed_n, P_status, location)
DRUG (D_id, warnings)
THERAPY (Th_id, P_id, date_st, date_end,)

made_of (Th_id, D_id, quantity, time_of_day)

Figure 11.5 The tailored database.

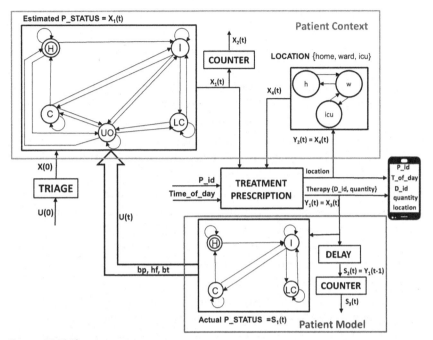

Figure 11.6 The system structure.

Listing 11.2 Querying the nurse's device.

```
select D_id, quantity, location
    from patient, therapy, made_of, D_id
    where P_id = "pp" AND time_of_day = "hh:mm"
```

In Fig. 11.6 the global system structure is represented showing the Moore state diagrams of the estimated and the actual patient status, and of the location respectively, as it will be detailed in Section 11.4.

11.4 The Boolean Control Network system model

We are now in a position to introduce the BCN models for our case study.

11.4.1 The patient context model

Let us first consider the patient context model. We assume as input vector the 3-dimensional vector $U(t)$, where:

$U_1(t)$ denotes the (*low*, *medium*, or *high*) value of the body temperature (*bt*) at time t;

$U_2(t)$ denotes the (*low, medium,* or *high*) value of the body pressure (*bp*) at time *t*;

$U_3(t)$ denotes the (*low, medium,* or *high*) value of the heart frequency (*hf*) at time *t*.

The corresponding canonical vector, $u(t)$, therefore belongs to $\mathscr{L}_{27}$, since each variable $U_i(\cdot)$, $i = 1, 2, 3$, can take three distinct values.

The state variable $X(t)$ is a 4-dimensional vector, where:

- $X_1(t)$ denotes the Estimated Patient status (in other words, the diagnosis) at time *t* with respect to a specific form of illness: it takes values in the set $\{H, C, UO, I, LC\}$.

- $X_2(t)$ represents a counter variable, that keeps track of how many consecutive times up to time *t* the estimated patient status has remained invariant. In other words, $X_2(t) = m$ if $X_1(t) = X_1(t-1) = \ldots = X_1(t-m+1)$, but $X_1(t-m+1) \neq X_1(t-m)$. To ensure that X_2 takes values in a finite set, and for the sake of simplicity,[1] we assume that we keep track until $X_2(t)$ reaches the value 3, and then we stop. This amounts to saying that $X_2(t)$ belongs to $\{1, 2, \geq 3\}$.

- $X_3(t)$ is the prescribed therapy at time *t*, belonging to a finite set, say $\{Th0, Th1, \ldots, Th5\}$, where $Th0$ means that the patient does not receive any drug.

- $X_4(t)$ is the prescribed location (*home, ward, ICU*) where the patient will get the therapy at time *t*.

The corresponding canonical representation, $x(t)$, under the previous assumptions will belong to $\mathscr{L}_{270}$, since $270 = 5 \times 3 \times 6 \times 3$. Finally, we assume as output of the Patient context the 2-dimensional vector $Y(t)$, where:

$Y_1(t)$ is the prescribed therapy at time *t*;

$Y_2(t)$ is the prescribed location (*home, ward, ICU*) where the patient will get the therapy at time *t*.

Clearly, $Y_1(t) = X_3(t)$ and $Y_2(t) = X_4(t)$. Moreover, the canonical representation of $Y(t)$, $y(t)$, belongs to $\mathscr{L}_{18}$, since 18 is the number of possible combinations of therapies and locations. Note, however, that the set of possible outputs can be significantly reduced: for instance, the location *home* is compatible only with the choice to dismiss the patient, after considering his/her health status, and with prescribed therapies such as $Th0$ (no drugs) or a light therapy (say, $Th1$). At the same time certain therapies can be administered only when the patient is in the *ICU*. So,

---

[1] All the numbers used in this context are, of course, arbitrary and meant to purely exemplify how to design the algorithm and to convert it into a BCN.

one may reasonably assume that a good number of the 18 output values are not realistic and hence can be removed, thus reducing the size of $y(\cdot)$.

It is worthwhile to introduce a few comments about the initial state $X(0)$ (or its canonical representation $x(0)$) and about the update of the state variables $X_i(t)$, $i = 1$,   2,   3,   4. The initial state can be regarded as the result of the triage process: when patients are admitted to the *Emergency Room*, a preliminary diagnosis is made, based on the three measures $U_1(0)$, $U_2(0)$, and $U_3(0)$, since there may be no previous history of the patient and the hospital admission requires a fast evaluation of the medical conditions of the patient. So, $X_1(0)$ may be a static function of $U(0)$. $X_3(0)$ is automatically set to *Th0*, while $X_2(0)$ is set to 1 and $X_4(0)$ to *home*.

We note that $X_1(t + 1)$ is naturally expressed as a logic function of $X_1(t), X_2(t), X_3(t), X_4(t)$ and $U(t)$, say $X_1(t + 1) = f_1(X(t), U(t))$. On the other hand, $X_2(t + 1)$ naturally depends on $X_2(t), X_1(t)$ and $X_1(t + 1)$, and, since we have just pointed out that $X_1(t + 1) = f_1(X(t), U(t))$, we can in turn express $X_2(t + 1)$ as $X_2(t + 1) = f_2(X(t), U(t))$. Similarly, $X_3(t + 1)$ and $X_4(t + 1)$ are functions of $X_1(t + 1)$, $X_2(t + 1)$, $X_3(t)$, $X_4(t)$, and $U(t)$, and hence can be expressed, in turn, as functions of $X_1(t), X_2(t), X_3(t), X_4(t)$, and $U(t)$. On the other hand, as we previously remarked, $Y_1(t) = X_3(t)$ and $Y_2(t) = X_4(t)$. This implies that

$$X(t + 1) = f(X(t), U(t)),$$

while

$$Y\big(t\big) = \begin{bmatrix} X_3(t) \\ X_4(t) \end{bmatrix},$$

and hence

$$x(t + 1) = L \ltimes u(t) \ltimes x(t),$$

$$y(t) = Mx(t), \cdots \rightarrow t \in \mathbb{Z}_+$$

for suitable choices of the logical matrices $L \in \mathscr{L}_{270 \times (27 \cdot 270)}$ and $M \in \mathscr{L}_{18 \times (27 \cdot 270)}$.

11.4.2 The patient model

At this point we consider the patient model. A reasonable choice of the patient state variables is the following one:

- $S_1(t)$ represents the actual patient status that takes values in the set $\{H, C, I, LC\}$. Note that this is a proper subset of the set where the

Estimated Patient status takes values, since of course the value UO in this case does not make sense.

- $S_2(t)$ represents the therapy that has been prescribed at time $t-1$, and hence it coincides with $Y_1(t-1)$.
- $S_3(t)$ is a counter variable that keeps track of how many consecutive times up to time t the therapy has remained invariant. In other words, $S_3(t) = m$ if $S_2(t) = S_2(t-1) = \ldots = S_2(t-m+1)$, but $S_2(t-m+1) \neq S_2(t-m)$. Also in this case, we put a bound on m and assume that $S_3(t)$ belongs to $\{1, 2, \geq 3\}$.
- Finally, $S_4(t)$ is the vector collecting the measures of the vital parameters at time $t-1$, namely $S_4(t) = U(t-1)$.

For the patient model the natural input is $Y(t)$ (in fact, $Y_1(t)$ could suffice), while the output is $U(t)$. Since $U(t)$ is the patient vital parameters at time t, it is reasonable to assume that these measures depend on their values at time $t-1$ (and hence on $S_4(t)$), on the patient status $S_1(t)$, the given therapy at time $t-1$, $S_2(t)$ (indeed it is not realistic to assume that the effect of the therapy is instantaneous), and on the duration of the therapy, namely, on $S_3(t)$.

Following a similar reasoning to the one adopted for the patient context model, we can claim that the patient model is described by the logic equations

$$S(t+1) = f_p(S(t), Y(t), U(t)),$$

$$U(t) = h_p(S(t)),$$

and hence by the BCN

$$s(t+1) = F \ltimes y(t) \ltimes u(t) \ltimes s(t),$$

$$u(t) = Hs(t), t \in \mathbb{Z}_+,$$

for suitable choices of the logical matrices $F \in \mathcal{L}_{1944 \times (18 \cdot 27 \cdot 1944)}$ and $H \in \mathcal{L}_{27 \times 1944}$, since $1944 = 4 \cdot 6 \cdot 3 \cdot 27$. So, to summarize, we have the following two models:

$$x(t+1) = L \ltimes u(t) \ltimes x(t), \tag{11.4}$$

$$y(t) = Mx(t), t \in \mathbb{Z}_+ \tag{11.5}$$

and

$$s(t+1) = F \ltimes y(t) \ltimes u(t) \ltimes s(t), \tag{11.6}$$

$$u(t) = Hs(t), t \in \mathbb{Z}_+ \qquad (11.7)$$

For the sake of simplicity, we will use the following notation: $270 = \dim x = : N_x$, $1944 = \dim s = : N_s$, $27 = \dim u = : N_u$ and $18 = \dim y = : N_y$. If we replace (11.7) and (11.5) in (11.6) and keep into account that

$$s(t) \ltimes s(t) = \Phi s(t),$$

where $\Phi \in \mathscr{L}_{N_s^2 \times N_s}$ is a logical matrix known as *power-reducing matrix* (Cheng et al., 2011), then (11.6) becomes

$$s(t+1) = F \ltimes M \ltimes x(t) \ltimes H \ltimes \Phi \ltimes s(t). \qquad (11.8)$$

At the same time, we can swap, namely, reverse the order of, the vector $x(t)$ and the vector $H \ltimes \Phi \ltimes s(t)$ by resorting to the swap matrix W of suitable size (Cheng et al., 2011), thus obtaining

$$s(t+1) = F \ltimes M \ltimes W \ltimes H \ltimes \Phi \ltimes s(t) \ltimes x(t) = A(s(t) \ltimes x(t)), \quad (11.9)$$

where

$$A := F \ltimes M \ltimes W \ltimes H \ltimes \Phi \in \mathscr{L}_{N_s \times N_s N_x}$$

Similarly, if we replace (11.7) in (11.4), we get:

$$x(t+1) = L \ltimes H \ltimes s(t) \ltimes x(t) = B(s(t) \ltimes x(t)), \qquad (11.10)$$

where

$$B := L \ltimes H \in \mathscr{L}_{N_x \times N_s N_x}.$$

Now, the overall model, keeping into account both the patient context and the patient model, becomes:

$$s(t+1) = A \ltimes s(t) \ltimes x(t), \qquad (11.11)$$

$$x(t+1) = B \ltimes s(t) \ltimes x(t). \qquad (11.12)$$

If we introduce the status of the overall system

$$v(t) := s(t) \ltimes x(t) \in \mathscr{L}_{N_s N_x},$$

we get

$$v(t+1) = (A \ltimes v(t)) \ltimes (B \ltimes v(t)).$$

It is a matter of elementary calculations to verify that once we denote by a_i the ith column of A and by b_j the jth column of B, the previous equation can be equivalently rewritten as

$$v(t+1) = Wv(t), \qquad (11.13)$$

where

$$W := \begin{bmatrix} a_1 \bowtie b_1 & a_2 & \bowtie & b_2 & \dots & a_{N_s N_x} \bowtie b_{N_s N_x} \end{bmatrix} \in \mathscr{L}_{N_s N_x \times N_s N_x}.$$

In addition, one can assume as system output

$$y(t) = Mx(t)$$

that can be rewritten as

$$y(t) = \Psi v(t), \tag{11.14}$$

where

$$\Psi := \begin{bmatrix} M & M & \dots & M \end{bmatrix} \in \mathscr{L}_{N_y \times N_s N_x}.$$

So, Eqs. (11.13) and (11.14) together describe a BN that models the overall closed-loop system.

11.5 Real-life properties and their mathematical formalization

In this section, we investigate the properties of the overall system, obtained by the feedback connection of the patient context and of the patient model, namely, the BN (11.13) and (11.14).

As stated in Section 11.2, we aim at providing general ideas about the mathematical properties of the system that have a clear practical relevance in this context, rather than checking those properties for a specific choice of the logical matrices involved in the system description. Thus we shall not provide numerical values for the quadruple of logical matrices (L, M, F, H), but *we shall show how to reduce our specific feedback system (or parts of it) to standard set-ups for which these properties have already been investigated.*

Thus the purpose of this section is to illustrate how issues regarding the correct functioning of the system, the possibility of identifying its "real state" from the available measurements, etc., that mathematically formalize the natural requirements on a closed-loop context-aware system describing a medical application (but not only!), can be easily addressed in the context of BCNs.

11.5.1 Identifiability of the patient status

A first question that is meaningful to pose is whether the Patient model is a good one, namely it will lead to the correct functioning of the overall system. To clarify what we mean when posing this question, we first need to better explain the perspective we have taken in modeling the patient. We have assumed that the patient is in a certain medical condition with respect to a specific medical problem. Thus the diagnosis pertains only to the level/seriousness of the patient's health condition, and not to the specific cause of the illness. Such a medical condition is revealed by the fact that patient vital parameters (*bp, bt, hf*), namely, the patient output $U(t)$, take values outside of the "normal range." The medical status is of course affected by the therapy Y and can be associated with different values of U, so the output measure $U(t)$ at time t together with the therapy $Y(t)$ (or $Y_1(t)$) do not allow to uniquely determine $S(t)$. In addition, some therapies may need some time to become effective (which is the reason why we introduced the state variable $S_2(t)$). On the other hand, a good (deterministic) model of the patient[2] necessarily imposes that the measured vital parameters are significant and hence allow physicians to determine the actual patient status after a finite number of observations. From a mathematical point of view, this amounts to assuming that the patient model (11.6) and (11.7) is reconstructable, namely, there exists $T \in \mathbb{Z}_+$ such that the knowledge of the signals $u(\cdot)$ and $y(\cdot)$ in $[0, \quad T \quad]$ allows to uniquely determine $s(T)$. Specifically, we have the following:

Definition 1 The BCN (11.6) and (11.7), with $s(t) \in \mathscr{L}_{N_s}, u(t) \in \mathscr{L}_{N_u}$ and $y(t) \in \mathscr{L}_{N_y}$, is said to be *reconstructable* if there exists $T \in \mathbb{Z}_+$ such that the knowledge of the input and output vectors in the discrete interval $\{0, \quad 1, \quad . \quad . \quad . \quad , \quad T\}$ allows to uniquely determine the final state $s(T)$.

It is worth noticing that the BCN (11.6) and (11.7) is different from the standard ones for which the observability and reconstructability problems have been addressed in the literature (see Fornasini & Valcher, 2013a; Laschov et al., 2013; Zhang & Zhang, 2016), since this BCN is intrinsically in a closed-loop condition, as the BCN output u(t) affects the state update at time $t + 1$. However, by replacing (11.7) in (11.6), and by using again the power-reducing matrix, we can obtain:

$$s(t \quad + \quad 1) = F \ltimes y(t) \ltimes H \quad \ltimes \Phi \ltimes s(t),$$

$$u(t) = Hs(t), t \in \mathbb{Z}_+,$$

---

[2] As previously mentioned, we have adopted a deterministic model and assumed that everything works according to statistics and well-settled procedures: therapies are designed according to specific protocols and statistically lead to the full recovery of the patient. This is the reason why the possibility that the patient dies is not contemplated.

which, in turn, can be rewritten as

$$s(t + 1) = \mathscr{F} \ltimes y(t) \ltimes s(t), \tag{11.15}$$

$$u(t) = Hs(t), t \in \mathbb{Z}_+, \tag{11.16}$$

where

$$\mathscr{F} := \begin{bmatrix} \mathscr{F}_1 & \mathscr{F}_2 & \cdots & \mathscr{F}_{N_s} \end{bmatrix}$$

and

$$\mathscr{F}_i := \begin{bmatrix} f_i \ltimes (H \ltimes \Phi)\delta_{N_s}^1 & \cdots & f_i \ltimes (H \ltimes \Phi)\delta_{N_s}^{N_s} \end{bmatrix},$$

where we have denoted by f_i the ith column of the matrix F. This allows to reduce the reconstructability problem for this specific BCN to a standard one, for which there are lots of results and algorithms (see Fornasini & Valcher, 2013a; Zhang & Johansson, 2020; Zhang & Zhang, 2016; Zhang et al., 2019).

Clearly, the matrices F and H must be properly selected to guarantee the reconstructability of the Patients' status. This means, in particular, that *the vital parameters to measure must be chosen in such a way that they are significant enough to allow to identify the actual medical conditions of the patient.*

From a less formal viewpoint, it is worth underlying that the reconstructability problem reduces to the problem of correctly identifying the state variable $s_1(t)$, since the definition of $s_i(t)$, $i = 1, 2, 3, 4$, allows to immediately deduce that such values can be uniquely determined from the variables $y_1(t)$ and $u(t)$. So, one could focus on a lower dimension model expressing $s_1(t + 1)$ in terms of $s_i(t)$, $i = 1, 2, 3, 4, u(t)$ and $y_1(t)$, where $s_i(t)$, $i = 1, 2, 3, 4$, $u(t)$ and $y_1(t)$ are known and address the reconstructability of $s_1(t)$ from $u(t)$, assuming $s_i(t)$, $i = 1, 2, 3, 4$, and $y_1(t)$ as inputs.

11.5.2 Correct diagnosis

Of course, once we have ensured that the patient model (11.6) and (11.7) is reconstructable, and hence we have properly chosen the vital parameters to measure to identify the patient status, the natural question arises: *Is the patient context correctly designed so that after a finite (and possibly small) number of steps T, the patient status $s_1(t)$ and the estimated patient status $x_1(t)$ coincide for every $t \geq T$?* This amounts to saying that the protocols to evaluate the Patient Status have been correctly designed.

To formalize this problem, we need to introduce a comparison variable, say $z(t)$. This variable takes the value δ_2^1 (namely the unitary or YES value) if $s_1(t) = x_1(t)$ and the value δ_2^2 (namely, the zero or NO value) otherwise. Keeping in mind that $S_1(t)$ takes values in $\{H, C,\ \ I,\ \ LC\}$ (and hence $s_1(t) \in \mathcal{L}_4$), while $X_1(t)$ takes values in $\{H, C, UO, I, LC\}$ (and hence $x_1(t) \in \mathcal{L}_5$), this leads to

$$z(t) = \begin{bmatrix} C_1 & C_2 & C_3 & C_4 \end{bmatrix} \ltimes s_1(t) \ltimes x_1(t),$$

where[3] $C_i \in \mathcal{L}_{2 \times 5}$ for every $i \in [1, 4]$. Moreover,
C_1 is the block whose first column is δ_2^1 while all the others are δ_2^2;
C_2 is the block whose second column is δ_2^1 while all the others are δ_2^2;
C_3 is the block whose fourth column is δ_2^1 while all the others are δ_2^2;
C_4 is the block whose fifth column is δ_2^1 while all the others are δ_2^2.

Clearly, $z(t)$ can also be expressed as a function of $s(t)$ and $x(t)$ and hence as a function of $v(t)$. This leads to

$$z(t) \quad = \quad \mathbb{C}v(t),$$

for a suitable $\mathbb{C} \in \mathcal{L}_{2 \times N_s N_x}$. Thus *the problem of understanding whether the system is designed to produce the correct diagnosis can be equivalently translated into the mathematical problem of determining whether for every initial condition, $v(0)$, the output trajectory of the system*

$$v(t\ +\ 1) \quad = \quad Wv(t) \tag{11.17}$$

$$z(t) \quad = \quad \mathbb{C}v(t) \tag{11.18}$$

eventually takes the value δ_2^1. In other words, we need to ensure that there exists $t \in \mathbb{Z}_+$ such that, for every $v(0) \in \mathcal{L}_{N_s N_x}$, the corresponding output trajectory $z(t), t \in \mathbb{Z}_+$, satisfies $z(t) = \delta_2^1$, for every $t \geq T$. Note that the idea is that once the seriousness level of the patient illness has been correctly diagnosed, this information will never be lost, even if the patient health status will change.

An alternative approach to this problem is to define the set of states

$$CD := \left\{ v(t) \in \mathcal{L}_{N_s N_x} : s_1(t) \ltimes x_1(t) \in \left\{ \delta_4^1 \ltimes \delta_5^1, \delta_4^2 \ltimes \delta_5^2, \delta_4^3 \ltimes \delta_5^4, \delta_4^4 \ltimes \delta_5^5 \right\} \right\},$$

---

[3] To improve the notation one could sort the set of values of the estimated patient's status as follows: $\{H, C,\ \ I,\ \ LC, UO\}$. In this way, each of the blocks C_i would have the ith column equal to δ_2^1 and all the remaining ones equal to δ_2^2.

that represent all possible situations where the estimated patient status $x_1(t)$ coincides with the patient status $s_1(t)$ (in other words, CD is the set of correct diagnoses) and to impose that such a set is *a global attractor* of the system. From a formal point of view, the set CD is a *global attractor of the BN* (11.17) if there exists $T \geq 0$ such that for every $v(0) \in \mathscr{L}_{N_s N_x}$, the corresponding state evolution $v(t)$, $t \in \mathbb{Z}_+$, of the BN (11.17) belongs to CD for every $t \geq T$.

This property can be easily checked (Cheng et al., 2011; Fornasini, & Valcher, 2013b) by simply evaluating that all rows of $W^{N_s N_x} \in \mathscr{L}_{N_s N_x} \times \mathscr{L}_{N_s N_x}$, the $N_s N_x$ power of W, are zero except for those whose indexes correspond to the canonical vectors in CD.

11.5.3 Successful therapies

As previously mentioned, when modeling the evolutions of the patient context and the patient model in a deterministic way, we are describing the evolution of the average case of a patient affected by a specific form of illness. Accordingly, as mentioned in Section 11.2.1, we are giving certain interpretations to the patient symptoms, as captured by the values of his/her vital parameters, and based on them we are applying well-settled medical protocols to prescribe therapies and locations where such therapies need to be administered. In this context it is clear that death is not contemplated, since this would correspond to assuming that a given medical protocol deterministically leads to the death of the patient and this does not make sense. Similarly, a protocol that deterministically leads to an equilibrium state where the Patient status is C, I or LC is not acceptable. In other words, *the only reasonable solution is to have designed the Patient Context in such a way that (1) the patient status is eventually H; (2) the estimated patient status is, in turn, H.*

Conditions (1) and (2) correspond to constraining *the global attractor of the system evolution to be a proper subset, say $\mathscr{H}$, of the set CD* we previously defined. Specifically, we define the set $\mathscr{H}$ as follows:

$$\mathscr{H} = \{v(t) \in \mathscr{L}_{N_s N_x} : s_1(t) \bowtie x_1(t) = \delta_4^1 \bowtie \delta_5^1\},$$

that represent all possible situations where the estimated patient status $x_1(t)$ is healthy and it coincides with the patient status $s_1(t)$ (in other words, $\mathscr{H}$ is the set of states corresponding to a healthy patient whose health status has been correctly identified), and *to impose that such a set is a global attractor of the system.*[4]

Also, in this case, it is possible to verify whether such a requirement is met by evaluating if all rows of $W^{N_s N_x} \in \mathscr{L}_{N_s N_x} \times \mathscr{L}_{N_s N_x}$, the $N_s N_x$ power of W, are zero except for those whose indexes correspond to the canonical vectors in $\mathscr{H}$.

11.6 Evaluation and conclusions

In this paper we introduced a novel methodology, which uses well-established systems theory tools, to formally assess some safety properties of feedback context-aware database systems. As a proof of concept, we have used an interesting case study, related to the evolution of the health status of a patient, to illustrate how a feedback context-aware system can be modeled by means of a BCN. Indeed, the patient is subjected to medical therapies and his/her vital parameters are not only the outcome of the therapies, but also the input based on which therapies are prescribed. By making use of a simplified and deterministic logical model, expressed in terms of BCNs/BNs, we have been able to illustrate how the most natural practical goals that the overall closed-loop system needs to achieve may be formalized, and hence investigated, by resorting to well-known systems theory concepts. Clearly, the given model can be improved and tailored to the specific needs, to account for more complicated algorithms and more exhaustive sets of data, but the core ideas have already been captured by the current model. Also, we have addressed what seemed to be the most natural targets in the specific context, but different or additional properties may be investigated, in case the same modeling technique is applied to describe closed-loop context-aware systems of different nature.

The use of a deterministic model of the patient health evolution, to plan therapies based on measured vital parameters, represents a first step

---

[4] Note that we are not introducing additional constraints, in particular we are assuming that the vital parameters u of the patient can change within the set of values compatible with a healthy status. Of course, one could further constrain the set $\mathscr{H}$ by assuming that the prescribed therapy is $Th0$, the patient is at home, and all the counters have reached the saturation level. Even in this case, we may regard as acceptable the existence of a limit cycle, since this would only correspond to oscillations of the values of the state variable s_4 within a small set of values that do not raise any concern. Clearly, one may impose also for s_4 and hence for u a prescribed desired value, and this would mean asking that the system has a single *equilibrium point* (the set $\mathscr{H}$ has cardinality one) which is a global attractor.

toward the design of an accurate algorithm to employ in the mobile device of a nurse.

A final question deals with the computational cost of the procedure and here the bad news comes. Many studies have established that verifying properties such as observability, controllability, and stabilizability of BCNs are NP-hard in the number of nodes (Weiss et al., 2018; Zhang & Johansson, 2020); however, in some cases, the computational complexity can be reduced (Lu et al., 2019; Zhao et al., 2016) and it will not exceed $O(N^2)$ with $N = 2^n$, where n is the number of state variables in a BCN (Zhu et al., 2019). Thus further research is needed to find meaningful modularizations of the BCN into sets of BCNs with a smaller number of state variables each, which can then be reassembled into the global system.

Furthermore, a probabilistic model, together with some warning system that advises the nurse of when different decisions are possible with different confidence levels, and hence there is the need for the immediate supervision of a specialist, is the target of future research.

Acknowledgment

Fabio A. Schreiber was partially supported by INAIL, RECKON Project.

References

Arcaini, P., Ricobene, E., & Scandurra, P. (2015). Formal design and verification of self-adaptive systems with decentralized control. *ACM Transactions on Autonomous and Adaptive Systems, 11*(4), article 25, 25:1−25:35.

Baskar, S., Shakeel, P. M., Kumar, R., Burhanuddin, M., & Sampath, R. (2020). A dynamic and interoperable communication framework for controlling the operations of wearable sensors in smart Healthcare applications. *Computer Communications, 149,* 17−26.

Bolchini, C., Curino, C., Orsi, G., Quintarelli, E., Rossato, R., Schreiber, F. A., & Tanca, L. (2009). And what can context do for data? *Communications of the ACM, 52* (11), 136−140.

Bolchini, C., Curino, C., Quintarelli, E., Schreiber, F. A., & Tanca, L. (2007a). A data-oriented survey of context models. *ACM SIGMOD Record, 36*(4), 19−26.

Bolchini, C., Quintarelli, E., & Tanca, L. (2007b). Carve: Context-aware automatic view definition over relational databases. *Information Systems, 38*(1), 45−67.

Cardozo, N., & Dusparic, I. (2020) Language abstractions and techniques for developing collective adaptive systems using context-oriented programming. In: *Proceedings of 5th eCAS workshop on engineering collective adaptive systems (ACSOS 2020).*

Castillejo, P., Martinez, J., Rodriguez-Molina, J., & Cuerva, A. (2013). Integration of wearable devices in a wireless sensor network for an e-health application. *IEEE Wireless Communications, 20*(4), 38−49.

Cheng, D., & Qi, H. (2009). Controllability and observability of Boolean Control Networks. *Automatica, 45*(7), 1659−1667.

Cheng, D., & Qi, H. (2010a). A linear representation of dynamics of Boolean Networks. *IEEE Transactions on Automatic Control, 55*(10), 2251−2258.

Cheng, D., & Qi, H. (2010b). State space analysis of Boolean Networks. *IEEE Transactions on Neural Networks, 21*(4), 584−594.

Cheng, D., Qi, H., & Li, Z. (2011). *Analysis and control of Boolean Networks.* London: Springer-Verlag.

Cherfia, T. A., Belala, F., & Barkaoui, K. (2014). Towards formal modeling and verification of context-aware systems. In: *Proceedings of the 8th international workshop on verification and evaluation of computer and communication systems,* Bejaïa, Algeria.

Dey, A. K. (2001). Understanding and using context. *Personal Ubiquitous Computing, 5*(1), 4−7.

Diao, Y., Hellerstein, J. l., Parekh, S., Griffith, R., Kaiser, G., & Phung, D. (2005). Self-managing systems: A control theory foundation. In: *Proceedings of the 12th IEEE international conference and workshops on the engineering of computer-based systems (ECBS'05)* (pp. 441−448).

Djoudi, B., Bouanaka, C., & Zeghib, N. (2016). A formal framework for context-aware systems specification and verification. *Journal of Systems and Software, 122C,* 445−462.

Filieri, A., Ghezzi, C., Leva, A., & Maggio, M. (2011) Self-adaptive software meets control theory: A preliminary approach supporting reliability requirements. In: *Proceedings of the 26th IEEE/ACM international conference on automated software engineering (ASE 2011)* (pp. 283−292).

Fornasini, E., & Valcher, M. E. (2013a). Observability, reconstructibility and state observers of Boolean Control Networks. *IEEE Transactions on Automatic Control, 58*(6), 1390−1401.

Fornasini, E., & Valcher, M. E. (2013b). On the periodic trajectories of Boolean Control Networks. *Automatica, 49,* 1506−1509.

Fornasini, E., & Valcher, M. E. (2016). Recent developments in Boolean Control Networks. *Journal of Control and Decision, 3*(1), 1−18.

Laschov, D., Margaliot, M., & Even, G. (2013). Observability of Boolean Networks: A graph-theoretic approach. *Automatica, 49*(8), 2351−2362.

Li, N., Tsigkanos, C., Jin, Z., Hu, Z., & Ghezzi, G. (2020). Early validation of cyberphysical space systems via multi-concerns integration. *Journal of Systems and Software, 170,* 110742.

Lu, J., Liu, R., Lou, J., & Liu, Y. (2019). Pinning stabilization of Boolean Control Networks via a minimum number of controllers. *IEEE Transactions on Cybernetics, 51* (1), 1−9.

Nzekwa, R., Rouvoy, R., & Seinturier, L. (2010) A flexible context stabilization approach for self-adaptive application. In: *Proceedings of the 2010 8th IEEE international conference on pervasive computing and communications workshops (PERCOM workshops)* (pp. 7−12), Mannheim, Germany.

Padovitz, A., Zaslavsky, A. B., Loke, S. W., & Burg, B. (2004). Stability in context-aware pervasive systems: A state-space modeling approach. In: *Proceedings of the 1st international workshop on ubiquitous computing* (pp. 129−138), Porto, Portugal.

Padovitz, A., Zaslavsky, A. B., Loke, S. W., & Burg, B. (2005). Maintaining continuous dependability in sensor-based context-aware pervasive computing systems. In: *Proceedings of the 38th annual Hawaii international conference on system sciences* (pp. 1−10), Big Island, Hawaii.

Schreiber, F. A., & Panigati, E. (2017). Context-aware self-adapting systems: a ground for the cooperation of data, software, and services. *International Journal of Next Generation Computing, 8*(1).

Schreiber, F. A., & Valcher, M. E. (2019). Formal assessment of some properties of context-aware systems. *International Journal of Next Generation Computing, 10*(3).

Schreiber, F. A., Tanca, L., Camplani, R., & Vigano, D. (2012). Pushing context-awareness down to the core: more flexibility for the PerLa language. In: *Electronic Proceedings of the 6th PersDB 2012 workshop (co-located with VLDB 2012)*, Istanbul, Turkey. https://schreiber.faculty.polimi.it/listpub.html.

Serral, E., Valderas, P., & Pelachano, V. (2010). Towards the model driven development of context-aware pervasive systems. *Pervasive and Mobile Computing, 6,* 254–280.

Shehzad, A., Ngo, H. Q., Pham, K. A., & Lee, S. (2004) Formal modeling in context aware systems. In: *Proceedings of the first international workshop on modeling and retrieval of context* (pp. 1–12), Ulm, Germany.

Sindico, A., & Grassi, V. (2009) Model driven development of context aware software systems. In: *Proceedings of the international workshop on context-oriented programming (COP 09)* (Article 7, pp. 1–5), Geneva, Italy.

Tran, M. H., Colman, A., Han, J., & Zhang, H. (2012). Modeling and verification of context- aware systems. In: *Proceedings of the 19th Asia-Pacific software engineering conference* (pp. 79–84), Hong Kong, China.

Wang, J.-s., Dong, X., & Zhou L. (2011). Formalizing the structure and behaviour of context-aware systems in bigraphs. In: *Proceedings of the first ACIS international symposium on software and network engineering* (pp. 89–94), Seoul, South Korea.

Weiss, E., Margaliot, M., & Even, G. (2018). Minimal controllability of conjunctive Boolean Networks is NP-complete. *Automatica, 92,* 56–62.

Zhang, K., & Zhang, L. (2016). Observability of Boolean Control Networks: A unified approach based on finite automata. *IEEE Transactions on Automatic Control, 61*(9), 2733–2738.

Zhang, Z., & Johansson, K. H. (2020). Efficient verification of observability and reconstructibility for large Boolean Control Networks with special structures. *IEEE Transactions on Automatic Control, 65*(12), 5144–5158.

Zhang, Z., Leifeld, T., & Zhang, P. (2019). Reconstructibility analysis and observer design for Boolean Control Networks. *IEEE Transactions on Control of Network Systems, 7*(1), 516–528.

Zhao, Y., Ghosh, B. K., & Cheng, D. (2016). Control of large-scale Boolean Networks via network aggregation. *IEEE Transactions on Neural Networks and Learning Systems, 27* (9), 1527–1536.

Zhu, Q., Liu, Y., Lu, J., & Cao, J. (2019). Further results on the controllability of Boolean Control Networks. *IEEE Transactions on Automatic Control, 64*(1), 440–442.

CHAPTER 12

Mental stress detection using a wearable device and heart rate variability monitoring

Christos Goumopoulos and Nikolaos G. Stergiopoulos
Department of Information and Communication Systems Engineering, University of the Aegean, Samos, Greece

12.1 Introduction

Mental stress or fatigue is a condition in which a person is driven after intense cognitive activity such as after prolonged work. The effects of mental stress on cognitive performance (Lorist et al., 2000), emotional state (Boksem et al., 2006), alert state (Lal & Craig, 2001), and physical performance (Marcora et al., 2009) have been substantially studied. A proliferation of studies have also enlightened the high risks in human health caused by continuous periods of intensive mental activity and revealed the association of mental stress to cardiovascular diseases (Esler et al., 2008), neurological disorders (Tanaka et al., 2014), diabetes (Kelly & Ismail, 2015), and burnout (Demerouti et al., 2002).

The competitive environment and challenges of contemporary business organizations as well as massive time pressure shape the everyday work of many people and place ever higher demands on the management of work tasks. Findings from research confirm the significant increase in mental stress (e.g., Calnan et al., 2001). As a result, negative effects on the successful completion of work tasks, on occupational safety, as well as on the health and well-being of employees can be observed. Fig. 12.1 illustrates that persistent mental stress in the workplace can have a negative impact on the performance in several areas. Therefore the ability to monitor mental stress in real time is of major importance.

Even though mental stress is a subjective feeling, to provide proper interventions, stress must be determined using a suitable measurement method. Mental stress can be recorded using various methods such as self-reports, observations, tests, or measurements. Since mental stress cannot

Edge-of-Things in Personalized Healthcare Support Systems
DOI: https://doi.org/10.1016/B978-0-323-90585-5.00011-4

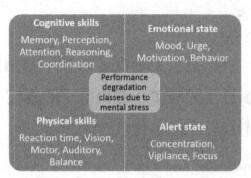

Figure 12.1 Performance degradation classes due to mental stress in workplace.

be measured directly, it must be determined using surrogate parameters. Mental stress influences various physiological parameters, such as heart rate, via the autonomic nervous system (Hjortskov et al., 2004). To assess the positive results of stress detection methods, the ground truth should be gathered from subjects, a process commonly achieved by using psychometric tests (Goumopoulos & Menti, 2019).

Electroencephalogram (EEG) is the primary physiological signal to monitor mental stress with high accuracy since the brain is the center of cognition. Direct measurement of the brain should, therefore, provide a faster measurement of mental stress than measurement of peripheral physiology (Alonso et al., 2015). EEG depicts the spontaneous and occasionally induced brain activity recorded through electrodes. The recording of brain waves correlates with the mental activity of a person, and their type and frequency are indicative of the level of stress experienced by the person. In addition, the location of the scalp where the wave activity is detected is also characteristic of the emotions and mental state of the person. The band category and frequency range of brain waves are very informative since beta waves (13–30 Hz) are associated with anxiety and stress, while alpha waves are considered indicative of relaxation (Hou et al., 2015). It has also been reported that the theta/beta brain wave ratio is analogous to the individual's stress levels. However, the invasive nature of the EEG sensor, since many electrodes should be attached to the human scalp, makes its application not quite practical in ubiquitous environments.

Another physiological signal that has been associated with the positive assessment of mental stress especially in the case of car drivers is electrooculogram (EOG) (Hsieh & Tai, 2013). EOG is a biosignal reflecting the electric field potential generated between the eye retina and cornea recording variations as a result of any eye activity (e.g., eye blinking and movement). Several variables have been suggested as suitable fatigue predictors including percent of time the eyes are closed (PERCLOS), blink frequency, duration and amplitude, eye-ball movement, and lid reopening delay (Hosseini et al., 2017). EOG signal collection is performed by placing electrodes near the eye causing a discomfort similar to the EEG case when applied in real-life scenarios. Eye-associated metrics can be acquired alternatively with glass-based or camera-based eye-tracking systems coupled with image processing techniques to realize mental stress detection systems (Zhang et al., 2018). Depending on the accuracy required and the operation scenario involved the installation of such a system may also be difficult, for example, when several cameras need to be installed opposite the subject or when privacy issues are concerned. Furthermore, system detection performance may depend on situational conditions such as illumination and background noise thus limiting its applicability.

To explore mental stress detection, the approach taken in this work is the use of a wearable device. This measurement method can be easily applied in real-life situations without interrupting the current task. Given the invasive nature of EEG and EOG sensors, the use of heart rate variability (HRV) was motivated since HRV can be determined using noninvasive methods. To do this, subjects are wearing a low-cost sensor that can record the electrocardiogram (ECG) signal and after a predefined measurement period the data can be evaluated. Undoubtedly, a low-budget wearable device cannot deliver physiological signals at the same quality level as signals which would be obtained from costly medical devices. Nevertheless, this was compensated by the requirement that it would be desirable to explore evidence of how well mental stress classification strategies could be realized in everyday environments.

This work presents, therefore, a methodology to detect mental stress using HRV features and training a classification model by employing machine learning algorithms such as NB (Naïve Bayes), SVM (support vector machine), and k-NN (k-nearest neighbors). An experimental protocol was designed to simulate mental stress in daily life by performing a quiz that includes memory tests, mathematical problems, and other problem solving and reasoning tests that collectively induce mental stress. The study took place in Samos Greece with the participation of 30 volunteer students in the University of the Aegean. HRV and subjective data were collected and analyzed. The results obtained from the trained machine learning models demonstrated a promising classification accuracy of up to 85% given the use of a methodologically selected set of HRV features.

12.2 Related work

There are several studies focusing on the detection of mental stress. A characteristic example is the research of Huang et al. (2018) which proposed a machine learning model that detects mental stress using HRV-based features with an accuracy of 75.5%. An experimental protocol similar to the approach described in this work was used to gather ECG signals from 35 volunteers using a portable single-channel ECG device. Various machine learning algorithms were applied and k-NN had demonstrated the best accuracy using an optimal number of three features.

By using a random forest (RF) classifier and an ECG signal combined with respiratory data, Zanetti et al. (2019) reported a mental stress detection accuracy of 76.5%. An experimental study with 17 participants was conducted and stress was induced by arithmetic tests and by playing a serious game that required intensive attention. The use of wearable devices to acquire the physiological data is considered by the researchers as an advantage of their approach. By adding more signals in terms of EEG and blood volume pulse the prediction accuracy was improved to 84.6% using logistic regression as the machine learning method.

The combination of different biomarkers including HRV, EEG, skin conductance, and respiratory signals with a deep learning approach was proposed for mental stress detection (Masood & Alghamdi, 2019). Data were collected in a laboratory setting from 24 participants to train a convolutional neural network (CNN). The proposed approach achieved an accuracy of 90% in distinguishing between rest and stress events. The study concluded that the contribution of the EEG signals was the most important in the classification model.

The EEG signal was also used by King et al. (2006) for estimating mental fatigue with a classification performance between 81.49% and 83.06% using artificial neural networks (ANNs) with a very short time window analysis of time-domain data that were processed into different bands. Other researchers have used the full spectrum from 1 to 30 Hz to estimate mental fatigue in a driving situation (Zhang et al., 2017). They were thus able to estimate mental fatigue in real time by means of a wearable brain—computer interface and by using SVM with an average accuracy of 90.70%.

Another study focused on mental stress detection by using ocular movement data gathered while viewing audio—visual material (Yamada & Kobayashi, 2018). In this research, mental stress was detected by collecting eye-tracking measures from 31 participants while they viewed movie clips. Using the iris of the eyes and simulating the fatigue state with the video projection, the researchers were able to detect if the candidate was in a fatigue state in the first 30 seconds of the video with very high accuracy (91.0%) using 10-fold cross-validation and SVM. Features used to detect cognitive fatigue were associated to blink frequency, amplitude and duration, percent of time the eyes are closed, eye-ball movement directions, and starring division features.

In the work of Laurent et al. (2013) mental fatigue was detected using a combination of ECG, EEG, and EOG signals. The EOG characteristics calculated were blink count, the average blink amplitude and duration as well as their standard deviation, and the interblink interval. The ECG characteristics calculated were *meanHR*, standard deviation for the RR interval, and spectral characteristics of the interpolated RR signal (*LF*, *HF*, *LF/HF* ratio). Eye blinks were calculated from the EEG characteristics. In this multimodal approach, EEG outperformed the other biomarkers in cognitive fatigue detection when used alone with a prediction performance varying between 80% and 94%.

In a different study, eye tracking and typing performance were used to detect mental fatigue (Bafna et al., 2021). A working-memory-based eye-typing experiment was designed and 19 healthy participants took part while data about typing performance and eye tracking were collected. The features of the prediction model included pupil diameter, eye-based height and blink frequency, and task difficulty. Mental fatigue was predicted based on subjective ratings and using RF regression for the creation of the prediction model. Additional machine learning methods used were bagging-based support vector regression, partial least squares regression, and adaBoost. Normalization of testing and training data was performed, and fivefold cross-validation was employed using 20% testing data and 80% training data split method for model performance comparison. Mean absolute error was used as the performance metric and the best performance with respect to baseline simulations was 1.057 with a minimum/ maximum testing error of 0.609/1.894.

Can et al. (2019) developed a mental stress detection approach that was tested in a real-life setting during a summer camp for students focusing on algorithm programming tasks. Physiological signals (HRV, skin conductance, skin temperature, and accelerometer) were collected from 21 participants for 9 days using high-quality wristbands and commodity smartwatches. The data collected were analyzed and processed for feature extraction, and then prediction models were created using different machine learning methods such as SVM, k-NN, linear regression, RF, and multilayer perceptron. In addition, linear discriminant analysis and principal components analysis were applied as dimensionality reduction methods. Other noise reduction methods that were applied at a preprocessing step before feature extraction included the use of percentage thresholds and interpolation. The effect of selecting different time window sizes (2−20 minutes) on the classification accuracy was also examined with respect to the feature extraction and a time window of 600 seconds was found to yield the best performance. The best accuracy (92.15%) was achieved using all the collected signals and the general model with the multilayer perceptron method. Correspondingly, when only the HRV signal was used the best classification accuracy was achieved with RF ranging from 84.67% using a low-cost smartwatch to 90.40% using a high data quality device.

Finally, a review study on mental health monitoring based on wearable devices and machine learning is provided by Garcia-Ceja et al. (2018). Diverse machine learning methods, including supervised, unsupervised, reinforcement, and transfer learning, have been adopted in mental stress detection and other related applications. The classification performance depends on the variety of the monitored stress signals, the stress test used, and the machine learning methods applied in combination with the number of classes defined. Combining mental stress detection with activity recognition can enhance the acquired context awareness and thus improve the detection performance. It is also shown that participants are unwilling to use intrusive devices for data collection since they are feeling uncomfortable with such equipment. On the other hand, the unobtrusive wearable devices that can be applied in everyday life come with a data measurement quality that is not equivalent to laboratory equipment. Despite the fact that smart wearable devices are available for everyday usage, their measurement quality and impact on stress detection performance have not been studied adequately.

12.3 Methodology

12.3.1 Electrocardiogram sensor

Data were collected using Zephyr HXM BT ECG sensor which is a wearable device. This device is a belt worn on the skin around the chest of the user. The ECG signal is measured at 250 Hz sampling rate. A data packet, however, is sent every 1 second by the sensor to a Bluetooth paired mobile device (Fig. 12.2). After the connection with the sensor is made, the mobile device (e.g., a smartphone) will receive information packets from the sensor with the structure shown in Fig. 12.2. Each packet stores 15 heartbeat timestamps. Each timestamp is a 16-bit integer representing the time of a heartbeat in milliseconds. The mobile phone stores the data in text format in a separate file for each participant in the experimental study.

Interbeat interval, measured typically in milliseconds, is the time between individual beats of the heart, otherwise known as the interval between two successive R-peaks of a typical ECG waveform (Fig. 12.3). HRV is obtained by identifying the times at which R-peaks of QRS complexes occur. The QRS complex is usually the most prominent feature of an ECG tracing, where the R peak occurs as the heart's ventricles contract. An R–R signal is then acquired by determining the peak-to-peak intervals. In the case of Zephyr sensor, an R–R interval is calculated

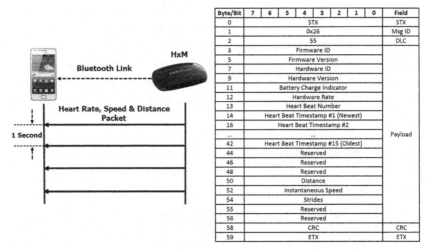

Figure 12.2 Communication protocol of the mobile device with Zephyr HXM-BT sensor (on the left) and the structure of the information packet sent by the sensor (on the right).

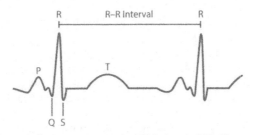

Figure 12.3 R–R interval component in the ECG signal. *ECG*, Electrocardiogram.

from the difference of two consecutive heartbeat timestamps. Thus, in total, 7 R—R intervals are calculated from each packet.

12.3.2 Experimental study

An experimental study was performed involving 30 volunteers whose ages range between 21 and 26 years (26 male and 4 female, mean 22.6 ± 1.9 years) to gather HRV and subjective data. None of the participants were under any serious medical condition or any medication treatment. Before the experiment took place, volunteers were asked to provide their demographic information and sign a typical consent form.

Before and after data collection, participants were also requested to fill a questionnaire to provide a subjective assessment of their mental stress. Questions were obtained from Chalder fatigue scale as shown in Table 12.1. The aim was to determine whether the perceived stress and the factors influencing the participants changed during the duration of the study. The questionnaire includes 15 items where every question concerns the level of fatigue of the test subject at that time and responses are given on a scale of 1 (low) to 5 (high). Participants with a fatigue score above 50, which indicates an increased fatigue level, had to be excluded from the study.

Table 12.1 Subjective self-report questions from the Chalder fatigue scale (Chalder et al., 1993).

Item	Question
1	Do you feel have problems with tiredness now?
2	Do you need to rest more now?
3	Do you feel sleepy or drowsy now?
4	Do you have problems starting things now?
5	Do you start things without difficulty but get weak as you go on now?
6	Do you start things without difficulty but get weak as you go on now?
7	Are you lacking in energy now?
8	Do you have less strength in your muscles now?
9	Do you feel weak now?
10	Do you have difficulty concentrating now?
11	Do you have problems thinking clearly now?
12	Do you make slips of tongue when speaking now?
13	Do you find it more difficult to find the correct word now?
14	How is your memory now?
15	Have you lost interest in the things you used to do now?

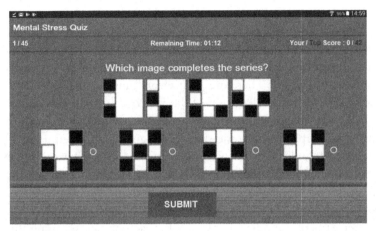

Figure 12.4 Mental stress quiz.

In the beginning of the experimental study, each participant rested listening to music for 10 minutes while wearing the sensor to collect rest state data.

Mental fatigue can be provoked in experimental studies using various methods proposed in the literature such as arithmetic tests, speech tasks, academic exams, mental pictures, memory games, trail-making tests, Stroop color-word disturbance tests, etc. (Castaldo et al., 2015). In this work a questionnaire was formed encapsulating characteristic types of cognitive load stressors in accordance with other similar studies (Huang et al., 2018). Thus the experiment was designed to induce mental fatigue through a mental stress quiz consisting of 45 questions. The administration of the quiz was performed through an Android application developed specifically for the purpose of this study (Fig. 12.4). The quiz includes math problems, memory tests, spatial questions, and logical inference problems which require an intense mental effort and collectively lead to a state of mental stress. The quiz takes approximately 50 minutes to complete and data collected at the last 10 minutes of the quiz were used to characterize the mental stress state.

12.3.3 Feature extraction

After the R−R signal is collected from the participants a preprocessing step is applied to correct principal errors before starting the feature extraction process. The output of this preprocessing step is an HRV signal which is called the N–N signal (N–N intervals indicate the time between two normal heartbeats). The automatic detection of R-peaks is prone to errors due to external factors. Even a few of such artificially inserted heartbeats can strongly influence the HRV feature extraction (Acharya et al., 2006). Therefore such errors must be identified and corrected in order for the calculation of the HRV features to be valid.

The R–R signal was first filtered to eliminate noise due to the movements of the Zephyr against the skin. The signal was filtered using a low/high pass filter with a cut-off frequency of 0.5/10 Hz. These frequencies have been established empirically to best preserve the position of the signal peaks while eliminating the maximum amount of motion artifacts.

Detecting and correcting incorrect R–R values was done automatically after gathering the R–R intervals for each participant using the artifact detection algorithm by Rand et al. (2007). The algorithm compares the differences of the surrounding R–R intervals with a threshold value calculated individually for each participant. The R–R intervals are marked as artifacts when exceeding the threshold value compared to the differences of the surrounding R–R intervals. An advantage of this algorithm is that it can be applied online without requiring the entire set of the recorded data.

HRV features have been calculated as characteristic markers of the N–N signal (Table 12.2). These features which are divided in time and

Table 12.2 Time and frequency domain HRV (heart rate variability) features.

Predictor	Description	Unit
meanNN	$\mu_{NN} = \frac{1}{N}\sum_{i=1}^{N} NN_i$	ms
medianNN	Median (NN)	ms
RMSSD	$\sqrt{\dfrac{\sum_{i=1}^{N}(\Delta NN_i)^2}{N}}$	ms
PNN50	$\frac{count(abs(\Delta NN) > 50)}{N}100\%$	%
meanHR	Mean heart rate	beats/min
TP	Total spectral power	ms^2
HF	High-frequency band (0.15–0.40 Hz) power	ms^2
LF	Low-frequency band (0.04–0.15 Hz) power	ms^2
VLF	Very-low-frequency band (0.0033–0.04 Hz) power	ms^2
LF/HF	LF and HF band powers ratio	–

frequency domain were calculated for each of the 60 collected data samples (30 for rest state and 30 for mental stress state). To produce valid results, time windows of five minutes on the preprocessed data were defined (Malik, 1996). The time-domain HRV parameters are based on methods from statistics. These methods assume a normal distribution of the data (N−N intervals). Concerning HRV frequency domain analysis, it is generally relied on power spectral density (PSD) analysis. PSD is performed on a series of discrete events, for example on a sequence of instantaneous heart rhythms, although this results to an irregularly sampled time signal. This sequence must be then interpolated to obtain it in a regular way as a function of time. The HRV features used in this study were calculated using Kubios software (Tarvainen et al., 2014).

The heart rate is subject to a physiological variability and reflects the interaction between the parasympathetic and the sympathetic nervous system (PNS and SNS) (Acharya et al., 2006). The sympathetic is the area in the brain that is responsible for stress control and the parasympathetic is responsible for relaxation. The predictors in Table 12.2 reflect the dynamic interactions between the SNS and the PNS on the cardiovascular system. Analysis of the frequency spectrum gives information about the level of sympathetic and parasympathetic activation in the heart autonomic control. PNS is thought to be a major contributor to the *HF* component of HRV. On the other hand, the origin of the *LF* component remains controversial as some consider it to reflect SNS activity, while others suggest that signifies both PNS and SNS activity. The *LF/HF* ratio is quite often used to characterize certain physiological or cognitive states. This could represent the balance of PNS and SNS activity.

Table 12.3 gives the statistics of HRV features in the mental stress and rest states. The Wilcoxon statistical test was applied to examine whether the variation of the HRV features between the two states is statistically significant. This test was selected since HRV features in the frequency domain are expected to follow a nonsymmetric distribution. All features demonstrate significant changes between the two states with the exception of *medianNN* and *LF/HF* which were excluded from further analysis.

Furthermore, the selected features should have as little correlation with each other as possible, while at the same time they should have as much correlation as possible with the target class. This is examined with the feature correlation and feature importance inspection procedures respectively described in the following sections.

Table 12.3 Descriptive statistics of HRV (heart rate variability) markers in the two states.

Predictor	Rest		Mental stress		P
	Mean	Stdev	Mean	Stdev	
meanNN	739.5	84.4	775.6	96.1	.001
medianNN	751.1	91.5	768.3	100.0	.118
RMSSD	36.9	20.0	46.6	20.6	.000
PNN50	13.6	12.6	21.1	14.5	.000
meanHR	82.0	8.8	78.5	9.9	.006
TP	1820.7	1952.1	2878.7	2330.2	.000
HF	572.8	916.4	956.0	1192.5	.000
LF	1196.1	1038.3	1830.1	1162.0	.000
VLF	51.0	75.3	91.6	104.2	.001
LF/HF	2.8	1.3	2.9	1.6	.826

12.3.4 Feature correlation inspection

By exploring the degree to which features are correlated to each other it is possible to identify redundant features. The exclusion of such features enables the creation of simpler and faster classification models when machine learning methods are applied to the training datasets.

In the context of this study the SPSS statistical tool was used to analyze the data and identify correlations, if any, between the features that were calculated. The existence of a correlation is demonstrated by the Pearson correlation coefficient r. Pearson r demonstrates the existence or not of an association between two variables and calculates the form of this correlation (i.e., positive or negative correlation) but also its intensity (i.e., level of statistical significance). This statistical criterion checks the null hypothesis that there is no correlation between two variables. The level of statistical significance (2-tailed probability), indicates whether and how much the relationship is important. The correlation is statistically significant if P (statistical significance level) is less than .05.

The correlations are shown in Table 12.4. This table contains the Pearson's r index for all possible combinations of variables involved in the analysis. The correlation of each variable with itself is 1. The data points count in all cases is 60 representing the size of input data and asterisks indicate statistically significant correlations.

Two variables are considered redundant when there is a very strong correlation of absolute value 0.8 or above and the correlation is also

Table 12.4 Feature correlation matrix.

	meanNN	PNN50	RMSSD	TP	LF	HF	VLF	HR
meanNN	1							
PNN50	0.627[b]	1						
RMSSD	0.535[b]	**0.962**[b]	1					
TP	0.461[b]	0.762[b]	0.791[b]	1				
LF	0.430[a]	0.687[b]	0.767[b]	0.728[b]	1			
HF	0.409[a]	0.654[b]	0.739[b]	0.749[b]	0.654[b]	1		
VLF	0.625[b]	0.706[b]	0.785[b]	0.661[b]	0.673[b]	0.685[b]	1	
HR	**−0.991**[b]	−0.667[a]	−0.565[a]	−0.420[a]	−0.397[a]	−0.439[a]	−0.672[b]	1

[a] $P < .05$.
[b] $P < .01$.

statistically significant ($P < .05$). In practical terms, between such a pair of variables, the variable that has a higher correlation with the rest of the variables was selected for exclusion. As shown in the correlation table, there are two pairs of features that can be considered for redundancy analysis: (meanNN, HR) and (PNN50, RMSSD). Features RMSSD and HR met the above criteria and were excluded.

12.3.5 Feature importance inspection

HRV features are used to create classification models with machine learning algorithms that distinguish the two states: rest and mental stress. At this point it is essential to consider the importance of each feature in terms of the target state to create simpler, more robust, and faster classification models (Foster et al., 2014). To this end, it is necessary to train a model for this purpose only. The RF machine learning algorithm (Breiman, 2001) utilized to inspect the importance of the HRV features and to cross-check these results with the accuracy that is achieved by each selected classification algorithm. RF classifiers encapsulate many decision trees, and as a result, they extract the class that is decided more often by the individual trees. In particular, each decision tree gives a categorization, a suggested dependent variable value. From the decision trees set each possible value of the dependent variable has been proposed specific times, which is also the classification basis for the RF algorithm. A key advantage of the RF classifier is its ability to effectively manage a large number of independent variables and the short execution time required.

The information that can assist in finding the most important features using a classifier model is mean decrease in impurity (MDI) also known as Gini importance and permutation importance also known as mean decrease in accuracy (MDA). MDI calculates the importance of each feature as the sum of the number of branches (split) in all trees containing the feature, divided by the number of data samples it divides. As for the MDA, it is essentially calculated by an algorithm that in each iteration replaces the values of a feature with random values and recalculates the accuracy of the model. When the target class is heavily dependent on a feature, then this will present high MDA values as changing its values randomly would have a strong effect on the model accuracy.

Table 12.5 presents the ranking of the qualified HRV features using the MDI metric which is considered stronger. The MDI/MDA scores were   calculated   using   the   Scikit-learn   library   and   the   Python

Table 12.5 Importance ranking of HRV (heart rate variability) features by MDI (mean decrease in impurity).

Rank	Predictor	MDI	MDA
1	TP	4.40	2.35
2	meanNN	4.12	3.83
3	LF	3.57	6.43
4	PNN50	3.31	4.22
5	HF	2.46	3.38
6	VLF	2.25	2.27

MDA, Mean decrease in accuracy.

programming language. Since both MDI and MDA scores are positive, all features are considered important for predicting accurately the target class. Therefore six variables will be explored in the training of the mental stress detection models.

12.3.6 Feature scaling

The scaling process is a preprocessing type action required in the proposed methodology. Scaling is essentially a transformation that is enforced to the available data so that the values of the independent variables end up having certain common characteristics, such as being within a certain range of values, having an average value 0, having the same standard deviation, and so on. The main reason to have the feature values on the same scale is that the model would not give more importance to a feature just because its value range is higher compared to other features. Another important reason for applying scaling, and more specifically the standardization technique, is to avoid the effects of concept drift. Essentially, for some features the distribution of values can change widely with each new dataset. Therefore, for each new dataset, standardization can be applied to minimize the negative impact on the performance of the model that may impose a change in the value distribution of a feature.

There are two main approaches that serve the scaling goal, namely, normalization and standardization. Normalization is useful when values in a certain interval are needed as it serves to rearrange data in a range of values between 0 and 1. To find this range the *min−max* method is applied to each feature, which is calculated by the formula:

$$X_{norm}^i = \frac{x^i - x_{min}}{x_{max} - x_{min}} \tag{12.1}$$

In the above formula x^i is the value of the specific feature, x_{min} is the feature minimum value and x_{max} is the feature maximum value in the dataset.

The standardization method is sometimes better and more practical in many machine learning algorithms, especially when linear models such as support vector machines with a linear kernel and logistic regression are used. This is due to the fact that these algorithms initialize weights to values equal to or close to 0. The standardization keeps the average of the features at 0 and the standard deviation at 1, having the form of the normal distribution, thus helping to better learn the weights. The standard scaler is calculated by the formula:

$$X_{std}^i = \frac{x^i - \mu_x}{\sigma_x} \qquad (12.2)$$

where μ_x is the feature mean and σ_x is the feature standard deviation in the given dataset.

The selection of the scaling technique depends heavily on the machine learning algorithm that will be used for training a classification model. For example, in the case of building the mental stress detection model with the k-NN algorithm, the normalization method was used to support the algorithm for accurate and smooth training. On the other hand, the standardization method was applied in the case of the SVM algorithm.

12.3.7 Machine learning methods

The NB, SVM, and k-NN machine learning methods were applied in the model training step. These three different approaches are presented briefly in the following subsections.

12.3.7.1 Naïve Bayes

The NB algorithm classifies a sample using the Bayes simplistic model, assuming that the variables are mutually independent, given the target class (Russel & Norvig, 2013). Suppose that there is a set of data D signifying an n-dimensional vector $X = (x_1, x_2, \ldots, x_n)$ which are the measurements in each sample of the n features. Assume that there are m categories $C_1, C_2, \ldots, C_m$ of a class C, which must be predicted.

To classify a sample X the probabilities $P(C_1 | X)$, $P(C_2 | X)$, $\ldots$, $P(C_m | X)$ are calculated, namely, the probability that the sample belongs to the category $C_1, C_2, \ldots, C_m$ respectively. Sample X is classified into that category whose probability $P(C|X)$ is the maximum. To determine the probability of a sample X associated to a class C_i, the Bayes theorem is used:

$$P(C_i/X) = \frac{P(X/C_i) \cdot P(C_i)}{P(X)} \quad (12.3)$$

$P(X)$ is the same in all categories, so the focus is only on the numerator $(X|C_i) \cdot (C_i)$. To calculate the probability $P(X|C_i)$ the simplistic assumption is made that the feature effect on a class C_i is independent of the other features values. Therefore the following formula is used:

$$P(X/C_i) = \prod_{k=1}^{n} P(X_k/C_i) = P(X_1/C_i)P(X_2/C_i)...P(X_n/C_i) \quad (12.4)$$

Learning the NB model works very well in a wide range of applications, as it is one of the most effective machine learning algorithms, which can be scaled efficiently into very large problems. Finally, data containing some noise does not affect the algorithm, thus providing answers to the classification problem.

12.3.7.2 Support vector machine

The SVM method tackles a classification problem by managing the separation of a pair of classes by a separator line (or a level or more generally a hyperplane in a multiclass problem). SVM is a vector learning method and in contrast to other classifiers, such as the NB algorithm, which locate any linear separator or look for the best potential linear separator based on a criterion, SVM is aiming to identify a decision threshold between classes (Wang, 2005). This limit should be at the maximum possible distance from any point of the training dataset, namely, the goal is to maximize the distance of the separator line from both classes at the same time. Suppose there is a learning machine $f(x_i, a)$ where a is the set of parameters of the function based on points, which consist of the subset of the data where the separator position is defined. These points are called support vectors. The aim is then to train the learning machine to determine the relation $x_i \rightarrow y_i$, that is to correctly classify x_i which have come from the same probability distribution $P(x_i, y_i)$ with those of the training set.

There are two quantities that are related to the performance of the learning machine, in terms of the associated error, the expected or actual risk, and the empirical risk. The former is defined by the formula:

$$R(a) = \int 1/2|y - f(x, a)| dP(x, y) \quad (12.5)$$

which is showing the deviation of the classification assessment $f(x_i, a)$ from the actual y to new points (outside the training set). Respectively, empirical risk is defined as the error rate on the whole training and is defined by the formula:

$$R_{emp}(\alpha) = \left(\frac{1}{2} \sum_{i=1}^{n} |y_i - f(x_i, \alpha)| \right) \qquad (12.6)$$

which is essentially the average of the classification error of the n points of the training set.

The goal of SVM is not to minimize the empirical risk but to minimize the upper limit of a generalization error. To achieve this goal, the decision limit of machine learning needs to have the optimal distance from the nearest training point.

12.3.7.3 k-Nearest neighbors
The last supervised learning algorithm is the k-NN algorithm, which is quite different in the process it follows for categorizing data, compared to the aforementioned algorithms (Cover & Hart, 1967). k-NN provides a quick and simple classification approach and thus consists of a common choice for addressing classification problems especially in cases where data distribution is not known a priori. k-NN is called the lazy algorithm because it memorizes the entire set of training in memory (Peterson, 2009). The advantage of this approach is that the classifier immediately classifies the new training data collected. An important disadvantage is that the introduction of new data for classification linearly increases computational complexity.

The basic idea of the k-NN algorithm for categorizing an element is that the properties of each specific element given as input to the algorithm should be similar to the properties of the other points, at a certain distance from it. This distance is also called neighborhood, hence the name of the algorithm. The steps that are followed for the implementation of the k-NN algorithm are the following: select the number of neighbors and the distance measure; find the neighbor that the item should be categorized; categorize the new item. To measure the similarity or the distance between points, a measure of distance $D(x_1, x_2)$ must be used. The k-NN classifier is usually based on the Euclidean distance between two data sets, the test set and the training set.

According to the k-NN algorithm, the constant parameter k is predetermined. The algorithm searches in the two-dimensional space the k

points (observations) that are closest to the new unknown observation. The classifier assigns the new observation to the class that has a majority among the k-NN. There are several approaches to finding the optimal number for k. However, there is typically no structured method for locating it. It must be discovered by various attempts through trial and error, assuming that the training data are unknown. Choosing very small values for k can be noisy and have a strong effect on the result. On the other hand, higher values of k will have looser decision limits, which means lower variance and increased systematic error (bias). It is also a computational expensive process.

12.3.7.4 Model evaluation methods and measures

For the evaluation of the proposed models the data used to construct the machine learning models cannot be also used for their evaluation. This is because the model that is created has the ability to remember the entire training set, so it will constantly classify any data from the training set correctly. However, the appearance of new data without the categorization tag helps to assess the prediction model performance. Specifically, the data collected are divided into two segments, where one segment is used to construct the machine learning model and is called training dataset, whereas the data that will be used to assess the effectiveness of the model is called the test dataset.

Various techniques are employed to validate classification models. The k-fold cross-validation method splits the initial dataset into k segments, which are named folds. These segments are different one to each other and are about the same size, where the classification model is repeatedly trained and tested k times. In each reiteration, one segment is maintained as the test dataset whereas the remaining segments are exploited for retraining the classification model. The value of k is usually 5 or 10.

The split method is the most popular approach for separating datasets into a training dataset and a test dataset. If, for example, the split method is 80%, this means that a random percentage of 80% of the dataset will be reserved for the training dataset and the remaining 20% for the test dataset.

The most basic measure used to evaluate the effectiveness of a classification model is *accuracy*. Other measures that are useful for evaluating the model are *precision*, *recall*, and *F-measure*. The classification accuracy given a dataset is the percentage of data instances that were correctly categorized in the training model whether the prediction is positive or negative. In

Table 12.6 Classification model evaluation measures.

Measure	Formula
Accuracy	$\dfrac{TP + TN}{TP + FN + FP + TN}$
Precision	$\dfrac{TP}{TP + FP}$
Sensitivity/Recall	$\dfrac{TP}{TP + FN}$
Specificity/Selectivity	$\dfrac{TN}{TN + FP}$
F-measure/F1-score	$2 \cdot \dfrac{Precision \cdot Sensitivity}{Precision + Sensitivity}$

TP/TN: Positive/Negative samples that have been correctly predicted by the model. FP/FN: Negative/Positive samples that have been incorrectly predicted as positive/negative.

other words, accuracy can be specified as the ratio of correct predictions of positive and negative instances to the total number of data instances. Precision is defined as the measure of effectiveness when our goal is to reduce false-positive categorized data. It is important that the model has high precision, which means that false-positive categorized data are avoided. Recall or *sensitivity* is defined as a measure of effectiveness when the goal is to identify all positive samples. From another point of view, *specificity* expresses the percentage of actual negative samples that have been correctly classified as negative. Many times the results of accuracy and recall do not give an accurate view of the classification results. Another measure that gives better results is the F-measure (*F1-score*), which combines precision and sensitivity as their harmonic mean. Table 12.6 summarizes the classification model evaluation measures.

12.3.8 Classification results

After separating the data into samples of 5-minute time windows and after scaling the extracted features, the training process of the classification models can start. As specified previously the development of the classifiers that use HRV features for mental stress detection is based on three different methods: NB, SVM, and *k*-NN. The adaptation of the classification model to the training dataset is associated with specifying values to a number of parameters that define the model. The behavior of most machine learning algorithms is also characterized by an additional set of parameters that must be defined before the training phase.

Parameter tuning for SVM involves a grid search approach to optimize the Gamma (G) and Cost (C) parameters under the selected radial basis function (rbf) kernel. G was varied from 10^{-4} to 10 and C was varied from 0.01 to 10^4. k-NN parameter tuning entails the selection of an optimal value for k parameter in a range between 3 and 12, using Euclidean as the distance metric. The Scikit-learn library in the Python programming environment was used for classifiers training and validation.

The classification performance was validated initially by the subjective data recorded in the beginning and at the end of the experimental protocol to assess the mental stress. Moreover, the classification models were validated with the leave-one-subject-out cross-validation strategy given the limited number of samples. According to this technique, the classifier is trained with a dataset from which the data samples of a subject have been removed. The data subset that has been excluded is then used as the test dataset to validate the classification model. The same procedure is reiterated for every subject in the dataset one after another and the classifier is resampled using the updated training dataset each time.

Given that six HRV features have been qualified from the relevance and redundancy analysis it is feasible to examine an exhaustive search strategy to identify the best features combination that achieves the higher detection performance. There are 63 combinations in total that must be tested starting from single variable models up to six variables models. Each case needs to be tested for each machine learning method. Table 12.7 summarizes for each method the best classification performance results of the models tested through cross-validation starting with models of a single variable and extending up to six variables. The prediction performance is reported in terms of the evaluation metrics discussed previously.

The best performance is achieved by SVM with a model of four HRV features (*TP, LF, HF, meanNN*): 85% accuracy, 80% precision, 88.9% sensitivity 81.8% specificity, and 84.2% F1-score. The SVM parameters used were $C = 0.5$ and $G = 0.5$. The feature synthesis of the model is in conformity with the feature importance analysis and the ranking of the features computed, as three of the variables are in the top list of the MDI-based ranking. This model attains also the highest F1-score and since this metric is a combination of precision and sensitivity it provides a performance assessment that is less biased.

On the single variable models the best accuracy is 66.7% and is realized by the NB method using the *TP* feature. However, the other two methods are close to that accuracy using the same predictor in the case of SVM or *PNN50* in the case of k-NN ($k = 3$). When two variables are used the best accuracy (71.7%) is delivered by k-NN ($k = 3$) by means of the *TP*, *PNN50* feature combination. Furthermore, a high precision (76.7%) is

Table 12.7 Classification performance results.

Method	# Variables	Features	Accuracy (%)	Precision (%)	Sensitivity (%)	Specificity (%)	F1 (%)
NB	1	TP	66.7	60.0	69.2	64.7	64.3
SVM		TP	65.0	56.7	68.0	62.9	61.8
k-NN		PNN50	63.3	56.7	65.4	61.8	60.7
NB	2	TP, PNN50	68.3	60.0	72.0	65.7	65.5
SVM		TP, meanNN	70.0	63.3	73.1	67.6	67.9
k-NN		TP, PNN50	71.7	76.7	69.7	74.1	73.0
NB	3	TP, LF, PNN50	70.0	73.3	68.8	71.4	71.0
SVM		TP, LF, meanNN	78.3	83.3	75.8	81.5	79.4
k-NN		TP, LF, PNN50	75.0	70.0	77.8	72.7	73.7
NB	4	TP, LF, HF, PNN50	66.7	60.0	69.2	64.7	64.3
SVM		TP, LF, HF, meanNN	**85.0**	80.0	88.9	81.8	84.2
k-NN		TP, LF, HF, PNN50	76.7	80.0	75.0	78.6	77.4
NB	5	TP, HF, VLF, PNN50, meanNN	61.7	56.7	63.0	60.6	59.6
SVM		TP, HF, LF, PNN50, meanNN	81.7	83.3	80.6	82.8	82.0
k-NN		TP, HF, LF, PNN50, meanNN	73.3	70.0	75.0	71.9	72.4
NB	6	TP, LF, HF, VLF, PNN50, meanNN	63.3	70.0	61.8	65.4	65.6
SVM			78.3	73.3	81.5	75.8	77.2
k-NN			68.3	73.3	66.7	70.4	69.8

k-NN, k-Nearest neighbors; NB, Naïve Bayes; SVM, support vector machine.

accomplished which is associated with the ability for reducing false positives. SVM outperforms the other two methods on the accuracy metric in the remainder of the distinct model dimensions: 78.3% on the three variables models (*TP*, *LF*, and *meanNN*), 81.7% on the five variables models (*TP*, *HF*, *LF*, *PNN50*, and *meanNN*), and 78.3% on the six variables models (*TP*, *LF*, *HF*, *VLF*, *PNN50*, and *meanNN*).

On the three variables models the second best performance is achieved by *k*-NN (*k* = 3) with 75.0% accuracy and a feature combination consisting of *TP*, *LF*, and *PNN50*. *k*-NN (*k* = 7) outperforms NB also in all other cases, namely, on the four variables models (*TP*, *LF*, *HF*, and *PNN50*) with 76.7% accuracy, on the five variables models (*TP*, *LF*, *HF*, *PNN50*, and *meanNN*) with 73.3% and on the six variables models with 68.3%.

By inspecting the contents of Table 12.7 it can be inferred that the classification performance follows an ascending course as the set of HRV features increases reaching a high plateau when the predictors are three or four. From another point of view, when the number of predictors further increases the performance degrades indicating an information disturbance due to the added variables and potentially an overfitting effect. It can also be observed that *TP*, which is the most important variable according to the feature importance inspection, is present in every combination of the features that result in high performance. On the other hand, *VLF*, which was assessed as the less important variable, does not occur in any combination of variables associated with high accuracies.

12.4 Discussion

Most of the research in the field of mental stress detection uses various biomarkers, such as EEG, ECG, and EOG or even a combination of these signals measured, however, with costly and intrusive medical devices, and as a consequence the detection accuracy is expected to be quite high. However, the proposed methodology here focuses on creating a model that is capable of detecting mental stress on everyday environments, which means the device that provides the data to the prediction model must be comfortable and easy to use on a daily basis as well as affordable for the typical user. In this context, a methodology that detects mental stress with an accuracy of 85% was presented based on a single biomarker and the SVM machine learning method. Table 12.8 summarizes characteristics of representative studies on mental stress detection.

Table 12.8 Characteristics of representative studies on mental stress detection.

Reference	Signals[a]	Stress test	Study participants	Duration[b]	Machine learning methods[c,d]	Accuracy
Huang et al. (2018)	ECG/HRV	Arithmetic, cognitive, memory tasks	35	54 min	SVM, NB, k-NN, LR	75.5%
Zanetti et al. (2019)	EEG, ECG, BVP, RESP	Arithmetic tests, serious game	17	38 mins	SVM, **LR**, **RF**	84.6%
Masood and Alghamdi (2019)	HRV, EEG, EDA, RESP	Memory search, color-word test, tracing mirror image, dual task, public speech	24	32 min	**CNN**	90%
King et al. (2006)	EEG	Driving task	55	–	**ANN**	81.49%– 83.06%
Zhang et al. (2017)	EEG	Driving task	10	120 min	**SVM**	90.70%
Yamada and Kobayashi (2018)	EOG	Calculation task, paced auditory serial attention test	31	17 min	**SVM**	91.0%
Laurent et al. (2013)	ECG, EEG EOG	Task-switching	13	20 min	**SVM**	94%
Bafna et al. (2021)	Eye tracking and typing performance	Eye-typing task	19	10 trials in 2 sessions	RT, PLSR, **RFR**, SVR	1.057[e]

Can et al. (2019)	ECG/HRV, EDA, ST, ACC	Algorithm programming tasks	21	9 days	PCA, LDA, SVM, k-NN, LR, RF, **MLP**	92.15%
This work	ECG/HRV	Arithmetic, reasoning, memory tasks	30	50 min	k-NN, **SVM**, NB	85%

[a] ACC, Accelerometer; BVP, blood volume pulse; ECG, electrocardiogram; EEG, electroencephalogram; EOG, electrooculogram; EDA, electrodermal activity; HRV, heart rate variability; RESP, respiration; ST, skin temperature.

[b] Experimental protocol mean time per participant.

[c] Machine learning algorithm achieving the best accuracy is indicated in bold.

[d] ANN, Artificial neural network; CNN, convolutional neural network; k-NN, k-nearest neighbors; LDA, linear discriminant analysis; LR, logistic regression; MLP, multilayer perceptron; NB, Naïve Bayes; PCA, principal components analysis; PLSR, partial least squares regression; RF, random forest; RFR, random forest regression; RT, regression trees; SVM, support vector machine; SVR, support vector regression.

[e] Performance in terms of mean absolute error.

The study carried out uses a structured processing chain that includes a data collection step using a wearable ECG sensor, a data preprocessing step (filtering and artifact reduction), a feature extraction step, feature selection steps (based on feature correlation and feature importance inspection), a classification model creation step by using three machine learning methods (NV, SVM, and k-NN) and finally a leave-one-subject-out cross-validation strategy for validating the predictive model. The criteria to select the specific machine learning algorithms were based on factors such as their applicability to the specific problem domain and their practicality for implementing the relevant models in a reasonable amount of time. Other factors include the complexity, scalability, interpretability, and accuracy of the generated model. All three algorithms have well-known characteristics that have qualified them as appropriate methods for detecting mental stress in several relevant studies (see Table 12.8). For example, SVM is simpler than using ANNs and CNNs and hence this facilitates convergence within a shorter training time. On the other hand, k-NN is resilient to noisy data that can emerge especially with the use of wearable devices and NV is a typical classifier for two or multiple classes achieving positive results on a variety of problems.

On the basis of the classification results presented *TP*, *LF*, *HF*, and *meanNN* are the fundamental HRV predictors for detecting the mental stress state. Involving less or more predictors affects the classification performance. Using only four HRV features provides the opportunity to collect and process the HRV data rapidly, which is imperative for the online monitoring of health status based on inexpensive wearable devices. Moreover, a limited number of predictors enables better interpretability of the accomplished results from a medical perspective and evades, to a degree, the pitfall of overfitting in the case of a limited dataset.

HRV is the variability of the interval between two successive heart beats, or in other terms the variability of the R−R interval. To characterize it, two types of methods are mainly employed, those in the time domain, and those in the frequency domain. In this study it was found that mental stress is accompanied by modulations of heart activity and in particular a decrease in heart rate (*meanHR*), resulting in an increase in the duration of the interbeat interval (*meanNN*). Therefore mental stress is accompanied by an increase of time-domain HRV features which is in accordance with relevant studies (Castaldo et al., 2015). In the frequency domain, mental stress is also characterized by an increase in the high-frequency power spectrum component (*HF*) as suggested also by other researchers (Zhao et al., 2012). Similarly, an increase in the power of its low-frequency component (*LF*) was recorded.

The preprocessing step of the HRV signal generally includes frequency filtering as discussed previously, with the purpose of limiting the impact of the surrounding electrical activity on the physiological signal. The data splitting must be carried out taking into account the minimum duration of a cycle and the number of iterations necessary to constitute a robust average of the features studied. In that respect a time window of 5 minutes was used.

An advantage of the ECG/HRV signal over other physiological signals such as EEG and EOG is its convenience for use in everyday measurements. On the other hand, HRV, although showing very reliable markers, these are indirect measures of mental state and are therefore slower than, for example, EEG markers. Thus the evaluation of the feasibility of using HRV measurements could be beneficial for applications that are attentive to the physiological state of the users on the corresponding time-frames (i.e., 5 minutes time windows). On the other hand, applications with very short time response requirements may have some lag effect as, for example, in cases where users should be immediately prevented from making quick and potentially erroneous decisions when high mental fatigue is detected.

Data labeling that is the assignment of correct mental state on the gathered sensor data represents another significant challenge for developing mental stress detection systems. The quality of data labeling affects the training and performance of machine learning models and acquiring such data demands a significant amount of time and effort. In the proposed methodology the ground truth of the mental stress state is ascertained through the Chalder fatigue scale where subjects are deemed to be fatigued if their score is above a predefined threshold. Although the administration of questionnaire-based assessments relies on introspection and is susceptible to bias, this risk was partially addressed by examining the trend of the HRV features and showing how the features change as the mental fatigue builds up throughout the cognitive test duration. A promising technique to alleviate the possible lack of labeled data in mental health monitoring applications, such as stress detection, is transfer learning (Garcia-Ceja et al., 2018). Existing knowledge from analogous domains is used to learn new information in cases where scarce annotated training data are available.

This study allowed the assessment of HRV markers with respect to a systematic estimate of the state of mental fatigue. The utility of the established predictors should be evaluated over longer and overlapping periods to allow online monitoring with more sound results than those obtained here. In addition, the relevance of other HRV predictors should be assessed (e.g., duration of complex QRS).

12.5 Conclusion

Prolonged mental stress arising in work environments has been linked with severe health-related and safety conditions thus triggering research and development efforts in recent years. The aim of this work was to develop a detection model for mental stress using a low-cost wearable device to collect successfully and unobtrusively physiological signals in everyday environments. Based on HRV data analysis and machine learning algorithms, conclusions were drawn about possible key features that operate as predictors to determine the mental stress state. Such awareness could be used to draw conclusions about the tendency for individuals to experience chronic stress in environments such as workplaces. With the help of such predictions, organizations can take appropriate measures to protect their employees from serious medical conditions and plan their tasks in a way to strike a balance between labor productivity and human well-being.

A future work direction could focus on a richer feature set comprising HRV and more composite predictors such as features calculated in the geometric domain. Moreover, the development of a more sophisticated machine learning approach in terms of ensemble models while exploiting dimensionality reduction techniques to achieve high diversity is considered. Additional probable future work is to attempt to further enrich the pool of base classifiers considering not only more classification algorithms but also several versions of the same approach with different parameter settings.

References

Acharya, U. R., Joseph, K. P., Kannathal, N., Lim, C. M., & Suri, J. S. (2006). Heart rate variability: A review. *Medical & Biological Engineering & Computing, 44*(12), 1031–1051.
Alonso, J. F., Romero, S., Ballester, M. R., Antonijoan, R. M., & Mañanas, M. A. (2015). Stress assessment based on EEG univariate features and functional connectivity measures. *Physiological Measurement, 36*(7), 1351.
Bafna, T., Bækgaard, P., & Hansen, J. P. (2021). Mental fatigue prediction during eye-typing. *PLoS One, 16*(2), e0246739.

Boksem, M. A., Meijman, T. F., & Lorist, M. M. (2006). Mental fatigue, motivation and action monitoring. *Biological Psychology, 72*(2), 123—132.

Breiman, L. (2001). Random forests. *Machine Learning, 45*(1), 5—32.

Calnan, M., Wainwright, D., Forsythe, M., Wall, B., & Almond, S. (2001). Mental health and stress in the workplace: The case of general practice in the UK. *Social Science & Medicine, 52*(4), 499—507.

Can, Y. S., Chalabianloo, N., Ekiz, D., & Ersoy, C. (2019). Continuous stress detection using wearable sensors in real life: Algorithmic programming contest case study. *Sensors, 19*(8), 1849.

Castaldo, R., Melillo, P., Bracale, U., Caserta, M., Triassi, M., & Pecchia, L. (2015). Acute mental stress assessment via short term HRV analysis in healthy adults: A systematic review with meta-analysis. *Biomedical Signal Processing and Control, 18,* 370—377.

Chalder, T., Berelowitz, G., Pawlikowska, T., Watts, L., Wessely, S., Wright, D., & Wallace, E. P. (1993). Development of a fatigue scale. *Journal of Psychosomatic Research, 37*(2), 147—153.

Cover, T., & Hart, P. (1967). Nearest neighbor pattern classification. *IEEE Transactions on Information Theory, 13*(1), 21—27.

Demerouti, E., Bakker, A., Nachreiner, F., & Ebbinghaus, M. (2002). From mental strain to burnout. *European Journal of Work and Organizational Psychology, 11*(4), 423—441.

Esler, M., Schwarz, R., & Alvarenga, M. (2008). Mental stress is a cause of cardiovascular diseases: From scepticism to certainty. *Stress Health, 24*(3), 175—180.

Foster, K. R., Koprowski, R., & Skufca, J. D. (2014). Machine learning, medical diagnosis, and biomedical engineering research-commentary. *Biomedical Engineering Online, 13*(1), 1—9.

Garcia-Ceja, E., Riegler, M., Nordgreen, T., Jakobsen, P., Oedegaard, K. J., & Tørresen, J. (2018). Mental health monitoring with multimodal sensing and machine learning: A survey. *Pervasive and Mobile Computing, 51,* 1—26.

Goumopoulos, C., & Menti, E. (2019). Stress detection in seniors using biosensors and psychometric tests. *Procedia Computer Science, 152,* 18—27.

Hjortskov, N., Rissén, D., Blangsted, A. K., Fallentin, N., Lundberg, U., & Søgaard, K. (2004). The effect of mental stress on heart rate variability and blood pressure during computer work. *European Journal of Applied Physiology, 92*(1), 84—89.

Hosseini, S. H., Bruno, J. L., Baker, J. M., Gundran, A., Harbott, L. K., Gerdes, J. C., & Reiss, A. L. (2017). Neural, physiological, and behavioral correlates of visuomotor cognitive load. *Scientific Reports, 7*(1), 1—9.

Hou, X., Liu, Y., Sourina, O., Tan, Y. R. E., Wang, L., & Mueller-Wittig, W. (2015) EEG based stress monitoring. In: *Proceedings of the 2015 international conference on systems, man, and cybernetics* (pp. 3110—3115). IEEE.

Hsieh, C. S., & Tai, C. C. (2013). An improved and portable eye-blink duration detection system to warn of driver fatigue. *Instrumentation Science & Technology, 41*(5), 429—444.

Huang, S., Li, J., Zhang, P., & Zhang, W. (2018). Detection of mental fatigue state with wearable ECG devices. *International Journal of Medical Informatics, 119,* 39—46.

Kelly, S. J., & Ismail, M. (2015). Stress and type 2 diabetes: A review of how stress contributes to the development of type 2 diabetes. *Annual Review of Public Health, 36,* 441—462.

King, L. M., Nguyen, H. T., & Lal, S. K. L. (2006). Early driver fatigue detection from electroencephalography signals using artificial neural networks. In: *Proceedings of the 2006 international conference of the IEEE engineering in medicine and biology society* (pp. 2187—2190). IEEE.

Lal, S. K., & Craig, A. (2001). A critical review of the psychophysiology of driver fatigue. *Biological Psychology, 55*(3), 173—194.

Laurent, F., Valderrama, M., Besserve, M., Guillard, M., Lachaux, J. P., Martinerie, J., & Florence, G. (2013). Multimodal information improves the rapid detection of mental fatigue. *Biomedical Signal Processing and Control, 8*(4), 400–408.

Lorist, M. M., Klein, M., Nieuwenhuis, S., De Jong, R., Mulder, G., & Meijman, T. F. (2000). Mental fatigue and task control: Planning and preparation. *Psychophysiology, 37* (5), 614–625.

Malik, M. (1996). Heart rate variability: Standards of measurement, physiological interpretation, and clinical use: Task force of the European Society of Cardiology and the North American Society for Pacing and Electrophysiology. *Annals of Noninvasive Electrocardiology, 1*(2), 151–181.

Marcora, S. M., Staiano, W., & Manning, V. (2009). Mental fatigue impairs physical performance in humans. *Journal of Applied Physiology (Bethesda, MD: 1985), 106*(3), 857–864.

Masood, K., & Alghamdi, M. A. (2019). Modeling mental stress using a deep learning framework. *IEEE Access, 7,* 68446–68454.

Peterson, L. E. (2009). K-nearest neighbor. *Scholarpedia, 4*(2), 1883.

Rand, J., Hoover, A., Fishel, S., Moss, J., Pappas, J., & Muth, E. (2007). Real-time correction of heart interbeat intervals. *IEEE Transactions on Bio-Medical Engineering, 54*(5), 946–950.

Russel, S., & Norvig, P. (2013). *Artificial intelligence: A modern approach.* London: Pearson Education Limited.

Tanaka, M., Ishii, A., & Watanabe, Y. (2014). Neural effects of mental fatigue caused by continuous attention load: A magnetoencephalography study. *Brain Research, 1561,* 60–66.

Tarvainen, M. P., Niskanen, J. P., Lipponen, J. A., Ranta-Aho, P. O., & Karjalainen, P. A. (2014). Kubios HRV—heart rate variability analysis software. *Computer Methods and Programs in Biomedicine, 113*(1), 210–220.

Wang, L. (Ed.), (2005). *Support vector machines: Theory and applications.* Springer Science & Business Media.

Yamada, Y., & Kobayashi, M. (2018). Detecting mental fatigue from eye-tracking data gathered while watching video: Evaluation in younger and older adults. *Artificial Intelligence in Medicine, 91,* 39–48.

Zanetti, M., Mizumoto, T., Faes, L., Fornaser, A., De Cecco, M., Maule, L., & Nollo, G. (2019). Multilevel assessment of mental stress via network physiology paradigm using consumer wearable devices. *Journal of Ambient Intelligence and Humanized Computing,* 1–10.

Zhang, L., Zhou, Q., Yin, Q., & Liu, Z. (2018). Assessment of Pilots Mental Fatigue Status with the Eye Movement Features. In: *Proceedings of the international conference on applied human factors and ergonomics* (pp. 146–155). Cham: Springer.

Zhang, X., Li, J., Liu, Y., Zhang, Z., Wang, Z., Luo, D., ... Wang, C. (2017). Design of a fatigue detection system for high-speed trains based on driver vigilance using a wireless wearable EEG. *Sensors, 17*(3), 486.

Zhao, C., Zhao, M., Liu, J., & Zheng, C. (2012). Electroencephalogram and electrocardiograph assessment of mental fatigue in a driving simulator. *Accident; Analysis and Prevention, 45,* 83–90.

CHAPTER 13

Knowledge discovery and presentation using social media analysis in health domain

Heba M. Wagih
Information Systems Department, The British University in Egypt, Cairo, Egypt

13.1 Introduction

The use of social networks has been evolving constantly and on a broad scale. Social network plays a vital role in many application domains where communities can share and exchange ideas, knowledge, and information. The use of social media has increased from 8% to 72% since 2005 in the United States and many social network applications exceeded millions of users, for example, the number of Facebook users has exceeded 1 billion users all over the globe (Bernhardt et al., 2014). Numerous problems in social networks are still under examination such as social network modeling and analysis, recommender systems in social networks, sentiment analysis, community detection, and other research challenges.

Social media has a notable role in the medical field where it has a great influence in improving the healthcare services and activities where it gives the opportunity for both patients and professionals to be highly engaged through their participation. Various social networks have been introduced into the medical field, among these social networks is the QuantiaMD. QuantiaMD is a very famous online platform for physicians, clinical members, and patients to interact. An important survey was introduced in Fogelson et al. (2013) that shows that more than 4000 physicians conducted the QuantiaMD and almost 65% use this site for professional reasons. Medicine 2.0 is another web-based personal health application that was introduced in (Eysenbach, 2008). Medicine 2.0 is designed to support patients, health specialists, and researchers using semantic web technologies and virtual reality. Medicine 2.0 added great value to the health domain as it promoted the interaction between researchers and different users from all categories (patients, physicians, decision-makers). PatientsLikeMe (PatientsLikeMe, 2017) is another popular web-based healthcare application

Edge-of-Things in Personalized Healthcare Support Systems
DOI: https://doi.org/10.1016/B978-0-323-90585-5.00012-6

with more than 500,000 members. This application is concerned with collecting and publishing medical information as well as enhancing the communication between different application participants. It provides real-time insights into a wide variety of diseases and symptoms thousands of diseases and conditions. Many benefits are granted by this application; first, it is free of charge where all participants can enjoy freely the health services offered by the application. The application facilitates the communication between patients having the same medical problems where patients can share their experiences, advice, and recommendations with other participants.

Medical applications have benefited from human behavior trajectory analysis where both the visited locations and time of visits were added to the users' trajectories. Medical applications that are supported with the human behavior trajectory analysis help patients to share information about their health condition as well as their locations. This would help in advising the nearest hospital or clinic for the patient in any case of emergencies. Also, patients can post all their visited locations via the check-in feature, and following the patients' trajectories could help in monitory and defining infected areas. In the last year the medical web applications have a remarkable strike in helping people to face COVID-19 where patients were able to report their location, their symptoms on daily basis, and if they need urgent medical help.

Machine learning, a branch of artificial intelligence, has a significant role in the healthcare domain and greatly supports both patients and medical staff in various ways. Machine learning is used in various fields (e.g., cancer detection) for medical decision support systems, development of medical care guidelines, and others. Different machine learning algorithms were introduced from simple ones as the K-nearest neighbor (KNN) to more complicated as random forests and boosting model. The random forest model is one of the most famous machine learning algorithms and it is based on the concept of decision trees. The main drawback of the random forest model is it tends to overfit the training data. Boosting is another machine learning algorithm that is based on the idea of combining weak classifiers into much stronger ones. A more detailed explanation for different machine learning models is presented in Stamp (2018). It is worth mentioning that clinical staff are responsible to prepare the medical data in such a way that makes the system easily identify patterns and inferences. Machine learning systems have shown equivalent performance compared to experienced physicians. Machine learning is an important driving force behind enhancing healthcare services.

Since healthcare systems use complex and comprehensive medical data, thus a need for analyzing and representing the medical concepts arises. Ontologies have a notable role in the healthcare domain for their ability to integrate and represent such data and knowledge. Earlier, ontologies were considered a branch of philosophy and with the technology arising era, the ontology was used to represent the world reality where it classifies any world domain into a set of categories that are connected through some object properties.

In Gruber (1993) the first formal definition for ontology was presented where the author described an ontology as a "formal, explicit specification of a shared Conceptualization." Ontology models support the data exchange separately from any application field. For better knowledge representation for ontologies, description logics are used. Description logics were introduced in Baader et al. (2003) and are the most famous knowledge representation formalism for semantic web. Description logics consist of two main components; TBox to represent concepts, relationships, and different constraints (e.g., cardinality) and the ABox to introduce assertions about individuals.

Finally, healthcare systems have noticed the importance of detecting patients' locations and analyzing the patients' activity trajectories. In this work, we introduce a new concept which is the patient activity trajectories which is a specific class of semantic trajectories as time dimension is added to the raw trajectory to represent all the visiting places by a patient. Detecting these places can have a great influence especially in the COVID-19 pandemic where analyzing these trajectories can help in identifying and locating all visited places and predicting other places that could have also been visited by the patient and thus taking all medical precautions that can help in controlling the virus propagation.

The objective of this chapter is to discover and present knowledge about COVID-19. The main contributions are summarized as follows:

1. developing an ontology-based model, coronology, that describes the main concepts and relations related to coronavirus;
2. presenting an SROIQ description logic to describe and reason about the concepts of coronology model; and
3. utilizing random forest machine learning model and Naïve Bayes algorithm to predict the infection rates from coronavirus.

The rest of this chapter is arranged as follows: Section 13.2 presents some related works to social media and its impact on the healthcare domain, healthcare applications, knowledge representation using ontology, and the role of machine learning algorithms in the health sector. Section 13.3 presented the coronology model, the underlying description logics, and some query formulation as well as the experimental results on three real case studies. Finally, Section 13.4 concludes and suggests directions for future work.

13.2 Related work

COVID-19 pandemic has caused major pressure on healthcare systems and applications worldwide. Many works were introduced lately to support the fight against this disease. Many research were introduced by the scientific community in various fields such as social media and its impact on coronavirus, detecting and analyzing human activity trajectories, ontology engineering and its roles in information representation, and using machine learning algorithms to predict the corona infection state for patients. In this work, we will shed the light on each of these areas and show its impact on the corona era.

Social media has been a great source of a wealth of information in the past years and many application domains have benefited it. The healthcare domain is one of the fields that has recently relied on social media where patients and healthcare professionals can interact and discuss health-related questions. Not only that but also healthcare applications are used for the detection and tracking of infectious diseases. Social media goes beyond asserting facts and describing events to saving lives and controlling the propagation of infectious diseases (Kotov, 2015). According to Fox (2011), it was stated that 61% of American adults use online health applications to search for medical information and 37% were able to retrieve or post health information online. Some works had been introduced to study the impact of social media in the health sector. In Rolls et al. (2016) the authors presented a survey on the use of social media by healthcare professionals in improving virtual interactions that promote professional networking and knowledge distribution. An important review was introduced about the use of social media in health and how it affects the relationship between patients and healthcare professionals. The authors presented some of the challenges that might affect the use of social media in healthcare as the difficulty in assuring the accuracy of health-related posts due to the exponential growth of the communication channels in healthcare applications (Smailhodzic et al., 2016).

Social media applications in the health domain have opened the opportunity for studying new challenges as detecting and analyzing patient mobility. By defining the location and the points of interest for a patient many medical services can be offered and here the role of recommendation systems arises. Social media applications in the health domain have opened the opportunity for studying new challenges as detecting and analyzing patient mobility. By defining the location and the points of interest for a patient many medical services can be offered and here the role of location recommendation systems arises. Location recommendation Systems have provided a better-customized user experience to healthcare users. Based on the patient's location, nearest medical centers or hospitals can be provided, professionals could be suggested, and medical instructions could be given. Also, by analyzing the patients' trajectories in case of COVID-19 infection, recommender systems could warn other users of the spread of the virus in certain zones. Many works have been introduced to study the importance of location-based recommender systems. Among which the work presented in Sarwat et al. (2012) where the authors introduced the Sindbad application, a location-based social network application that is composed of three main components—the location-aware news feed, the location-aware news ranking, and the location-aware recommendation. Three data categories were fed to the system, namely, spatial messages, user profiles, and spatial ratings. Based on these components, the system provided users with geo-tagged messages recommended by his or her friends to suggest a particular place within a certain area.

More contributions were introduced in the healthcare domain, among which is the development of medical ontologies. Ontologies are shared vocabulary that represents concepts and relations in a specific field of interest. A surveillance ontology for COVID-19 was introduced in McGagh et al. (2021). Their ontology consists of 32 classes that facilitate the monitoring of COVID-19 cases especially for patients who suffer from respiratory conditions. According to Helmy et al. (2015), the authors introduced health, food, and user's profile ontologies for custom-based information retrieval. They integrated different ontologies (e.g., disease ontology and nutrition ontology) from food and medical domains as well as developing new ontologies such as body function and body part ontologies to suggest the best health recommendations based on the patients' health state.

Among the recent contributions in the healthcare domain is the utilization of machine learning algorithms in enhancing the prediction of several diseases. An algorithm of a stacked ensemble built using different baseline models was presented in Jain et al. (2021). The experiments were performed using the proposed model and other machine learning models, like support vector machine (SVM), Naïve Bayes, KNN, Gradient boosting, and others to analyze and predict the dataset. The results of the experiments proved that the proposed model has a higher accuracy rate than other machine learning models. Bullard et al. (2020) used the diagnostic samples to perform their investigation based on predicting SARS-CoV-2 infection. They adopted the diagnostic samples dataset and performed several experiments to predict whether a person is suffering from SARS-CoV-2. Another important contribution was proposed in Khalil et al. (2020). The authors presented a real-time system that performs sentiment prediction on tweets about the coronavirus pandemic. The author examined several machine learning models (KNN, logistic regression, random forest, SVM, and decision tree) to find the model with the higher accuracy rate in the coronavirus sentiment analysis prediction and then utilizes it in real-time applications. The proposed system is developed in two stages: develop an offline sentiment analysis and model an online prediction. The authors used n-gram and TF-ID feature extraction models, to extract the main features in the dataset. The authors performed several experiments, and the random forest model has achieved the highest performance.

Following the above contributions and driven by the impact of social media in the healthcare domain, we introduce an ontology model for representing the medical domain using the Protege Ontology Development tool along with its underlying description logics. We performed deep analysis on several datasets to uncover all correlations and hidden patterns that can affect the propagation of coronavirus. Finally, we used the random forest machine learning and Naïve Bayes algorithm model to predict coronavirus pandemic.

13.3 Coronology: proposed ontology for representing COVID-19 information

Coronology is the proposed ontology that shows the concepts' categorization that represents the COVID-19 domain, also the restrictions and axioms of these concepts. The proposed ontology consists of six main entities: patients, regions, regions population, routes, tests, and cases. Each entity is represented by an OWL annotation that defines its relevant properties. The proposed schema model is the initial model for our work and many enhancements are expected in the future such as introducing disjointness between some classes, introducing some inverse and transitive relations, and introducing Boolean combination of classes. Since such features are not introduced in the RDFs, we favored using the OWL language as the language for our ontology model. In our schema model, we introduce 6 main entities, 32 dataproperty, and 5 objectproperty. Dataproperty and objectproperty are used to present the different types of relations used in our model. Dataproperty relation presents the relation between an entity and its attributes, while the objectproperty relation presents the relation between two entities. The main entities in the proposed model are *PatientInformation*, *Region*, *RegionPopulation*, *Test*, *Case*, and *PointOfInterest*.

PatientInformation' attributes include demographic features (age, gender) as well as other relevant features (e.g., contactNo, ID, and SymptomOnsetDate). The attributes are presented using five dataproperty relations, namely, hasgender, has Age, hasID, hasContactNumber. Since each patient lives in a certain region, thus there is an objectproperty relation between *PatientInformation* entity and *Region* entity named liveIn. *Region* is defined by seven attributes, namely, CityName, CountryName, Longitude, Latitude, MaxTemp, MinTemp, and WindSpeed. The attributes are presented using seven dataproperty relations, namely, hasCityName, hasCountryName, hasLongitude, hasLatitude, hasMaxTemp, hasMinTemp, and hasWindSpeed. The population of each region is stored in *RegionPopulation* entity, and a relation named hasPopulation of type objectproperty defines the relationship between these two entities. RegionPopulation is defined by four attributes, namely, ElementaryCount, SchoolCount, UniversityCount, and ElderlyCount. The attributes are presented using four dataproperty relations, namely, hasElementaryCount, hasSchoolCount, hasUniversityCount, and hasElderlyCount. Since each patient can visit different locations which are represented by their longitude and latitude coordinates in the *PointOfInterest* entity. Thus identifying these locations can help in detecting the behavior of the patient movements which sequentially helps in controlling the spread of the disease by examining people who visited these places and may also close these places for some time. The visited locations are presented by the entity *PointOfInterest*. The relation between the *PatientInformation* and *PointOfInterest* entities is presented by the objectproperty named hasCheckedIn.

The *PointOfInterest* entity is defined by five attributes, namely, Location, Latitude, Longitude, Date, and Time. The attributes are presented using five dataproperty relations, namely, hasLocation, hasLatitude, hasLongitude, atDate, and atTime. Each patient can perform several tests, and if the test is positive, then the patient is considered a case and all information regarding the patient as the diagnosed disease (strain type), infection causes, and current medical situation are all stored in *Test* and *Case* entities. An objectproperty named hasTest represents the relation between entity *PatientInformation* and *Test*. The *Test* entity is defined by three attributes, namely, Date, Time, and TestResult. The attributes are presented using three dataproperty relations, namely, atDate, atTime, and hasTestResult. An objectproperty named isArchived is used to represent the relation between *Test* and *Case* entities. The last entity in the proposed model is *Case* where the *Case* entity is defined by eight attributes, namely, CaseID, Disease, InfectionCause, MedicalState, Released, Deceased, IsolatedFrom, and IsolatedTill. The attributes are presented using eight dataproperty relations, namely, hasCaseID, hasDisease, hasInfectionCause, hasMedicalState, isReleased, isDeceased, isIsolatedFrom, and sIsolatedTill. Fig. 13.1 shows a complete schema describing each of these entities with its corresponding attributes.

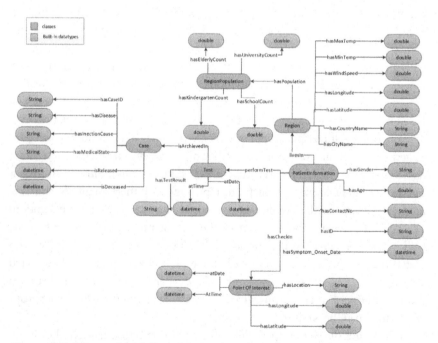

Figure 13.1 Schema description of the proposed ontology model.

The following set of axioms formally describes the proposed model in description logics. The description logic used was presented in Horrocks et al. (2006).

Axiom 1: PatientInformation $\sqsubseteq$ $\exists$hasgender.string $\prod$ $\exists$hasAge.double $\prod$ $\exists$hasID.string $\prod$ $\exists$hasContactNumber.string $\prod$ $\exists$hasSymptom_Onset_date. datetime $\prod$ $\exists$PerformTest.Test $\prod$ $\exists$liveIn.Region $\prod$ $\exists$hasCheckedIn. PointOfInterest

- hasgender: A DataProperty of Domain Class PatientInformation and Range DataType string
- hasAge: A DataProperty of Domain Class PatientInformation and Range DataType double
- hasID: A DataProperty of Domain Class PatientInformation and Range DataType string
- hasContactNumber: A DataProperty of Domain Class PatientInformation and Range DataType string
- hasRace: A DataProperty of Domain Class PatientInformation and Range DataType string
- hasSymptom_Onset_date: A DataProperty of Domain Class PatientInformation and Range DataType datetime
- PerformTest: An ObjectProperty of Domain Class PatientInformation and Range Class Test
- liveIn: An ObjectProperty of Domain Class PatientInformation and Range Class Region
- hasCheckedIn: An ObjectProperty of Domain Class PatientInformation and Range Class PointOfInterest

Axiom 2: Region $\sqsubseteq$ $\exists$hasCityName.string $\prod$ $\exists$ hasCountryName.string $\prod$ $\exists$hasLongitude.double $\prod$ $\exists$hasLatitude.double $\prod$ $\exists$hasMaxTemp.double $\prod$ $\exists$hasMinTemp.double $\prod$ $\exists$hasWindSpeed.double $\prod$ $\exists$hasPopulation. RegionPopulation

- hasCityName: A DataProperty of Domain Class Region and Range DataType String
- hasCountryName: A DataProperty of Domain Class Region and Range DataType String
- hasLongitude: A DataProperty of Domain Class Region and Range DataType double
- hasLatitude: A DataProperty of Domain Class Region and Range DataType double
- hasMaxTemp: A DataProperty of Domain Class Region and Range DataType double

- hasMinTemp: A DataProperty of Domain Class Region and Range DataType double
- hasWindSpeed: A DataProperty of Domain Class Region and Range DataType double
- hasPopulation: An ObjectProperty of Domain Class Region and Range Class RegionPopulation

Axiom 3: RegionPopulation $\sqsubseteq$ $\exists$hasElementaryCount.double $\prod$ $\exists$ hasSchoolCount.double $\prod$ $\exists$hasUniversityCount.double $\prod$ $\exists$hasElderlyCount. double

- hasElementaryCount: A DataProperty of Domain Class RegionPopulation and Range DataType double
- hasSchoolCount: A DataProperty of Domain Class RegionPopulation and Range DataType double
- hasUniversityCount: A DataProperty of Domain Class RegionPopulation and Range DataType double
- hasElderlyCount: A DataProperty of Domain Class RegionPopulation and Range DataType double

Axiom 4: PointOfInterest $\sqsubseteq$ $\exists$hasLocation.String $\prod$ $\exists$hasLatitude. double $\prod$ $\exists$hasLongitude.double $\prod$ $\exists$atDate.datetime $\prod$ $\exists$atTime. datetime

- hasLocation: A DataProperty of Domain Class PointOfInterest and Range DataType String
- hasLongitude: A DataProperty of Domain Class PointOfInterest and Range DataType double
- hasLatitude: A DataProperty of Domain Class PointOfInterest and Range DataType double
- atDate: A DataProperty of Domain Class PointOfInterest and Range DataType datetime
- atTime: A DataProperty of Domain Class PointOfInterest and Range DataType datetime

Axiom 5: Test $\sqsubseteq$ $\exists$hasTestResult.String $\prod$ $\exists$atDate.datetime $\prod$ $\exists$atTime.datetime $\prod$ $\exists$isArchievedIn.Case

- hasTestResult: A DataProperty of Domain Class Test and Range DataType String
- atDate: A DataProperty of Domain Class Test and Range DataType datetime
- atTime: A DataProperty of Domain Class Test and Range DataType datetime

- isArchievedIn: An ObjectProperty of Domain Class Test and Range Class Case

  Axiom 6: Case $\sqsubseteq$ $\exists$hasCaseID.String $\prod$ $\exists$ hasDisease.String $\prod$ $\exists$ hasInfectionCause.String
  $\prod$ $\exists$ hasMedicalState.String $\prod$ $\exists$ isReleased.datetime $\prod$ $\exists$ isDeceased.datetime $\prod$ $\exists$ isIsolatedFrom.datetime $\prod$ $\exists$ sIsolatedTill. datetime

- hasCaseID: A DataProperty of Domain Class Case and Range DataType String
- hasDisease: A DataProperty of Domain Class Case and Range DataType String
- hasInfectionCause: A DataProperty of Domain Class Case and Range DataType String
- hasMedicalState: An ObjectProperty of Domain Case Test and Range DataType String
- isReleased: A DataProperty of Domain Class Case and Range DataType datetime
- isDeceased: An ObjectProperty of Domain Case Test and Range DataType datetime
- isIsolatedFrom: A DataProperty of Domain Class Case and Range DataType datetime
- isIsolatedTill: A DataProperty of Domain Class Case and Range DataType datetime

13.3.1 Analyzing patients' dynamics

As confirmed by experts in the medical domain, COVID-19 is highly spread among people especially in close distances, thus analyzing patients' trajectories could have a significant role in controlling the rapid spread of this virus. Patients' visited locations can be detected by analyzing patients' dynamics from data collected from social networking and hence taking extra precautions in these locations.

Consider the following example, assume that a group of patients have performed the COVID-19 test and the results turned to be positive. Thus we need to create a Query (Query1) that retrieves the test results for all patients who were tested for the coronavirus. As a result of Query1, it is crucial to analyze all the routes that were taken by the infected patients to identify areas that could have been infected and this is achieved through Query2. For providing better medical services, it is important to determine the cause of infection and the current medical state of the infected patients, and such information is retrieved by Query3. SPARQL Query Language introduced in (Kollia et al., 2011) is used in Protégé 5.0 beta (Musen, 2015) to query the proposed ontology and ensure the validity of our ontology model through expressing real-world queries.

Query1: *Retrieve the test result for all tested Patient.*
SPARQL Query:
>SELECT? Patient?Test
>WHERE? Patient onto:performTest?Test.

Query2: *Retrieve all the locations that were visited by infected Patients.*
SPARQL Query:
>SELECT? Patient? PointsofInterests
>WHERE? Patient onto:Check-In? PointsofInterests.

Query3: *Retrieve all the medical information related to infected Patients.*
SPARQL Query:
>SELECT? Patient?Test?Case
>WHERE? Patient onto:performTest?Test.
>?Test onto:isArchievedIn?Case.

13.3.2 Case study

To better explain the proposed approach and highlight its benefits, we used several datasets for COVID-19 cases for some different places over the globe as South Korea and the State of Tennessee which is a state in the Southeastern region of the United States. The South Korean dataset was provided by the Korean Centers for Disease Control & Prevention (Kaggle.com, 2021). The datasets consist of information that were collected for four months, namely, March, April, May, and June in 2020. It consists of five main files which are cases, patients, routes, regions, and time. The second dataset for the state of Tennessee and was provided by the Tennessee Department of Health (Tn.gov, 2021). The dataset is updated on daily basis and it consists of nine main files that describe the age, gender, counties, and daily cases information for new infections. The third dataset was provided by the Israeli Ministry of Health website (Data.gov.il, 2021) and it was published for public use. The dataset consists of initial records for all citizens who were tested for COVID-19. The dataset includes the basic clinical symptoms as if the patients suffer from fever, cough, or headache, also the gender and age were included.

Analyzing these files could help in detecting the highly infected groups of people, discovering the routes taken by suspected cases, determining red zones (places of high infection cases) to prevent people from visiting, and hence help in controlling the COVID-19 propagation. Fig. 13.2 presents a generalized system architecture for proposed model.

We started our experiments by first matching the used datasets with the proposed ontology to prove the capability of the proposed ontology to present corona information. In each of the following figures, we will discuss the matched part in our ontology. We started our investigation by analyzing the South Korea dataset from the gender perspective as shown in Fig. 13.3. It was shown that although the infection percentage is obviously higher in the female gender however their resistance to this virus is noticeable higher than the male gender, this can be returned to the fact that females' immune system better responds to infections rather than that in males. In this experiment, we compare the gender against the population, and this is represented in coronology using three entities which are PatientInformation (gender attribute), Region (country and city attributes), and RegionPopulation (ElementaryCount, SchoolCount, UniversityCount, and ElderlyCount attributes).

In Fig. 13.4, we considered different age groups starting from newborn to 80s. We found that the resistance of middle ages is low to this virus and the death rate tends to increase in older ages which are suspected due to the presence of other medical health issues that older ages usually suffer from. In this experiment, we compare the gender against the population, and this is

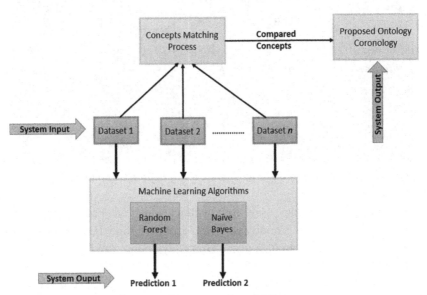

Figure 13.2 System architecture for proposed model Korea.

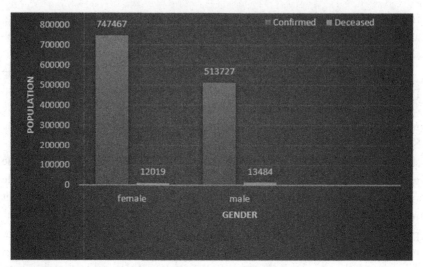

Figure 13.3 Gender infection percentage in South Korea.

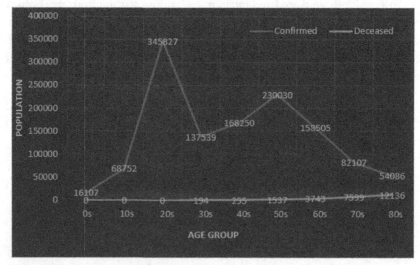

Figure 13.4 Infection percentage in different age groups in South Korea.

represented in coronology using three entities which are PatientInformation (age attribute), Region (country and city attributes), and RegionPopulation (ElementaryCount, SchoolCount, UniversityCount, and ElderlyCount attributes).

The higher the temperature and humidity have a clear role in affecting virus viability (Chan et al., 2011), on studying the given dataset we found that cities with lower average temperatures as Seoul and Incheon as shown higher infection rates among their citizens, however, the propagation of the virus and in turn the number of confirmed and deceased cases increased from March to June as shown in Fig. 13.5 which proves the high survival strength of the COVID-19 even in high temperatures. In this experiment, we compare the month against the population, and this is represented in coronology using two entities which are Region (country, city, MinTemp, and MaxTemp attributes) and RegionPopulation (ElementaryCount, SchoolCount, UniversityCount, and ElderlyCount attributes).

Besides the temperature effect in the frequency of virus propagation, other factors were considered in our investigation such as contacts with other patients, overseas flow, and others. On analyzing the dataset, we found that 31% of infection cases were due to contact with other patients, 16% from overseas overflows, 16% were unable to identify the cause and the rest were due to visiting commonplaces as religious places, malls, clubs, healthcare centers, and workplaces. Adding all these factors together draws a clearer picture of the most infected cities and the reasons for infection. Table 13.1 shows the percentage of isolated, released, and deceased cases in four of the major cities in South Korea, namely, Seoul, Busan, Incheon, and Daegu from March till June.

To relate Table 13.1 to our ontological schema, four entities are used, namely, *PatientInformation*, *Region*, *Test*, and *Case*. For example, suppose

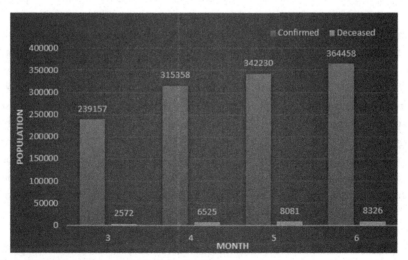

Figure 13.5 Cases state per month in South Korea.

Table 13.1 Cases state in different cities in South Korea.

City	% Isolated	% Discharged	% Deceased
Seoul	6.195547	19.07067	0.135528
Busan	0.71636	2.14908	0.058083
Incheon	2.187803	0.077444	0.387222
Daegu	4.782188	1.858664	0

we have Patient "X" with ID "123," all the personal information can be stored in the entity *PatientInformation*. Since Table 13.1 presents information for four cities in South Korea, then information about the patient's residential area is stored in entity *Region* using the data properties hasCountry and hasCity. If the patient performed the check test, then the test results, date, and time will be stored in entity *Test* and then the test is archived in entity *Case* which keeps the information about the patient medical condition, the infection cause, and the date and time of whether the patient is released or deceased. If the patient is isolated before being released, then the time in isolation will be stored using two data properties, namely, isIsolatedFrom and isIsolatedTill.

Moving to the second dataset for the state of Tennessee, we found that the results were almost similar to those of South Korea, for the gender dimension, infection among females was 53% among the population to 47% in males which is a noticeable higher percentage. Similar to Fig. 13.3, in this experiment, we compare the gender against the population, and this is represented in coronology using three entities which are PatientInformation (gender attribute), Region (country and city attributes), and RegionPopulation (ElementaryCount, SchoolCount, UniversityCount, and ElderlyCount attributes).

The other dimension that was considered is the age group which also shows similar results to that in South Korea where the experiments show that the resistance of the age group from 20 to 50 is lower to the virus infection than other age groups as shown in Fig. 13.6. Similar to Fig. 13.4, in this experiment, we compare the age against the population, and this is represented in coronology using three entities which are PatientInformation (age attribute), Region (country and city attributes), and RegionPopulation (ElementaryCount, SchoolCount, UniversityCount, and ElderlyCount attributes). The other dimension that was considered is the age group. The age group shows similar results to that in South Korea where the experiments show that the resistance of the age group from 20 to 50 is lower to the virus infection than other age groups.

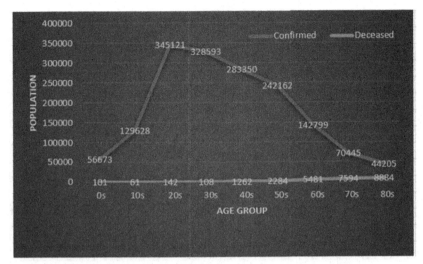

Figure 13.6 Infection percentage in different age groups in Tennessee.

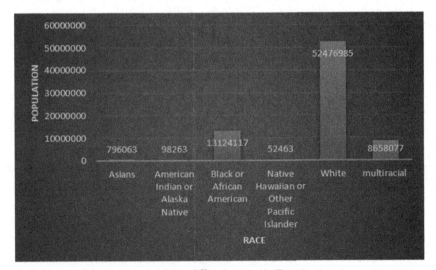

Figure 13.7 Infection percentage in different races in Tennessee.

Since the United States is known for the presence of a diversity of races thus races are considered an important dimension to be studied and examine the effect of COVID-19 on different races. We found that White races are more fragile to resist the virus in opposite to Asians or American Indians as shown in Fig. 13.7. Such observation sheds the light on how genes can have an important role in the resistance to some viruses. In this experiment, we

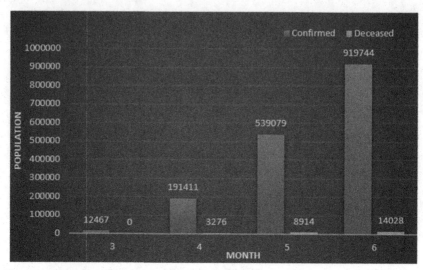

Figure 13.8 Cases state per month in Tennessee.

compare the race against the population, and this is represented in coronology using three entities which are PatientInformation (race attribute), Region (country and city attributes), and RegionPopulation (ElementaryCount, SchoolCount, UniversityCount, and ElderlyCount attributes).

Finally, for the temperature, although the temperature gets higher from March to June, however, this does not affect the virus spread among citizens as shown in Fig. 13.8. Similar to Fig. 13.5, in this experiment, we compare the month against the population, and this is represented in coronology using two entities which are Region (country, city, MinTemp, and MaxTemp attributes), and RegionPopulation (ElementaryCount, SchoolCount, UniversityCount, and ElderlyCount attributes).

The dataset introduced in Data.gov.il, (2021) presents more detailed features regarding tested cases where more symptoms (e.g., cough, headache, and fever) were analyzed and added for further studies. In our experiments, we uncovered the correlation between every pair of features to provide more insights about the COVID-19 characteristics as shown in Fig. 13.9. Correlation between features is performed using Pearson correlation coefficient. The experiments show a noticeable positive correlation between the three symptoms: headache, sore throat, and shortness of breath. Also, it was noticed a positive correlation between the age group above 60 and the shortness of breath. Another observation that was noticed regarding the gender is that males are subject to a higher infection rate of 53% when comparing to females with a 47% infection rate.

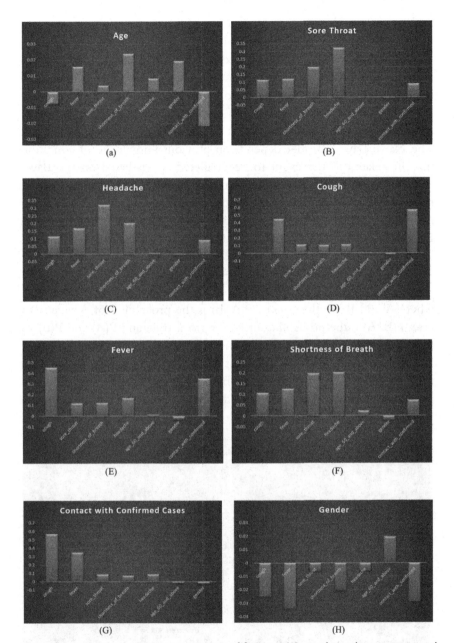

Figure 13.9 Correlation between every pair of features, (A) correlation between age and all other features, (B) correlation between sore throat and all other features, (C) correlation between headache and all other features, (D) correlation between cough and all other features, (E) correlation between fever and all other features, (F) correlation between shortness of breath and all other features, (G) correlation between contact with confirmed cases and all other features, (H) correlation between gender and all other features.

Such observations have an important role in defining categories of people who might be at risk when facing COVID-19.

We also considered the performed experiments to determine the possibility to distinguish COVID-19 patients through their set of symptoms. In our experiments we used two famous machine learning algorithms, the Naïve Bayes classification algorithm and the random forest classification algorithm, to classify the new patient as a positive or negative corona case. The Naïve Bayes classifier depends on previous knowledge of conditions that are related to the event to be predicted and is based on the Bayes' theorem. The Naïve Bayes classifier can quickly and efficiently predict multiple classes of datasets. One of the major benefits of Naïve Bayes in my study is that it is better adapted for categorical input variables. Eq. (13.1) presents the mathematical formulation for the Bayes' theorem.

$$P(A|B) = \frac{P(B|A) \cdot P(A)}{P(B)} \tag{13.1}$$

where A and B are the events, $P(A|B)$ is the probability of A given B is true, $P(B|A)$ is the probability of B given A is true, and $P(A)$ and $P(B)$ are the independent probabilities of A and B.

Random forest is one of the most powerful classifiers that is built using many decision tree models. Its strength is in its capability in executing both regression and classification models. It performs efficiently on massive

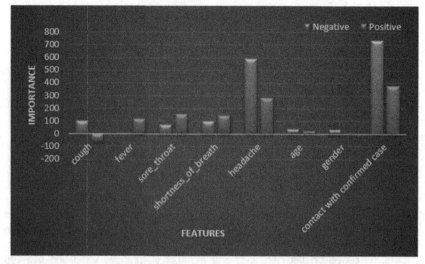

Figure 13.10 Importance of different features in COVID-19 prediction.

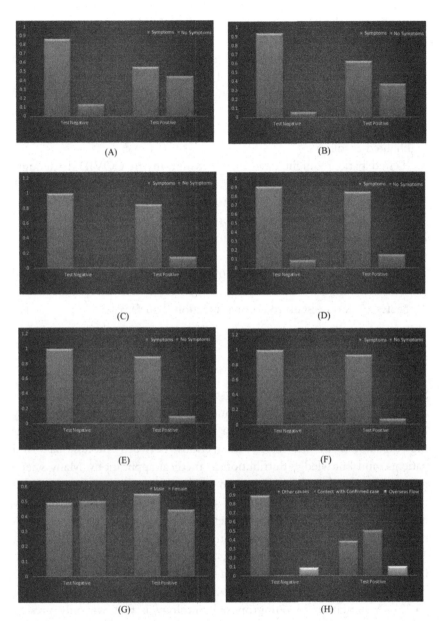

Figure 13.11 Impact of different features in the test results, (A) cough, (B) fever, (C) sore throat, (D) shortness of breath, (E) headache, (F) age, (G) gender, (H) infection cause.

datasets where the dataset is split into smaller chunks and the random forest algorithm runs in parallel.

We used the dataset introduced in Data.gov.il. and we applied both algorithms on the 200 + K records. The random forest algorithm shows a slightly higher accuracy 96.83% compared to the Naïve Bayes algorithm 95.3%. In Figs. 13.10 and 13.11, we examined the effect of each feature in predicting the test results (positive, negative). Interesting observations were highlighted as follows:

1. Tested patients might have similar symptoms to COVID-19 (cough, fever, sore throat, headache), however, their tests' results are negative which is an indicator that symptoms are not enough to decide if the patient has been infected with coronavirus or not.
2. Most of the positive cases were in contact with other confirmed positive patients.
3. Patients whose ages are above 60 are at higher risk to get infected with COVID-19.
4. Males are less resistant to corona infection than females.

13.4 Conclusion

Social media has shown remarkable strides in the past few years in various domains. The healthcare sector is one of the major fields that are highly impacted by the use of social media by considering many factors such as the capabilities of the users, interactions between professionals and patients, and knowledge distribution in medical applications. Many scientific works had been introduced to study the empowerment of social media in the health sector. In this chapter, we introduce an ontology model that represents the main concepts and relations related to coronavirus. Machine learning algorithms, namely, Naïve Bayes and random forest are used to provide more insights, predictions, and recommendations. Different experiments were conducted, and random forest has shown a better accuracy rate of 96.83% compared to the Naïve Bayes algorithm of 95.3%. A limitation in our proposed ontology is that we only present two-dimensional spatial space, thus we plan in our future work to enhance the version of our ontology by integrating the three-dimensional spatial space. We also consider in our future work to compare more machine learning algorithms as boosting models to figure out the best model for our datasets.

References

Baader, F., Calvanese, D., McGuinness, D. L., Nardi, D., & Patel-Schneider, P. F. (2003). *The description logic handbook: Theory, implementation, and applications.* Cambridge University Press.

Bernhardt, M., Alber, J., & Gold, R. S. (2014). A social media primer for professionals: digital do's and don'ts. *Health Promotion Practice, 15*(2), 168−172.

Bullard, J., Dust, K., Funk, D., Strong, J. E., Alexander, D., Garnett, L., Boodman, C., Bello, A., Hedley, A., Schiffman, Z., Doan, K., Bastien, N., Li, Y., Van Caeseele, P. G., & Poliquin, G. (2020). Predicting infectious SARS-CoV-2 from diagnostic samples. *Clinical Infectious Diseases: An Official Publication of the Infectious Diseases Society of America.*

Chan, K. H., Peiris, J. S., Lam, S. Y., Poon, L. L., Yuen, K. Y., & Seto, W. H. (2011). The effects of temperature and relative humidity on the viability of the SARS coronavirus. *Advances in Virology.*

Data.gov.il. (2021). Corona data. <https://data.gov.il/dataset/covid-19>.

Eysenbach, G. (2008). Medicine 2.0: Social networking, collaboration, participation, apomediation, and openness. *Journal of Medical Internet Research, 10*(3), e22. Available from https://doi.org/10.2196/jmir.1030.

Fogelson, N. S., Rubin, Z. A., & Ault, K. A. (2013). Beyond likes and tweets: An in-depth look at the physician social media landscape. *Clinical Obstetrics and Gynecology, 56*(3), 495−508.

Fox, S. (2011). *The social life of health information.* In *Pew Internet.* (8). Washington, DC: American Life Project. Available from https://www.pewresearch.org/internet/2011/05/12/the-social-life-of-health-information-2011/.

Gruber, T. R. (1993). A translation approach to portable ontology specifications. *Knowledge Acquisition, 5*(2), 199−220, Special issue: Current issues in knowledge modeling archive.

Helmy, T., Al-Nazer, A., Al-Bukhitan, S., & Iqbal, A. (2015). Health, food and user's profile ontologies for personalized information retrieval. *Procedia Computer Science, 52,* 1071−1076.

Horrocks, I., Kutz, O., & Sattler, U. (2006). The even more irresistible {SROIQ}. In: *Tenth international conference on principles of knowledge representation and reasoning* (pp. 57−67), Lake District of the United Kingdom, June 2−5, 2006.

Jain, N., Jhunthra, S., Garg, H., Gupta, V., Mohan, S., Ahmadian, A., Salahshour, S., & Ferrara, M. (2021). Prediction modelling of COVID using machine learning methods from B-cell dataset. *Results in Physics, 21.*

Kaggle.com. (2021). *[NeurIPS 2020] Data science for COVID-19 (DS4C).* <https://www.kaggle.com/kimjihoo/coronavirusdataset>.

Khalil, A. M., Zhang, X., Saleh, H., Younis, E. M. G., Sahal, R., & Ali, A. A. (2020). Predicting coronavirus pandemic in real-time using machine learning and Big Data streaming system. *Complexity, 2020,* 1−10.

Kollia, I., Glimm, B., & Horrocks, I. (2011) SPARQL query answering over OWL ontologies. In: *Proceedings of the 8th extended semantic web conference on the semantic web: Research and applications* (pp. 382−396), Heraklion, Crete, Greece.

Kotov, A. (2015). *Social media analytics for healthcare.* Healthcare Data Analytics.

McGagh, D., Liyanage, H., Delusignan, S., & Williams, J. (2021). *COVID-19 surveillance ontology.* Available at: https://bioportal.bioontology.org/ontologies/COVID19.

Musen, M. A. (2015). *The Protégé project: A look back and a look forward. AI matters.* Association of Computing Machinery Specific Interest Group in Artificial Intelligence. Available at: http://protégé.stanford.edu.

PatientsLikeMe. (2017). <http://www.patientslikeme.com>.

Rolls, K., Hansen, M., Jackson, D., & Elliott, D. (2016). How health care professionals use social media to create virtual communities: An integrative review. *Journal of Medical Internet Research, 18.*

Sarwat, M., Bao, J., Eldawy, A., Levandoski, J. J., Magdy, A., & Mokbel, M. F. (2012). Sindbad: A location-based social networking system. In: *Proceedings of the ACM SIGMOD international conference on management of data, SIGMOD 2012* (pp. 649–652), Scottsdale, AZ, USA, May 20–24, 2012.

Smailhodzic, E., Hooijsma, W., Boonstra, A., & Langley, D. J. (2016). Social media use in healthcare: A systematic review of effects on patients and on their relationship with healthcare professionals. *BMC Health Services Research, 16,* 442.

Stamp, M. (2018). A survey of machine learning algorithms and their application in information security. In: *Guide to vulnerability analysis for computer networks and systems – An artificial intelligence approach.*

Tn.gov. (2021). <https://www.tn.gov/health/cedep/ncov/data/downloadable-datasets.html>.

CHAPTER 14

Computationally efficient integrity verification for shared data in cloud storage

M.B. Smithamol and N. Sruthi
Department of Computer Science and Engineering, LBS College of Engineering, Kasaragod, India

14.1 Introduction

Business and organizations use cloud services as they provide measured and flexible services. However, conducting business in the cloud induces new security threats as confidential files and classified data are exposed to an environment where the user has no control over the data (Xu et al., 2013; Xu et al., 2020). To offer security, cloud service providers (CSPs) implement some security mechanisms for their platform and also monitor all data interactions with cloud storage applications and automatically encrypting sensitive data before movement (Ahmed & Litchfield, 2018; Zissis & Lekkas, 2012). Still, unauthorized users may try to access the stored data and data may get corrupted or deleted which is critical in health-care data (Subramanian & Jeyaraj, 2018; Alsmadi & Prybutok, 2018).

The data owner chooses a third-party auditor (TPA) for unbiased data auditing since the CSP is a semitrusted entity. Data auditing is a major requirement in sensitive data uploading because data has value only if it is accurate. Basically, any data auditing scheme requires that data owners have to generate some kind of signatures for the data before outsourcing to the cloud server (Worku et al., 2014; Rasheed, 2014). Generally, data auditing schemes uses message authentication code and digital signatures (Basu et al., 2018). Many approaches have been proposed like provable data possession (PDP) (Ateniese et al., 2007; Ateniese et al., 2008; Ren et al., 2015) and proof of retrievability (POR) (Shacham & Waters, 2008; Zhang et al., 2019) which supports remote data integrity verification and also to recover the original data even when some part of the data is

Edge-of-Things in Personalized Healthcare Support Systems
DOI: https://doi.org/10.1016/B978-0-323-90585-5.00013-8

315

corrupted. The batch integrity auditing seems to enhance auditing efficiency (Jin et al., 2016) with increased computation cost and few schemes like (Yuan & Yu, 2015) both read and write permission were given on shared data.

Data auditing schemes study the vulnerabilities involved in sensitive data outsourcing to evaluate the security of proposed method. The major attacks listed are replace attacks, collusion attacks, replay attacks, and forge attacks (Xu et al., 2020; El Ghoubach et al., 2019; Zou et al., 2020). In Jiang et al. (2015) a technique that involves vector commitment and verifier-local revocation group signatures was used to overcome collusion attacks. Many techniques (Liu et al., 2015; Shen et al., 2017; Li et al., 2016; Tian et al., 2015) were later proposed to support data dynamics, public integrity auditing, and less communication or computational audit cost. Yet these schemes suffer from the burden of complex certificate management. As a result, identity-based cryptography (Wang et al., 2013; Hu et al., 2018) was introduced. The identity of user, serves as the public key and key generation center (KGC), issues secret key based on the user identity, consequently there is no need of certification. Identity-based cryptography faces key escrow problem as all the secret keys are generated by it cannot be trusted and consequently another scheme called certificateless cryptography was popular. Certificateless scheme solves both the problem of certificate management and key escrow problem (Wu et al., 2020). In certificateless cryptography (Yuan & Yu, 2015), a partial key is generated by KGC based on user's identity and the private key is a combination of partial key and a secret value chosen by user, thereby gets rid off certificate management and key escrow problem. However, the tag (signature) generation algorithm has to generate more tags proportional to the number of data blocks in the input file. Therefore it demands efficient auditing methods that can perform integrity checking regardless of the file size. The proposed scheme CL-IVS (Certificate-Less Integrity Verification for Shared data) enables the user to perform efficient remote data integrity by reducing the tag generation overhead per block. Major contributions in this work are as follows:

1. The chapter presents a novel remote data auditing scheme CL-IVS which implements certificateless cryptography with a new tag generation scheme. This scheme enables user to verify data integrity regardless of the file size.

2. The proposed certificateless public auditing scheme relieves the user from tag computation overhead and also allows auditing of multiple blocks.

3. The security and performance analysis proves the strength of the proposed remote data auditing scheme and its practical implementation in real-life application.

The rest of this paper is organized as follows: Section 14.2 briefs the important literature. Section 14.3 gives the preliminaries and the proposed system model. Section 14.4 gives in-depth security analysis and Section 14.5 provides the performance analysis. Finally, we conclude the chapter in Section 14.6.

14.2 Related works

Over the years, issues of data auditing in cloud have been examined in many POR and PDP schemes. Jin et al. (2016) explain the concept of dynamic PDP scheme which allows users to modify the stored data. Also, an index switcher was used in tag computation to separate block index and tag index. To achieve multiple writes to data, Yuan and Yu (2015) used polynomial-based authentication tags as aggregation of authentication tags is possible. The authors used a proxy-based authentication tag update technique which allowed secure delegation of user revocation. However, it does not give importance to the data secrecy of group users. Jiang et al. (2015) detect the collusion attack in the existing scheme. The technique involves vector commitment and verifier-local revocation group signatures which will prevent the collusion of cloud and revoked users by giving permission to data owners take part in the user revocation. Liu et al. (2015) presented a public auditing scheme using proxy-based code regeneration in cloud storage. The proposed scheme suffers from storage overhead and has high bandwidth cost in the repair phase. Shen et al. (2017) introduced a method that supports data dynamics and handling potential disputes. The scheme used a doubly linked information table which was stored with the TPA as proof to solve disputes. This scheme also provides integrity proof of constant size and predetermined verification operations. However, the lower block size is not considered because it increases the cost resulting in a large number of exponentiations. A privacy-preserving scheme (Li et al., 2016) has been proposed for low-performance end devices in cloud. This method generates online/offline tags and stores them with the file in the CSP. Signatures are used to validate these tags. TPA maintains a Merkle Hash Tree which stores the hash values of the online tags making auditing faster but has more storage overhead. The protocol proposed in Tian et al. (2015) uses Dynamic Hash Table for public auditing with reduced computational cost. Batch auditing is made possible by using aggregate signatures. Although the method supports these features it requires more bandwidth. Also, these schemes have some advantages, they all suffer from the overhead of complex certificate management.

Wang et al. (2013) use identity-based remote data integrity checking protocol which uses user identity as the public key. This scheme also uses aggregately signatures supporting multiple updations. The following methods also support identity-based data auditing. Wang et al. (2016) make use of data outsourcing with comprehensive auditing using user identity that permits users to assign dedicated proxies with the job of data upload to the cloud storage server which are verified by the TPA. However, the storage cost is more compared to existing schemes. In Wang et al. (2018), the authors proposed a data auditing scheme that supports global and sampling block-less verification. This approach also provides batch auditing and provides nonrepudiation using index logic table. Even though computational overhead is removed, the storage overhead at the client-side remains the same. In Hu et al.'s scheme (Hu et al., 2018) homomorphic authenticators generated by group signatures take care of multiple tag updates and batch auditing. Shen et al. (2018) provide a method that protects the sensitive information and only other information is published. This scheme mainly focuses on electronic health records. Sanitizer converts hidden data blocks to a common acceptable format to ease further computation process and then uploads the file protecting the sensitive information from illegal access. In the above mentioned schemes certificate management is eliminated with the use of identity-based cryptography (IDC), however, these schemes have poor signature verification process. Also, IDC has key escrow problem because KGC is creating secret keys of all users and cannot be fully trusted.

To overcome key escrow problem a new scheme called certificateless cryptography was proposed in Li et al. (2005), which provides secure user revocation and batch updates by using certificateless signature and proxy signature schemes. However, the scheme is not secure against replace attacks. In Liu et al. (2007), the authors use certificateless signature scheme that was self-generated but has not mentioned how to secure the data. In Huang et al. (2007) the authors proved the security of their scheme against type-I and type-II adversaries. However, the scheme produces signatures with a larger length. In Wang et al. (2013), the authors have given a modified scheme of Huang et al. (2007) which requires more tag updates; another remote data integrity protocol in Li et al. (2018) provides efficient tag updates and user revocation in certificateless cryptography. Also, certificateless PDP scheme was used for broad area networks in He et al. (2017) and He et al. (2015), but the security and performance analyses given in He et al. (2017) show that the proposed scheme could improve security at the cost of increased computation and communication costs.

As the certificateless scheme shows better performance compared to that of PKI (public key infrastructure), the proposed data integrity method uses certificateless approach. The major goal of proposed work is to reduce significantly the tag generation cost and storage overhead at the client side.

14.3 Proposed work CL-IVS

The proposed method CL-IVS follows the same steps for setup and key generation as given in He et al. (2015). Let ID_U be the identity of the user. Suppose that the input file F is split into m blocks, represented as $F = \{f_1, f_2, \ldots, f_m\}$. The detailed construction for the scheme is demonstrated as shown in Fig. 14.1:

Setup: Given the input bilinear map e, the setup algorithm generates the master key and the public parameters.

Algorithm 1 SETUP
  Output: Public parameters $P = \left(q, G_1, G_2, g, e, h, \overline{h}, H, P_{pub}\right)$
  Initialization
1. KGC chooses a master private key randomly $l, l \in Z_q$ and stores this key secretly
2. Chooses additive cyclic group G_1, multiplicative group G_2 of the same prime order q
3. KGC chooses the bilinear map: $e: G_1 \times G_1 \to G_2$ and three secure hash functions $h, \overline{h}, H:\{0, 1\}^* \to G_1$
4. Setup computation
5. Computes the public key $P_{pub} = l.g$ where g is the generator of G_1
6. KGC computes the final parameters, $params = \{q, G_1, G_2, g, e, h, \overline{h}, H, P_{pub}\}$

PartialPrivateKey: A new user has to get the partial key from KGC by sending request message to KGC. Upon receiving the request KGC generates user partial private key as follows:

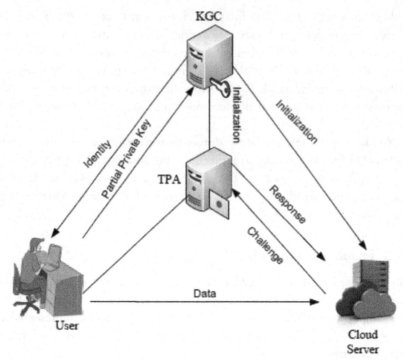

Figure 14.1 CL-IVS system model.

Algorithm 2: PartialPRivateKeyGeneration
   Output: Partial private key, $PPK_{Ui} = (SK_{1i}, SK_{2i})$
   Initialization
1. Let U_i be the ith user with identity ID_{Ui} randomly selects the secret key $x_i \in Z_q$ and computes the partial public key $P_i = x_i.g$
2. Then, the user registers with KGC by providing user credentials, user identity and P_i
   Partial private key computation
3. KGC chooses a random number $k_i \in Z_q$
4. Computes the first component of partial key, $SK_{1i} = k_i.g$
5. Computes the second component of partial key, $SK_{2i} = (k_i + h(ID_{Ui}, SK_{1i}).l) mod q$
6. Returns the partial private key, $PPK_{Ui} = (SK_{1i}, SK_{2i})$

User Key Generation: During user key generation phase both user private key and public key are computed.

Algorithm 3: UserKeyGeneration
   Output: User key pair (SK_U, PK_U)
   1. Verifies the partial key by checking whether $SK_{2i}.g = SK_{1i} + h(ID_{Ui}, SK_{1i}).P_{pub}$ holds true.
   2. User key generation phase
   3. Computes user private key, $SK_U = (SK_{2i}, x_i)$
   4. Computes user public key, $PK_U = (SK_{1i}, P_i)$.

Proposed tag generation algorithm: The file F is split into n blocks, represented as $F = \{f_1, f_2, \ldots, f_n\}$. Using the symmetric key, each block is encrypted $E = \{e_1, e_2, \ldots, e_n\}$. These n encrypted blocks are then grouped as pairs $D = \{(e_1, e_2), (e_3, e_4), \ldots, (e_{n-1}, e_n)\}$ or $D = \{t_1, t_2, \ldots, t_d\}$, where $t_k = (e_i, e_{i+1}), i \in \{1, 2, \ldots, n\}$ and $1 \le k \le d, d = n/2$. User U_1 with identity ID_{Ui}, private key $SK_{Ui} = (SK_{2i}, x_{Ui})$ and public key $PK_{Ui} = (SK_{1i}, P_{Ui})$, runs this algorithm to generate tag which is used to check the integrity of each pair of blocks.

Algorithm 4: TagGeneration
   Output: tag_id for each pair (e_i, e_{i+1})
   Initialization
   1. Let the block be t_k where $1 \le k \le d, d = n/2$.
      Tag Generation Phase
   2. $ID_{tk} = id_{ei}||id_{e(i+1)}$
   3. Computes the identity for each block as $f_{ID} = \bar{h}(ID_{Ui}, PK_{Ui}, P_{pub})$ and $\sigma = H(P_{pub})$
   4. Computes the tag for each data block as $\theta_k = (SK_{2i} + f_{ID}.x_{Ui})||H(ID\_t_k) + t_k.\sigma)$
   5. Uploads the file and tag (t_k, θ_k, ID_{tk})

As the tag generation is reduced by half of the number of block in F, CL-IVS is efficient compared to that of He et al. (2017).
   Challenge: Data integrity verification is very much essential for the data owner as the CSP is a semitrusted entity. Upon receiving audit request from the data owner, TPA generates the set $\varphi = \{m_1, m_2, \ldots, m_c\}$ randomly by choosing the index of the block from $\omega = \{1, 2, \ldots, d\}$. In this way a finite number, c number of blocks are added in φ.
   1. For every element $m_i \in \varphi$ generate a random number $w_j \in Z^*_q$.
   2. Send audit challenge, $Chall = \{(m_j, w_j)\}j \in \varphi$ to the CSP. Here m_j indicates the index of the block chosen for verification.

Proof Generation: After receiving the challenge message, *chall*, the CSP generates proof as follows:

1. Computes c element set $J = \sum_{j=1}^{c} w_j \cdot \theta_{ij}$ and $I = \sum_{j=1}^{c} w_j \cdot t_k mod\ q$
2. Send $\{J, I\}$ to TPA as proof.

Proof Verification: The TPA runs this algorithm to check the integrity of data by verifying the proof $\{J, I\}$ send by cloud server.

1. TPA computes $f_{ID} = \bar{h}\left(ID_{Ui}, PK_{Ui}, P_{pub}\right)$ and $\sigma = H(\frac{r}{t}P_{pub})$
2. TPA checks whether equation $e(J, P_i) = e(I \cdot \sigma + \sum_{j=1} w_j \cdot H(f_{ID}) + h \cdot P_{pub})$ holds.

14.4 Security analysis of CL-IVS

In this section we present the security analysis of the proposed scheme, CL-IVS. To perform the security analysis, CL-IVS has adopted the security model provided in He et al. (2015) to show that the proposed scheme is also secure against both type-I and type-II adversaries. Here the three hash functions h, $\bar{h}$, and H are considered as three random oracles.

Lemma 1: If the Computational Diffie-Hellman (CDH) assumption holds true then the proposed scheme CL-IVS is secure against the Type-I Adversary in the random oracle model.

Proof: Type-I adversary A_1 does not have access to master key. Assume that the type-I adversary A_1 could guess the master key with a probability ϵ and $\mathbb{C}$ is the challenger to compute the hardness of CDH with a substantial probability value ϵ. Given a CDH problem instance consisting of $(P, Q_1 = x \cdot P, Q_2 = y \cdot P)$, $\mathbb{C}$ sets $P_{pub} \leftarrow Q_1$ and returns the system parameters params $= \{q, G_1, G_2, P, e, h, H, P_{pub}\}$ to A_1. Based on the security game simulated as, $\mathbb{C}$ picks a user identity randomly as the identity for challenge, ID_U and needs to consider eight different types of the queries given below.

1. *Query,h*: While processing the query with key $\{ID_U, T_U\}$, $\mathbb{C}$ verifies whether a tuple $\{ID_U, T_U, h_U\}$ is valid and satisfies all constraints. If so, $\mathbb{C}$ returns h_u to A_1; otherwise, $\mathbb{C}$ generates a random number $\in \mathbb{Z}_q^*$, stores $\{ID_U, T_U, h_U\}$ in the associated list L_h, and returns h_u to A_1.
2. *Query,h*: While processing the query with the key $\{ID_U, T_U\}$, $\mathbb{C}$ verifies whether a tuple $\{ID_U, pk_U, P_{pub}, k_U\}$, is included in $L_{\bar{h}}$. If so, $\mathbb{C}$

returns k_U to A_1; otherwise, $\mathbb{C}$ generates a random number $k_U \in \mathbb{Z}_q^*$, stores $\{ID_U, pk_U, P_{pub}, k_U\}$ in $L_{\bar{h}}$, and returns k_U to A_1.

3. *Query,H:* While processing a query with message id_r, $\mathbb{C}$ verifies the membership of the tuple $\{id_r, c_r, z_r, Z_r\}$, in L_H. If present then $\mathbb{C}$ returns $H(id_r) = Z_r - t_r \cdot Q$ to A_1; else, $\mathbb{C}$ picks $b_r \in \{0, 1\}$ a random bit satisfying $Pr[b_r = 0]$ which is λ, where $\lambda \in \{0, 1\}$. If $b_r = 0$, $\mathbb{C}$ computes $Z_r = z_r \cdot P$; else $\mathbb{C}$computes $Z_r = z_r \cdot Q_2$, where $z_r \in \mathbb{Z}_q^*$ is the random number chosen by $\mathbb{C}$. If $b_r = 0$ Finally, $\mathbb{C}$ stores $\{id_r, c_r, z_r, Z_r\}$ in L_H and returns $H(id_r) = Z_r - t_r \cdot Q$ to A_1.

4. *Create-User:* $\mathbb{C}$ receives a query with key ID_r, and verifies the membership of the tuple $\{ID_U, s_U, T_U, x_U, P_U\}$, in L_K. If so, $\mathbb{C}$ returns $pk_U = \{T_U, P_U\}$ to A_1. If $ID_U = ID_O$, $\mathbb{C}$ generates two random numbers $t_U, x_U \in \mathbb{Z}_q^*$, sets $s_U \leftarrow \perp$, computes $T_U = t_U \cdot P$ and $P_U = x_U \cdot P$, and stores $\{ID_U, s_U, T_U, x_U, P_U\}$ in L_K; otherwise $(ID_U \neq ID_O)$, $\mathbb{C}$ generates three random numbers $s_U, w_U, x_U \in \mathbb{Z}_q^*$, computes $T_U = s_U \cdot P - w_U \cdot P_{pub}$ and $P_U = x_U \cdot P$, and stores $\{ID_U, s_U, T_U, x_U, P_U\}$ and $\{m_U, w_U\}$ in L_K and L_h, respectively, where $m = \{ID_U, T_U\}$. Finally, $\mathbb{C}$ returns $pk_U = \{T_U, P_U\}$ to A_1.

5. *Partial-Private-Key:* At the time of processing the query with key message ID_U, the challenger $\mathbb{C}$ verifies identity constraint that if ID_U and ID_O are equal. If it satisfies $ID_U = ID_O$, then the challenger $\mathbb{C}$ aborts the game; or else, $\mathbb{C}$ looks up L_K for tuple $\{ID_U, s_U, T_U, x_U, P_U\}$ and returns $\{s_U, T_U\}$ to A_1.

6. *Public-Key-Replacement:* During the time of executing a user query with key ID_U and $pk_U' = \{T_U', P_U'\}$(chosen by A_1), $\mathbb{C}$ looks up L_K for tuple $\{ID_U, s_U, T_U, x_U, P_U\}$ and replaces $pk_U = \{T_U, P_U\}$ with pk_U'.

7. *User-Secret-Value:* While processing a query with identity ID_U, $\mathbb{C}$ searches the list L_K for tuple $\{ID_U, s_U, T_U, x_U, P_U\}$ and returns x_U to A_1.

8. *Tag-Generation:* A query with a user's identity ID_U is processed and data t_r with identity id_r, $\mathbb{C}$makes a query H with id_r and gets tuple $\{id_r, c_r, z_r, Z_r\}$. If $c_r = 0$, $\mathbb{C}$ computes $S_r = z_r \cdot (T_U + h_U \cdot P_{pub} + k_U \cdot P)$ and returns $\{t_r, id_r, S_r\}$ to A_1;otherwise $(c_r = 1)$, $\mathbb{C}$ aborts the game.

Finally, A_1 outputs a proof $\{m^*, S^*\}$ of a subset $J^* = \{k_1^*, k_2^*, \ldots, k_{c*}^*\}$ corresponding to the owner's identity ID_O^* as his/her forgery. If $ID_O^* \neq ID_O$ or $c_{r_j^*} = 0 (j = 1, 2, \ldots, c^*)$, $\mathbb{C}$ rejects the game; if not, $\mathbb{C}$ initiates search operation in the tables L_K and L_H for tuples $\{ID_O, s_O, T_O, x_O, P_O\}$ and

$\left\{ id_{r_j^*}, c_{r_j^*}, z_{r_j^*}, Z_{r_j^*} \right\}$, respectively, where $j = 1, 2, \ldots, c^*$. $\mathbb{C}$ could get

$$
\begin{aligned}
e(J*, P) &= e\left(m* \cdot Q + \sum_{j=1}^{c*} w_j \cdot H\left(id_{r_j^*} \right), T_O + h_O \cdot P_{pub} + k_O \cdot P_O \right) \\
&= e\left(\sum_{j=1}^{c*} w_j \cdot Z_{r_j^*}, T_O + h_O \cdot P_{pub} + k_O \cdot P_O \right) \quad (14.2) \\
&= e\left(\sum_{j=1}^{c*} w_j \cdot Z_{r_j^*}, T_O + k_O \cdot P_O \right) \times e\left(\sum_{j=1}^{c*} w_j \cdot Z_{r_j^*}, h_O \cdot P_{pub} \right)
\end{aligned}
$$

Based on oracle model definition on hash function h given in Pointcheval and Stern (2000), $\mathbb{C}$ is able to produce another proof $\left\{ m^*, S^* \right\}$ satisfying

$$
e(S^{*\prime}, P) = e\left(\sum_{j=1}^{c^*} w_j \cdot Z_{r_j^*}, T_O + k_O \cdot P_O \right) \times e\left(\sum_{j=1}^{c^*} w_j \cdot Z_{r_j^*}, h_O' \cdot P_{pub} \right) \quad (14.3)
$$

From Eqs. (14.2) and (14.3) the challenger $\mathbb{C}$ is able to derive

$$
\begin{aligned}
e\left(J* - J^{*\prime}, P \right) &= e\left(\sum_{j=1}^{c*} w_j \cdot Z_{r_j^*}, (h_O - h_O') \cdot P_{pub} \right) \\
&= e\left(\left(\sum_{r_j^* \in J*, c_{r_j^*}=1} w_j \cdot Z_{r_j^*} \right) \cdot (h_O - h_O') \cdot Q_1, P \right) \times e\left((h_O - {}'_O) \left(\sum_{r_j^* \in J*, c_{r_j^*}=1} w_j \cdot Z_{r_j^*} \right) \cdot x \cdot y \cdot P, P \right) \quad (14.4) \\
&= e\left(S* - S^{*\prime} - \left(\sum_{r_j^* \in J*, c_{r_j^*}=1} w_j \cdot Z_{r_j^*} \right) \cdot (h_O - h_O') \cdot Q_1, P \right) \\
&= e\left((h_O - h_O') \left(\sum_{r_j^* \in J*, c_{r_j^*}=1} w_j \cdot Z_{r_j^*} \right) \cdot x \cdot y \cdot P, P \right)
\end{aligned}
$$

Now, for the CDH problem, $\mathbb{C}$ outputs $\Psi^{-1} \cdot \Phi$ where Ψ^{-1} is the inversion of Ψ such that $\Psi^{-1} \cdot \Psi \equiv 1 mod q$, Ψ, and $\Psi = \left(S^* - S^{*\prime} - \left(\sum_{r_j^* \in J*, c_{r_j^*}=1} w_j \cdot Z_{r_j^*} \right) \right) \cdot (h_O - h_O') \cdot Q_1$.

Analysis: Here, the probability that $\mathbb{C}$ could solve the given CDH problem by using Å_1 is computed. As per the simulated game, mainly the following three events are considered in connection with the success of the challenger.

1. E_1: *The* query *Partial-Private-Key* is processed by $\mathbb{C}$.
2. E_2: Å_1 generates valid proof of a subset $J^* = \left\{ k_1^*, k_2^*, \ldots, k_{c*}^* \right\}$ with at least one $c_{r_j^*} = 1 (j = 1, 2, \ldots, c^*)$.
3. E_3: After event E_2 happens, we have $ID_O^* = ID_O$.

From the process of the game, we could get $Pr[E_1] \geq \left(1 - (1/q_h) \right)^{q_{ppk}}$, $Pr[E_2 | E_1] \geq \left(1 - \lambda^* \right) \epsilon$, and $Pr[E_3 | E_1 \wedge E_2] \geq (1/q_h)$, with respect to the number of h queries q_h and q_{ppk}, the number of Partial-Private-key queries. Now we can derive the probability that C solves the CDH problem as in the following equation:

$$\Pr[E_1 \wedge E_2 \wedge E_3] = \Pr[E_1]\Pr[E_2|E_1]\Pr[E_3|E_1 \wedge E_2] \quad \geq 1/(q_h(1 - \lambda^{c*})(1 - 1/q_h)^{q_{ppk}}.\epsilon)$$

(14.5)

From the above statements it has been concluded that $\mathbb{C}$ could solve the CDH problem with a nonnegligible probability ϵ. This is a contradiction because the hardness of CDH has been theoretically proved. Hence, security CL-IVS is proved in the random oracle model against Type-I Adversary.

Lemma 2: If the CDH assumption holds true then the proposed scheme CL-IVS is secure against the Type-II Adversary security game within the random oracle model.

Proof: Let us the consider the case of Type-II Adversary $\mathbb{A}_2$. In this case the algorithm which simulates the game assumes that the probability of $\mathbb{A}_2$ could win the game is ϵ. Holding these assumptions true, we could construct a challenge $\mathbb{C}$ as in Lemma 1, to solve the CDH problem. With a problem state $(P, Q_1 = x \cdot P, Q_2 = y \cdot P)$ of the CDH problem, $\mathbb{C}$ finds public key $P_{pub} = s \cdot P$, and returns system parameters params $= \{q, G_1, G_2, g, e, h, H, P_{pub}\}$ and the master key to $\mathbb{A}_2$ where $s \in \mathbb{Z}_q^*$ as the chosen random number and is the master key. Given identity ID_O as a challenge identity $\mathbb{C}$ simulates the game as follows:

1. *Create-User*: While processing a query with message ID_r, $\mathbb{C}$ verifies the inclusion of the respective tuple $\{ID_U, s_U, T_U, x_U, P_U\}$, in L_K. If satisfied then the challenger $\mathbb{C}$ returns $pk_U = \{T_U, P_U\}$ to $\mathbb{A}_2$. If $ID_U = ID_O$, $\mathbb{C}$ generates a random numbers $t_U \in \mathbb{Z}_q^*$, computes $T_U = t_U \cdot P$, $h_U = h(ID_U, T_U)$ and $s_U = t_U + s \cdot h_U \bmod q$, and sets $P_U = Q_1$; otherwise $(ID_U \neq ID_O)$, $\mathbb{C}$ generates two random numbers $t_U, x_U \in \mathbb{Z}_q^*$ and computes $T_U = t_U \cdot P$, $h_U = h(ID_U, T_U)$, $s_U = t_U + s \cdot h_U \bmod q$ and $P_U = x_U \cdot P$. $\mathbb{C}$ stores $\{ID_U, s_U, T_U, x_U, P_U\}$ in L_K returns $pk_U = \{T_U, P_U\}$ to $\mathbb{A}_2$.

2. *Partial-Private-Key*: Consider a query with message key ID_U, then the challenger $\mathbb{C}$ searches the associated list L_K for the answer $\{ID_U, s_U, T_U, x_U, P_U\}$ and returns $\{s_U, T_U\}$ to $\mathbb{A}_2$.

3. *User-Secret-Value:* In the case of processing a query with message key ID_U, $\mathbb{C}$ verifies the Boolean expression that whether ID_U and ID_O are equal. If *truethatis* $ID_U = ID_O$, $\mathbb{C}$ terminates the game; otherwise, $\mathbb{C}$ begin searching for the respective tuple $\{ID_U, s_U, T_U, x_U, P_U\}$ in the list L_K and returns x_U to $\mathbb{A}_2$.

4. In the end, $\mathbb{A}_2$ evaluates and gives a proof $\{m^*, S^*\}$ of a subset $J^* = \left\{k_1^*, k_2^*, \ldots, k_{c^*}^*\right\}$ for the owner's identity ID_O^* as his/her forgery. If $ID_O^* \neq ID_O$ or $c_{j^*} = 0 (j = 1, 2, \ldots, c^*)$, $\mathbb{C}$ terminates the game; otherwise, $\mathbb{C}$ starts searching in the lists L_K and L_H for tuples $\{ID_O, s_O, T_O, x_O, P_O\}$ and $\left\{id_{r_j^*}, c_{r_j^*}, z_{r_j^*}, Z_{r_j^*}\right\}$, respectively, where $j = 1, 2, \ldots, c^*$. Eventually, $\mathbb{C}$ could deduce,

$$
\begin{aligned}
e(J*, Pi) &= e\left(m* \cdot Q + \sum_{j=1}^{c*} w_j \cdot H\left(id_{r_j^*}\right), T_O + h_O \cdot P_{pub} + k_O \cdot P_O\right) \\
&= e\left(\sum_{j=1}^{c*} w_j \cdot Z_{r_j^*}, T_O + h_O \cdot P_{pub} + k_O \cdot P_O\right) \\
&= e\left(\sum_{j=1}^{c*} w_j \cdot Z_{r_j^*}, T_O + k_O \cdot P_O\right) \times e\left(\sum_{j=1}^{c*} w_j \cdot Z_{r_j^*}, h_O \cdot P_{pub}\right).
\end{aligned} \tag{14.6}
$$

Now the solution to the CDH problem is $\Psi^{-1} \cdot \Phi$, such that $\Psi^{-1} \cdot \Psi \equiv 1 \bmod q$, $\Psi = \left((k_O - k_O')\left(\sum_{r_j^* \in J^*, c_{r_j^*}=1} w_j \cdot Z_{r_j^*}\right)\right)$, and $\Psi = \left(S^* - S^{*'} - \left(\sum_{r_j^* \in J^*, c_{r_j^*}=1} w_j \cdot Z_{r_j^*}\right) \cdot (k_O - k_O') \cdot Q_1\right)$

5. *Analysis:* Here, we compute the probability of solving CDH problem using $\mathbb{A}_2$ as its main subfunction. Consequently, we should examine the events E_1, E_2, and E_3 associated with the success of $\mathbb{C}$.

1. E_1: The Challenger continues the processing of the query, *User-Secret-Value* query.
2. E_2: $\mathbb{A}_2$ computes and gives the valid proof of a subset $J^* = \left\{k_1^*, k_2^*, \ldots, k_{c^*}^*\right\}$ with at least one $c_{j^*} = 1 (j = 1, 2, \ldots, c^*)$.
3. E_3: After the completion of E_2 we have $ID_O^* = ID_O$.

Following the conclusions of the above game simulation algorithm, eventually we could get $Pr[E_1] \geq \left(1 - (1/q_h)\right)^{q_{usv}}$, $Pr[E_2|E_1] \geq \left(1 - \lambda^{c^*}\right)\epsilon$, and $Pr[E_3|E_1 \wedge E_2] \geq (1/q_h)$, with respect to the cardinality of the h set, q_h and secret-value query set, q_{usv}. The probability of challenger to find the answer to CDH can be given as:

$$
Pr[E_1 \wedge E_2 \wedge E_3] = Pr[E_1]Pr[E_2|E_1]Pr[E_3|E_1 \wedge E_2] \geq 1/(q_h(1 - \lambda^{c^*})(1-1/q_h)^{q_{usv}}.\epsilon) \tag{14.7}
$$

From the above equation, we conclude that with a probability ϵ, $\mathbb{C}$ could find an answer to the CDH problem. This is indeed a contradiction to the hardness of the CDH which has been already proved. Hence, the proposed scheme is secure against the *type-II* adversary in the random oracle model.

Theorem 1: Holding hardness of CDH in the random oracle security model the proposed scheme CL-IVS is secure against any type of polynomial time adversary algorithms.

Proof with Lemma 1 and Lemma 2 we arrive at the above statement.

14.5 Performance analysis

The efficiency of the proposed scheme is evaluated based on the following parameters:

1. Computation cost
2. Communication cost

The proposed algorithm CL-IVS is implemented using JPBC and derived the running time of related operations. In this section, we compare the proposed scheme CL-IVS with He et al. (2015) for performance analysis. Also, we compare the computation time of our scheme CL-IVS with (Wang et al., 2013) computation mentioned in scheme (He et al., 2015).

1. Let T_p be the computational complexity involved in pairing operation, T_{pm} be the complexity of point multiplication in an elliptic curve, T_{pa} be the running time of an elliptic-curve point addition operation, T_H running time of a hash-to-point operation, and T_h running time of a general hash function. The experimental results on the platform (a personal computer with an Intel Core i3−6006U 2.00 GHz CPU with 4GB RAM and running Windows 10 operating system) show that $T_p, T_{pm}, T_{pa}, \ T_H$ and T_h are 0.078s,0.893s,0.023s,0.155s and 0.032s respectively. Let c be cardinality of the audit challenge message set. The computation costs for proof generation and proof verification of existing He et al.'s scheme (Jiang et al., 2015) and proposed scheme are analyzed and shown in Fig. 14.2.

2. In He et al. (2015) the auditing challenge is denoted as $\left\{ \left(i_j, w_j \right) \right\}_{j \in J}$ with c number of elements. The auditing scheme requires c number of elliptic curve point multiplications in pairing operations and $(c - 1)$ elliptic curve point addition operation in paring for proof generation.

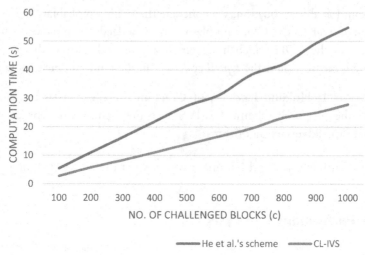

Figure 14.2 Proof generation analysis.

Hence, the time complexity in terms of the pairing operations for the scheme (He et al., 2015) is $c \times T_{pm} + (c - 1) \times T_{pa}$. In CL-IVS auditing challenge is denoted as $\left\{ \left(k_j, w_j \right) \right\}_{j \in J}$ where J as $c/2$ elements compared with existing scheme (He et al., 2015). Consequently, CL-IVS has $c/2$ point multiplications and $(c/2 - 1)$ point additions. Therefore the total running time of CL-IVS is $c/2 \times T_{pm} + (c/2 - 1) \times T_{pa}$. Fig. 14.2 depicts comparison of the total running time taken for proof generation in He et al. (2015) and proposed scheme CL-IVS.

In the proof verification phase of algorithm used in He et al. (2015), TPA must execute two pairing operations, $(c + 2)$ elliptic curve multiplications alongside $(c + 1)$ hash-to-point operations and two general hash operations so as to verify the proof. The total running time of He et al.'s scheme (He et al., 2015) is $2 \times T_p + (c + 3) \times T_{pm} + (c + 2) \times T_{pa} + (c + 1) \times T_H + 2 \times T_h$. However, CL-IVS significantly reduces curve point multiplications and curve point additions. That is, $\left(c/(2 + 3) \right)$ point multiplications, $(c/2 + 2)$ point additions with $(c/2 + 1)$ and with point operations mapping into hash functions used giving the total running time as $2 \times T_p + (c/2 + 3) \times T_{pm} + (c/2 + 2) \times T_{pa} + (c/2 + 1) \times T_H + 2 \times T_h$. Fig. 14.3 depicts the entire time complexity analysis for proof generation by the prevailing scheme and proposed scheme. The total running time combining proof generation and proof verification of He et al.'s scheme (He et al., 2015), Wang et al.'s scheme (Wang et al., 2018) and proposed scheme is analyzed in Table 14.1.

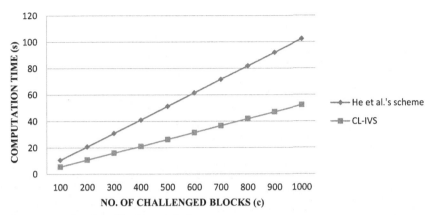

Figure 14.3 Proof verification analysis.

CL-IVS reduces the time complexity of proof generation algorithm since the number of elements in the auditing challenge set is half as compared to that of existing algorithms (He et al., 2015; Wang et al., 2018). The complexity is mainly computed by quantifying the bilinear pairing operations, point multiplications, point additions, and hash-to-point operations. The reduction is mainly in point multiplications and point addition with hash to point operations by half because CL-IVS computes tags for two blocks at a time.

14.5.1 Theoretical analysis

The comparison of running time is shown in Table 14.1. Based on the comparisons we deduce that the proposed scheme CL-IVS has lower computation time. Fig. 14.4 depicts the total running time taken for proof generation and proof verification by the existing scheme (He et al., 2015) and proposed scheme CL-IVS.

The analysis shows that CL-IVS requires fewer computation costs for proof generation and proof verification than He et al.'s scheme (He et al., 2015). The proposed scheme not only requires low computation cost but also the number of blocks being verified is more compared to the existing scheme. Fig. 14.4 shows that for the same number of challenged blocks the number of blocks getting verified is more for CL-IVS than He et al.'s scheme (He et al., 2015). This shows the efficiency of the proposed scheme CL-IVS.

Table 14.1 Comparison of computation cost.

Scheme	ProofGen	ProofVerify	Total
Wang et al.'s scheme	$c \times T_{pm} + (c-1) \times T_{pa}$	$3 \times T_p + (2 \times (c+1)) \times T_{pm} + (2 \times c + 1) \times T_{pa} + c \times T_H$	$3 \times T_p + (3 \times (c+1)) \times T_{pm} + (3 \times c + 1) \times T_{pa} + c \times T_H$
He et al.'s scheme	$c \times T_{pm} + (c-1) \times T_{pa}$	$2 \times T_p + (c+3) \times T_{pm} + (c+2) \times T_{pa} + (c+1) \times T_H + 2 \times T_h$	$2 \times T_p + (2 \times c + 3) \times T_{pm} + (2 \times c + 1) \times T_{pa} + (c+1) \times T_H + 2 \times T_h$
CL–IVS	$\frac{c}{2} \times T_{pm} + (\frac{c}{2} - 1) \times T_{pa}$	$2 \times T_p + (\frac{c}{2} + 3) \times T_{pm} + (\frac{c}{2} + 2) \times T_{pa} + (\frac{c}{2} + 1) \times T_H + 2 \times T_h$	$2 \times T_p + (2 \times \frac{c}{2} + 3) \times T_{pm} + (2 \times \frac{c}{2} + 1) \times T_{pa} + (\frac{c}{2} + 1) \times T_H + 2 \times T_h$

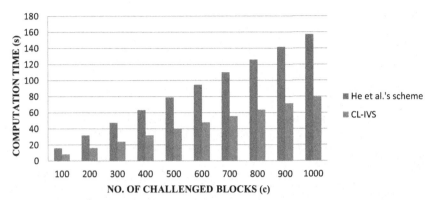

Figure 14.4 Overall computation analysis.

3. Communication cost: To achieve integrity checking the TPA sends audit challenge message and in response the cloud server sends proof message. In He et al. (2015), the auditing challenge is $\left\{ \left(i_j, w_j \right) \right\}_{j \in J}$ where $i_j \in J, w_j \in Z_q^*, J = \{i_1, i_2, \ldots, i_c\}$ is a subset of $\Omega = \{1, 2, \ldots, n\}$ and proof $\{m, s\}$ where $S = \sum_{j=1}^{c} w_j \cdot S_{i_j}$ and $m = \sum_{j=1}^{c} w_j \cdot m_{i_j} mod\ q$. In the proposed scheme CL-IVS the auditing challenge is $\left\{ \left(k_j, w_j \right) \right\}_{j \in J}$ where $r_j \in J, w_j \in Z_q^*, J = \{k_1, k_2, \ldots, k_c\}$ is a subset of $\Omega = \left\{ 1, 2, \ldots, n/2 \right\}$ and proof $\{m, s\}$ where $S = \sum_{j=1}^{c} w_j \cdot S_{r_j}$ and $m = \sum_{j=1}^{c} w_j \cdot t_{r_j} mod\ q$ When the existing scheme chooses c elements from n elements, the proposed scheme CL-IVS chooses c elements from $\frac{n}{2}$ elements, which significantly reduces the communication cost of CL-IVS compared to that of He et al.'s scheme (He et al., 2015).

14.6 Conclusion and future works

In this paper, we proposed a certificateless s remote data integrity scheme, CL-IVS for secure data outsourcing. However, previous auditing schemes suffer from a high computation cost problem with an increased number of file blocks. The proposed work CL-IVS efficiently verifies remote data integrity with lower computational and communication cost compared to that of existing works. The in-depth security framework and performance analysis illustrate the efficiency of the proposed scheme CL-IVS.

In future, CL-IVS could be improved by enabling sensitive data sharing like health records by modifying the tag generation algorithm.

References

Ahmed, M., & Litchfield, A. T. (2018). Taxonomy for identification of security issues in cloud computing environments. *Journal of Computer Information Systems, 58*(1), 79−88.
Alsmadi, D., & Prybutok, V. (2018). Sharing and storage behavior via cloud computing: Security and privacy in research and practice. *Computers in Human Behavior, 85,* 218−226.
Ateniese, G., Burns, R., Curtmola, R., Herring, J., Kissner, L., Peterson, Z., & Song, D. (2007, October). Provable data possession at untrusted stores. In: *Proceedings of the 14th ACM conference on computer and communications security* (pp. 598−609).
Ateniese, G., Di Pietro, R., Mancini, L. V., & Tsudik, G. (2008, September). Scalable and efficient provable data possession. In: *Proceedings of the 4th international conference on Security and privacy in communication networks* (pp. 1−10).
Basu, S., Bardhan, A., Gupta, K., Saha, P., Pal, M., Bose, M., & Sarkar, P. (2018, January). Cloud computing security challenges & solutions—A survey. In: *2018 IEEE 8th annual computing and communication workshop and conference (CCWC)* (pp. 347−356). IEEE.
El Ghoubach, I., Abbou, R. B., & Mrabti, F. (2019). A secure and efficient remote data auditing scheme for cloud storage. *Journal of King Saud University-Computer and Information Sciences.*
He, D., Kumar, N., Zeadally, S., & Wang, H. (2017). Certificateless provable data possession scheme for cloud-based smart grid data management systems. *IEEE Transactions on Industrial Informatics, 14*(3), 1232−1241.
He, D., Zeadally, S., & Wu, L. (2015). Certificateless public auditing scheme for cloud-assisted wireless body area networks. *IEEE Systems Journal, 12*(1), 64−73.
Hu, A., Jiang, R., & Bhargava, B. (2018). Identity-preserving public integrity checking with dynamic groups for cloud storage. *IEEE Transactions on Services Computing, 14.*
Huang, X., Mu, Y., Susilo, W., Wong, D. S., & Wu, W. (2007). *Certificateless signature revisited. Australasian conference on information security and privacy* (pp. 308−322). Berlin, Heidelberg: Springer.
Jiang, T., Chen, X., & Ma, J. (2015). Public integrity auditing for shared dynamic cloud data with group user revocation. *IEEE Transactions on Computers, 65*(8), 2363−2373.
Jin, H., Jiang, H., & Zhou, K. (2016). Dynamic and public auditing with fair arbitration for cloud data. *IEEE Transactions on Cloud Computing, 6*(3), 680−693.
Li, J., Yan, H., & Zhang, Y. (2018). Certificateless public integrity checking of group shared data on cloud storage. *IEEE Transactions on Services Computing, 14.*
Li, J., Zhang, L., Liu, J. K., Qian, H., & Dong, Z. (2016). Privacy-preserving public auditing protocol for low-performance end devices in cloud. *IEEE Transactions on Information Forensics and Security, 11*(11), 2572−2583.
Li, X. X., Chen, K. F., & Sun, L. (2005). Certificateless signature and proxy signature schemes from bilinear pairings. *Lithuanian Mathematical Journal, 45*(1), 76−83.
Liu, J., Huang, K., Rong, H., Wang, H., & Xian, M. (2015). Privacy-preserving public auditing for regenerating-code-based cloud storage. *IEEE Transactions on Information Forensics and Security, 10*(7), 1513−1528.
Liu, J. K., Au, M. H., & Susilo, W. (2007, March). Self-generated-certificate public key cryptography and certificateless signature/encryption scheme in the standard model. In: *Proceedings of the 2nd ACM symposium on Information, computer and communications security* (pp. 273−283).
Pointcheval, D., & Stern, J. (2000). Security arguments for digital signatures and blind signatures. *Journal of Cryptology, 13*(3), 361−396.

Rasheed, H. (2014). Data and infrastructure security auditing in cloud computing environments. *International Journal of Information Management, 34*(3), 364—368.

Ren, Y. J., Shen, J., Wang, J., Han, J., & Lee, S. Y. (2015). Mutual verifiable provable data auditing in public cloud storage. *Journal of Internet Technology, 16*(2), 317—323.

Shacham, H., & Waters, B. (2008). Compact proofs of retrievability. In: *International conference on the theory and application of cryptology and information security* (pp. 90—107).

Shen, J., Shen, J., Chen, X., Huang, X., & Susilo, W. (2017). An efficient public auditing protocol with novel dynamic structure for cloud data. *IEEE Transactions on Information Forensics and Security, 12*(10), 2402—2415.

Shen, W., Qin, J., Yu, J., Hao, R., & Hu, J. (2018). Enabling identity-based integrity auditing and data sharing with sensitive information hiding for secure cloud storage. *IEEE Transactions on Information Forensics and Security, 14*(2), 331—346.

Subramanian, N., & Jeyaraj, A. (2018). Recent security challenges in cloud computing. *Computers & Electrical Engineering, 71*, 28—42.

Tian, H., Chen, Y., Chang, C. C., Jiang, H., Huang, Y., Chen, Y., & Liu, J. (2015). Dynamic-hash-table based public auditing for secure cloud storage. *IEEE Transactions on Services Computing, 10*(5), 701—714.

Wang, B., Li, B., Li, H., & Li, F. (2013, October). Certificateless public auditing for data integrity in the cloud. In: *2013 IEEE conference on communications and network security (CNS)* (pp. 136—144). IEEE.

Wang, F., Xu, L., Wang, H., & Chen, Z. (2018). Identity-based non-repudiable dynamic provable data possession in cloud storage. *Computers & Electrical Engineering, 69*, 521—533.

Wang, H., Wu, Q., Qin, B., & Domingo-Ferrer, J. (2013). Identity-based remote data possession checking in public clouds. *IET Information Security, 8*(2), 114—121.

Wang, Y., Wu, Q., Qin, B., Shi, W., Deng, R. H., & Hu, J. (2016). Identity-based data outsourcing with comprehensive auditing in clouds. *IEEE Transactions on Information Forensics and Security, 12*(4), 940—952.

Worku, S. G., Xu, C., Zhao, J., & He, X. (2014). Secure and efficient privacy-preserving public auditing scheme for cloud storage. *Computers & Electrical Engineering, 40*(5), 1703—1713.

Wu, G., Zhang, F., Shen, L., Guo, F., & Susilo, W. (2020). Certificateless aggregate signature scheme secure against fully chosen-key attacks. *Information Sciences, 514*, 288—301.

Xu, Q., Shen, H., Sang, Y., & Tian, H. (2013, December). Privacy-preserving ranked fuzzy keyword search over encrypted cloud data. In: *2013 International conference on parallel and distributed computing, applications and technologies* (pp. 239—245). IEEE.

Xu, Y., Sun, S., Cui, J., & Zhong, H. (2020). Intrusion-resilient public cloud auditing scheme with authenticator update. *Information Sciences, 512*, 616—628.

Xu, Z., He, D., Wang, H., Vijayakumar, P., & Choo, K. K. R. (2020). A novel proxy-oriented public auditing scheme for cloud-based medical cyber physical systems. *Journal of Information Security and Applications, 51*, 102453.

Yuan, J., & Yu, S. (2015). Public integrity auditing for dynamic data sharing with multiuser modification. *IEEE Transactions on Information Forensics and Security, 10*(8), 1717—1726.

Zhang, J., Wang, B., He, D., & Wang, X. A. (2019). Improved secure fuzzy auditing protocol for cloud data storage. *Soft Computing, 23*(10), 3411—3422.

Zissis, D., & Lekkas, D. (2012). Addressing cloud computing security issues. *Future Generation Computer Systems, 28*(3), 583—592.

Zou, X., Deng, X., Wu, T. Y., & Chen, C. M. (2020). *A collusion attack on identity-based public auditing scheme via blockchain. Advances in intelligent information hiding and multimedia signal processing* (pp. 97—105). Singapore: Springer.

Further reading

Li, S., Zhong, H., & Cui, J. (2016, December). Public auditing scheme for cloud-based wireless body area network. In: *Proceedings of the 9th international conference on utility and cloud computing* (pp. 375–381).

Zhang, J., Wang, B., Wang, X. A., Wang, H., & Xiao, S. (2020). New group user based privacy preserving cloud auditing protocol. *Future Generation Computer Systems, 106,* 585–594.

CHAPTER 15

Intelligent analysis of multimedia healthcare data using natural language processing and deep-learning techniques

Rohit Kumar Bondugula[1], Siba K. Udgata[1], Nashrah Rahman[2] and Kaushik Bhargav Sivangi[1]
[1]School of Computer and Information Sciences, University of Hyderabad, Hyderabad, India
[2]Faculty of Engineering and Technology, Jamia Millia Islamia, New Delhi, India

15.1 Introduction

With increasing digitalization in health care, natural language processing (NLP) has also marked its place in this field. NLP can be used as effective clinical decision support (CDS) and can help solve many problems. Healthcare databases are growing exponentially, and this data can further add value with the use of text analytics and NLP to improve the outcomes of reports of patients. It can be helpful in streamlining operations and can also help in managing regulatory compliance. Moreover, physicians invest much time in determining the patient's conditions through the chart notes. Those notes are further unstructured, which has to be dealt with by some NLP methods for processing. Without NLP, a large amount of data goes into a not usable format for the extraction. One of the key applications of NLP in health care is information extraction from clinical notes. NLP techniques can help process the free-text notes and can pull out the required conditions from the text. These techniques are also helpful in extracting data from radiology reports. So, NLP algorithms help visualize the clinical notes, reports of the patients, lab reports, pathology reports, MRI scans, CT scans, etc. Speech recognition can also be an added feature to the domain.

Many applications revolve around NLP in the healthcare domain, where extracted data can be used in many decision-making applications. These techniques, in turn, are useful in reducing the cost and time of the analysis, reducing the error rates, which is improving the efficiency of the processes in health care.

Edge-of-Things in Personalized Healthcare Support Systems
DOI: https://doi.org/10.1016/B978-0-323-90585-5.00014-X

This chapter focuses on NLP, healthcare/clinical analytics, the preprocessing techniques revolving around NLP, text representations, types of models, and their implementation in the healthcare domain. Further, we will also discuss the results and evaluations of the models implemented in various implemented applications in health care.

15.1.1 Deep learning

Deep learning is a subdomain of machine learning which takes on learning the representations from the data. It consists of successive hidden layers of representations and extracts more comprehensive and complex features as layers progress. The models on deep learning consist of neural networks containing the latent variables. Due to its supremacy over ever-increasing data, there has been a rise in the application of deep learning in various fields in recent years. They have been essentially used in applications such as but not limited to image recognition, NLP, medical data analysis, and bioinformatics. Specifically, due to the significant increase in textual data and records, there has been a development in traditional NLP methods to move toward utilizing the potential of deep learning. Techniques such as recurrent neural networks (RNNs) and long short-term memory networks have proven effective compared to previous methods. One of the objectives of deep learning is to convert unstructured data into structured data to make predictions more accurately when compared to other machine algorithms.

15.1.2 Natural language processing

NLP is an artificial intelligence method which establishes communication between machines and humans using natural language. It is useful for machines to understand, analyze, and derive meaning from the human language in a developed manner. Using NLP, one can structure and organize the data to perform tasks such as automating the summarization, named entity recognition (NER), translation, sentiment analysis, speech recognition, and segmentation. In addition, NLP helps the machines to understand how the beings speak. NLP is used mainly in text mining, automated question answering, and machine translation.

15.1.2.1 Overview of healthcare/clinical analytics and relating to natural language processing

Healthcare analytics is a process of reviewing, analyzing the current and historical data for predicting the trends, improving the reach, and managing the spread of diseases. It is a method for improvising the quality of diagnosis, patient facilities, and managing the resources. It is actively coming out to rise in the research field in recent years. It focuses on developing analytical solutions that increase the efficiency of the application or product or the initiative by the medical practitioners. This study explores the area of research in clinical and nonclinical support with different designs, developments, and implementations.

Similarly, another term is interchangeably used with healthcare analytics but can further be described in a better way. Clinical analytics is a more focused area among many information technology companies, and they are looking to prioritize the health of their employees. Especially domains with hazardous working environments are working to incorporate such IT solutions. Furthermore, talking in general, the companies are looking for an efficient solution that can provide insurance to their employee's health and safety.

This field uses the real-time clinical data of the users to generate some valuable insights related to the tagging done for the application. This is useful in making intelligent decisions and forecasting the related variables or outcomes designed for the application. These kinds of applications are led by clinical analytics and, when combined with business intelligence and later with NLP, can lead to more efficient, cost-effective results, with better operativity of the application, which can be designed with an interactive user interface. Furthermore, talking about the organizations, this analysis can also help increase their productivity by monitoring their employee's health, which in turn causes them to generate increased revenues. The cost savings are on the part of the analysis that the machine learns on the implementation side. This also reduces the medical and manual errors for analyzing the reports/clinical notes. Recently, this has also led to the increased use of electronic health records (EHRs) in many organizations.

So, relating to the healthcare domain/clinical analytics, the methods used in this evaluation process are similar to NLP, which aims to produce the results using some computational approaches to the given or required prognosis or outputs. Furthermore, the performance of the algorithm or method can further be evaluated using some statistical terms like z-score, performance metrics, accuracy, precision, and recall.

There are two different types of evaluation metrics on a border picture, namely extrinsic and intrinsic evaluation metrics. Extrinsic evaluation metrics focus on the component's contribution to the performance of a complete application, which often involves the participation of a human in the loop. Furthermore, intrinsic evaluations measure the performance of an NLP component on its defined subtask, usually against a defined standard in a reproducible laboratory setting (Resnik et al., 2006). So, the validation of the outcomes or the prognosis can further metrics, and Clinical NLP can be evaluated using extrinsic or intrinsic evaluation metrics. As NLP's objective is to get automated solutions to the problem statement, the evaluation metric can be intrinsic and extrinsic. Restricting to clinical NLP, the evaluation is mainly focused on intrinsic metrics. The intrinsic metrics are a computational process of development evaluated on documents, words, sentences of the input data, with different attributes like sections of the input documents, content, named entity like symptoms and diseases (which we will discuss in the following sections of this chapter). These intrinsic metrics are valuable, but the information will be informative when it comes to a higher-level task like patient data or new data. For instance, the current state-of-the-art that is achieved in medical concept classification is >80% F-score, close to human agreement on the same task. However, suppose such a system was to be deployed in clinical practice. In that case, any error rate >0%, about misclassification of a drug or a history of severe allergy, might be seen as unacceptable (Velupillai et al., 2016).

15.1.2.2 Natural language processing approaches and their applications in health care

As we know that the healthcare domain is also fast growing in terms of technologies, the data is also increasing, the importance of data is being realized, and information is being collected from EHRs, sensors, etc. However, the collected data is 80% unstructured, and the rest 20% is structured, which is helpful, while the rest 80% goes to waste. Thus there is a high need to utilize the data, analyze it, and make it useful for doctors to make efficient decisions. This brings us to using NLP on the unstructured data, which helps make it into the structured form using some automated processes and algorithms.

The block diagram in Fig. 15.1 represents the workflow of the data processing, which is first in an unstructured form and later converted into the structured format with the help of the NLP techniques. Then the structured data is worked upon using the machine learning of deep-learning models for its evaluation, prediction, and forecasting of the requirements.

Figure 15.1 Diagram of the workflow of the data processing.

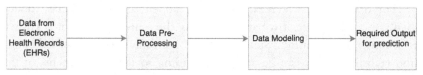

Figure 15.2 Brief procedure on how the data from EHRs can be analyzed. *EHRs, Electronic health records.*

Taking briefly about the data generated in the healthcare domain, 80% of the data is in an unstructured format. So, as discussed above, the data must be in a structured layout for the computations to be done. Because data needs to be preprocessed to be appropriately visualized for prediction or analytics purposes, there are many techniques for cleaning and processing the data. Different methods are being implemented for the preprocessing of the data to be done. These preprocessing techniques are followed by text representation techniques where the formatted data is passed to some algorithms that will convert them into a format that the machine understands. The machine understands the numerical representations. So, many methods are helpful in the process.

Fig. 15.2 is the image that briefly explains the process. First, let us discuss the data preprocessing and the text representation techniques more in detail with the practical applications of the methods in the healthcare domain. Moreover, later, we will further be studying some approaches and their applications in the healthcare domain.

15.1.2.3 Data preprocessing techniques
Different types of preprocessing techniques applied to the data can help convert the unstructured data to structured sets of data.

1. Tokenization

Tokenization is the process of breaking down raw or large texts into small chunks. These small chunks are known as tokens. The tokens are useful in having a better understanding of the context, and further will be helpful in the development of the particular NLP models. It helps in understanding the meaning of the data through the analysis of the sequence of the words. Here a bag of words is collected and then taken forward for further preprocessing. Multiple libraries help in tokenizing the data like NLTK, Gensim, and Keras.

One clinical note having data about a patient is taken as an example; the input and output of the tokenization implementation are shown in Fig. 15.3.

This method of tokenization can be added to the sentences, documents, etc. Moreover, if we are tokenizing the sentences, then it is known as sentence tokenization. Similarly, there are different kinds of tokenization techniques that can be applied according to the requirements of the problem statement in health care.

The above code snippets in Fig. 15.3 and the one present in the other section represented in Fig. 15.4A and B were executed in Google Colaboratory, a product by Google Research where python code can be written and executed.

2. Stop words removal

Stop words are the words in the data that do not add meaning to the sentence like "is," "to," "into," "under," and "numbers." After removing such words or characters, it will not affect the meaning. This also helps in removing the unnecessary words adding more meaning, and helping reduce the noise of the data.

3. Stemming and lemmatization

Stemming is the process of decreasing or reducing the inflection of words. In other words, it is a technique of extracting the base form of the words from the data. It is done by the removal of affixes from the words. The removal of the affixes is also valid, even if it does not make any meaningful word.

6 Intelligent Analysis of Multimedia Healthcare data using Natural Language Processing and Deep Learning Techniques

```
from nltk.tokenize import word_tokenize
word_tokenize("the patient should be readmitted to the hospital due to the criticality and severity of the infection")

['the',
 'patient',
 'should',
 'be',
 'readmitted',
 'to',
 'the',
 'hospital',
 'due',
 'to',
 'the',
 'criticality',
 'and',
 'severity',
 'of',
 'the',
 'infection']
```

Figure 15.3 Input and output for tokenization.

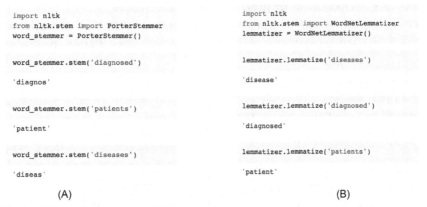

Figure 15.4 (A) Stemming and (B) lemmatization.

The library, NLTK, offers a class named PorterStemmer, with the help of which we can implement stemming and apply Porter Stemmer algorithms for the word to be stemmed.

For example, let us import PorterStemmer from NLTK first and look in some clinical notes if we have the following bag of words like diagnosed, patients, diseases.

In the output, we get the stemmed words like diagnosis, patient, disease, respectively as shown in Fig. 15.4A. Moreover, we can also notice that the stemmed words are not meaningful.

The usage of stemmed words reduces the sizes of the indexes stored and, in turn, increases the accuracy of retrieval.

Lemmatization: Lemmatization is also a method like the above-discussed method, stemming. It is a technique that switches any kind of a word to its base root mode. It can also be said to group different inflected forms of words into the root form, having the same meaning. The output of lemmatization is called lemma. Here, it will always get a valid word.

NLTK has a class WordNetLemmatizer, which helps in the implementation of the Lemmatization process.

For example, in the healthcare domain, see Fig. 15.4B for the same bag of words used for stemming.

4. PoS Tagger

PoS here stands for part-of-speech, and PoS tagger is a kind of classification that involves tagging. It is a technique of automatically assigning a description to the tokens. Here, the words are tagged according to the PoS used, including nouns, verbs, adverbs, adjectives, etc.

15.1.2.3.1 Utilization of preprocessing techniques

Using the above data preprocessing techniques, clinical NLP can help get the initial dataset gathered from various means like EHRs, pathology reports, and nursing reports. These techniques can help in improving the EHRs, risk mitigation, clinical predictive analytics in health care. For example, the text from social media can be used to analyze a person's emotions, sentiments, and anxiety levels.

EHR systems have allowed automated document classification to become an important research field of clinical predictive analytics. This helped to leverage the utility of narrative clinical notes (Charles et al., 2013). In the mentioned citation the researchers have used bag-of-words representation, which directly identified and normalized lexical variants from the unstructured text content, as the baseline of clinical feature representation. The research performed by Chen et al. (2021) was based on information retrieval, NER, literature-based discovery, and question answering on the upcoming urgent need due to the COVID-19 pandemic. They have also described the related work that can address various aspects like emotional and sentiment analysis, topic modeling, caseload forecasting, and misinformation detection. For the preprocessing of the data chunk collected, they have applied stemming and lemmatization. They have observed that using these techniques helped transform each word into its primary form or its root form. This utilized the use of domain-specific dictionaries of synonyms (Chen et al., 2021). Collecting the data, further, the models were applied for their required research.

In Chen et al., (2021), for another perspective of the objective, which was question answering, they used the QA engines categorized into three groups. In the first part, case folding, lemmatization, and stop word removal techniques helped them with sentence simplification. This caused all the complex and lengthy sentences to shorten and become much more straightforward.

Using the above techniques, many clinical questions can be addressed using NER. When applied with the preprocessing techniques and training over the models, the data collected can predict the early symptoms recorded from the EHRs. They also observed that the symptoms of COVID-19 were strongly associated with anosmia, fever, and chills.

During the research by Samuel et al., in which the team was working on the classification of the tweets based on COVID-19 public sentiment analysis. They have used different preprocessing techniques, like stop word removal, tokenization, PoS tagging, stemming, and lemmatization.

The redundant words would deprive the potential opportunities of the data analysis and prediction. So, therefore the above techniques helped to improve the accuracy and efficiency of the model proposed.

15.1.2.4 Levels of natural language processing in health care

Let us discuss the general steps which are followed in NLP and their applications in health care. First, some of the analyses are done as part of the preprocessing steps, including lexical analysis, syntactic analysis, semantic analysis, disclosure integration, and pragmatic analysis. Fig. 15.5 depicts the process flow diagram of the implementation of NLP in health care:

1. *Lexical analysis*: It involves the identification and analysis of the structure of words in our data. Then, it divides whole data into words, sentences, paragraphs, or phrases in a particular language. The process of tokenization does this. As discussed, tokenization is a process of breaking the content into small chunks of data. Here, a small chunk of data is known as lexeme. Typical examples of lexemes in health care include congestive heart failure and diabetes mellitus (Friedman & Johnson, 2006).

2. *Syntactic analysis*: Syntactic analysis is also known as syntactic parsing. It is a process in which the input is converted to a hierarchical structure

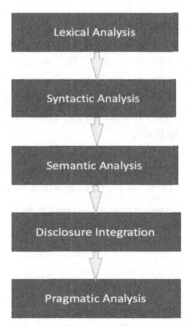

Figure 15.5 Process flow diagram of the implementation levels of NLP in health care.

representing the meaning in a sentence. It is related to the establishment of a relationship between the words in a sentence. It ensures if the tokens are in a series of a particular language.

3. *Semantic analysis*: Semantic analysis interprets the exact meaning from the extracted text. The text gets checked for it is in a valid form and if it is meaningful. The meaning is checked for all words, sentences, paragraphs, and phrases. Adding, there are many meanings or interpretations to a single word. This can be explained by taking an example, suppose we have two sentences, and the meaning of the word "discharge" stands to be different in both: "The doctor prescribed some medicines on discharge. The medicines were given as the patient was getting a translucent-yellowish discharge"

4. *Discourse integration*: This process of discourse integration brings out the meaning of a sentence just before it.

5. *Pragmatic analysis*: Pragmatic analysis is how the combining of the sentences takes place to get a discourse of that of documents and paragraphs. For instance, mass in a mammography report denotes breast mass which is a form of breast cancer, mass in a radiological report of the chest denotes mass in the lung. In contrast, mass in a religious journal denotes a ceremony (Friedman & Johnson, 2006).

15.1.2.5 Text representations

In the previous sections, we discussed how the data is preprocessed. In this section, we will discuss some of the methods taken up after the data preprocessing for the machine to understand the text of the data given. The machine can only understand once the words are vectorized. Fig. 15.6 represents the classification of the methods.

1. Frequency-based methods
   a. *Count Vectors*

   In this technique the vectors for the machine interpretation are built on the frequency or the number of occurrences of a word

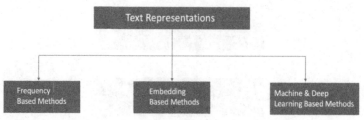

Figure 15.6 Diagrammatic view of the classification of text representations.

present in the bag of words. Sklearn is a framework that provides us with an inbuilt module, "CountVectorizer," to perform count vectors. So, the CountVectorizer counts the occurrences of tokens generated and builds a sparse matrix out of it. Taking the case of the clinical notes prepared by the doctor, in which the frequency of the words extracted into the bag of words is calculated. Moreover, this gives us a vector of the frequency of the words in the data.

b. *Term frequency—inverse document frequency (TF-IDF)*

This method is a statistical measure to determine how a word is important to a document in a set of collections of documents. The importance increases with the number of times the word appears in the document but is offset by the frequency of that particular word in collecting the documents or corpus. This method, TF-IDF, has efficiently been implemented in Kovacs and Kardkovacs (2014). In this work, the researchers have demonstrated a technique to determine the frequent sequential patterns occurring in healthcare databases. With sequential data mining, the patterns and TF-IDF implementations are applied in many real-life scenarios like real-time recommendation systems, healthcare service optimizations, and fraud detection. In addition, the technique of TF-IDF is used to boost the performance of the sequence mining algorithms.

2. Embedding-based methods

a. Word2Vec

Word embeddings have nowadays become the most used method and popular among the implementations in NLP. These concepts revolve around various traditional approaches like one-hot encoding, which captures the high dimensionality and inefficiency. The basic idea of the implementation should be, each word is judged and characterized based on the words accompanying it or are surrounded with. So, word embedding techniques combine the words with similar contexts in the new space-created model.

Similarly, word2vec has gained popularity, and embeddings have become the basic step in many NLP tasks. It finds its major application in transfer learning, where embedding is learned from many unlabeled data, then the embedded outputs are used in supervised tasks. Within the context of medical data, recent examples have shown that transfer learning works very well for imaging tasks (Beam & Kohane, 2016; Gulshan et al., 2016) due in large part to the availability of pretrained computer vision models (He et al., 2016; Simonyan & Zisserman, 2014; Szegedy et al., 2016) that were pretrained on the ImageNet database (Deng et al., 2009).

Initially, the word2vec approach consists of the models and algorithms like a continuous bag of words (CBOW) and skip-gram models. CBOW models help predict the probability of a required output word when its context is already defined in the window. At the same time, SG models help in predicting the surrounding context with the required output word.

Machine learning can be implemented in almost all domains in health care. As in health care, much data can be used for pretraining the resources for practical implementations. This representation of words as numerical vectors based on the embeddings or the context of the data is seen as the de facto method of analyzing the text with machine learning.

Similarly, for clinical text corpus, the vector embeddings can be created with intrinsic and extrinsic evaluation parameters, including the methods' limitations. Adding, the word embeddings can quickly learn on the input data from EHRs.

In Dudchenko and Kopanitsa (2019), the researchers have compared the embedding techniques over the electronic medical records. The model consisted of a multilayer perceptron and convolutional neural networks. The initial processing was pipelined with the embedding-based architecture with preliminary lemmatization. The highest F-score was achieved in their model with the word2vec implementation itself.

In Bai et al. (2018), word2vec was used as a vector representation for helping in the exploratory analysis and predictive modeling of the EHR data, which helps in knowing the details of the patterns of care and health outcomes and analyzing them. The EHR here contained an unstructured form of the data of the clinical notes, which helped the team evaluate more in-depth about the condition and treatments of the patients.

There is another research implementation (Ghosh et al., 2016), where the research workers characterize diseases from an unstructured set of data with the help of the word2vec approach. In this, the real-time sourced data from news, social media can be augmented with traditional disease surveillance. Word2vec is used here to model the diseases and constituent attributes as word embedding from the health map of the news corpus. These word embeddings are useful in automatically creating disease taxonomies. Moreover, in turn, it helps in evaluating the model against the human-annotated taxonomies. The accuracy of the model was compared with many word2vec methods. One of the methods, namely dis2vec, outperformed the traditional representations and was able to classify the attributes into various classes of diseases like endemic emerging and rare.

Classical methods of word2vec were applied to determine various outputs to linguistic tasks with considerate accuracy. Unfortunately, the algorithms were designed in an unsupervised fashion which does not help in generating good enough embedding for the healthcare domain where the uncovered relationships are treated with greater importance than the nondomain ones. The embedding part was designed to find interesting characterizations of diseases. In addition to the emerging, endemic, and rare attributes, this kind of classification can also be done and analyzed over different geographical regions.

b. *Doc2Vec*

Sentiment analysis has been one of the fastest-growing research areas. Sentiment analysis utilizes NLP to analyze the sentiment/opinion of a person's statement. Lately, sentiment analysis has not only been limited to online product reviews but has been shifted to but is not limited to analyzing social media texts from Facebook, Twitter, stock market analysis, disasters, and healthcare industries (Chen & Sokolova, 2018). Chen & Sokolova, (2018) have explored the oppositeness of Word2Vec and Doc2Vec methods for sentiment analysis of clinical data, where they have briefly explained the methods. Concretely, the application of these methods for clinical discharge summaries of patients has been studied. Clinical records are usually written in an objective way in which the text is descriptive rather than opinionated. Most opinions on patients' health are indirectly started subtly. This results in fewer sentiment terms present in the text. In addition, clinical records contain an excess of medical terms, which leads to the improper judgment of the degree of severity of underlying diseases. It is important to identify and address bias while dealing with patient care. This results in indiscrimination against afflictions and receives quality care. They have performed unsupervised sentiment analysis on a set of clinical discharge summaries written by health physicians and nurses. They identified three diseases: hypertension, diabetes, obesity, a higher number of discharge summaries and performed analysis. In addition, the set differences of the three chosen diseases are considered, which leaves with 12 subsets on which experiments can be done. Word2Vec develops word embeddings, that is, the words are represented numerically in the form of vectors. Word embeddings are made based on the context of the occurrence of words in the text in which the words which have synonymous meanings are grouped. They followed the continuous bag of models (CBOW) (Chen & Sokolova, 2018) architecture for the word2vec, which takes the context as input and predicts a specific word.

15.2 Deep-learning-based methods

The following sections discuss the overview and development of Deep Learning based methods employed for healthcare applications.

An RNN is a neural network model which is repeated over time. Concretely, the parameters of the network are shared across different time steps via self-loops. Compared to a standard feed-forward neural network, which maps an input vector to an output, RNN maps an input sequence into a sequence. Fig. 15.7 illustrates a standard RNN architecture.

Compared to traditional approaches, RNNs have shown to have better flexibility and efficiency. For example, RNN can handle varying length sequences as input which a regular neural network cannot achieve. Furthermore, due to parameters sharing over time steps (different input sequences), they have shown superiority over standard networks. Due to these advantages, recent applications of NLP in the healthcare domain have employed RNN models.

Long short-term memory (LSTM) is a particular class of RNN that can handle long-term dependencies (Soutner & Müller, 2013), avoiding the vanishing gradient problem. The vanishing gradient problem has been

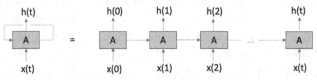

Figure 15.7 Standard RNN architecture (Hochreiter & Schmidhuber, 1997).

overcome by LSTM units by including additional gates into the standard RNN unit, namely input and forget gates. Within LSTM, there is a self-loop memory cell that allows gradients to flow through long sequences. The memory cell is used to manage the flow of information from one cell to the other. LSTM captures the input tokens from a given sentence in a distributed representation in the form of continuous values. In addition, LSTM cells contain other states such as cell state Ci and hidden state hi. The memory cell consists of 3 units, namely forget unit Fi, input unit Ii and output unit Oi, and their associated weights. These units allow the LSTM cell to pass the information from one timestep to another, and they constrict the amount of information flow. Fig. 15.8 describes an LSTM unit.

Attention mechanism enables to highlight relevant features of the input text sequences dynamically. As a result, attention has become a useful mechanism to extract superior results (Bahdanau et al., 2014). Since the standard LSTM could not detect the specific features which play an essential role in determining the outcome, a new attention-based LSTM architecture is proposed, as shown in Fig. 15.9.

15.2.1 Diagnosis and prognosis predictions and decisions

The evolution of deep NLP algorithms such as RNNs, LSTMs, and transformer networks has enabled an ample amount of applications in healthcare and clinical diagnostics. Choi et al. (2016) have leveraged the idea of temporal modeling from RNN and developed a tool named Doctor-AI applied on EHR consisting of time stamps and data from 260k patients by

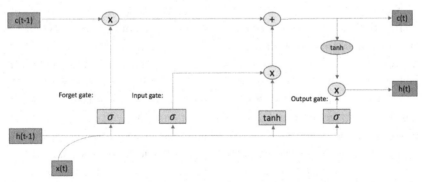

Figure 15.8 A standard LSTM cell containing additional gates for control over the flow of gradients (Hochreiter & Schmidhuber, 1997).

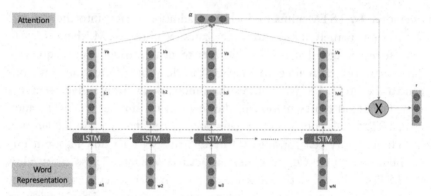

Figure 15.9 Attention mechanism with LSTM architecture.

2128 physicians. This helped to predict the diagnosis and medication for subsequent visits of the patient. The importance lies in how the model was able to identify the medical history of various patients and understand the inference over subsequent visits and diagnoses. This was a superior development over traditional approaches and has achieved a 79% recall value. This kind of evaluation has been an important part of the health-care domain. Furthermore, there is a growing amount of information that is being digitized which seeks human expertise. NLP algorithms can help patients with personalized recommendations based on previous visits and medical diagnosis history. Therefore combining the static information present in the EHRs and dynamic data consisting of a history of diagnosis over time has been used extensively by deep NLP algorithms to determine the prognosis and detect endpoints.

Esteban et al. have developed an RNN model for predicting clinical events related to this methodology. They have used a database consisting of patients who underwent kidney transplantation. Based on each patient's collected EHR, they developed the model that could predict when the endpoint could occur. Concretely, the endpoints are the rejection of the kidney, death of the patients due to complications, and kidney loss. In addition, they have also used the model to predict the next visit of the patient. On similar lines, Yang et al. (2016) have utilized the RNN framework to develop a CDS system. Often, while diagnosing patients, physicians need to consider an extended range of data involving the prognosis, making the decisions increasingly complex. In this case, RNNs have shown their superiority in predicting these decisions; this helps the physician with mutually dependent decisions.

15.2.2 Mortality rate prediction

Mortality prediction for patients in intensive care units (ICUs) is an important task and requires extensive monitoring. ICU admits the most severely afflicted patients who require comprehensive supervision. Therefore it is crucial to identify patients at risk of dying and provide relevant intercession to save a life. Therefore mortality prediction becomes an important task in ICUs that admits people are suffering from life-threatening injuries and diseases (Karunarathna, 2018). The purpose of this task is not only limited to identifying people with high-risk factors but also the decision to save ICU beds for patients in need (El-Rashidy et al., 2020).

Aczon et al. (2017) developed an RNN model to track the patient encounters in the pediatric intensive care unit (PICU). They extracted data from electronic medical records (EMRs) of over 12000 patients who were admitted to PICU over 10 years. The inputs given to the RNN model consist of laboratory and scan results, physiological observations, and prescribed drugs. The model then generates dynamic mortality prediction at user-specified times. Over the years, the adoption of EMRs has facilitated the access to health data and other related variables of several patients. Responding to this ever-growing data, deep NLP models have shown their supremacy to forecast the patient's risks and conditions, enabling easy decision making. Although the existing methods have shown higher accuracy of ICU mortality prediction, there is still a requirement for interpretability. Ge et al. (2018) developed an interpretable mortality prediction model based on RNN consisting of LSTM units. In this, interpretability has been provided as a scoring factor. In addition, they even analyzed nonsequential features from EMRs, such as patient demographics, prognosis, and diagnosis. These nonsequential features provide added benefits to understanding the patient's population and infer the appropriate long-term diagnosis. This helps add relevant information to the static and helps to achieve better results in mortality prediction and to identify the features that are strongly associated with the mortality rate (Fig. 15.10).

The interpretability of a deep NLP model is crucial in health care because the models' decisions can result in survival or death (Esteva et al., 2019). This ensures the physicians relevant and acceptable information for quick and actionable decisions. Ho et al. (2019) employed learned binary masks (LBMs) to identify the pertinent features across the RNN model.

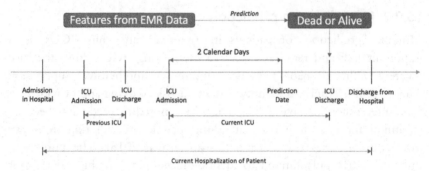

Figure 15.10 The sequential view for the prediction of mortality rate.

Table 15.1 Summary of proposed models for mortality rate prediction (MRP) and their test accuracy.

Summary of proposed models for MRP	Test accuracy
RNN	0.930
Interpret RNN-LSTM	0.760
RNN-LBM	0.939

RNN, Recurrent neural network.

LBM has shown the RNN model contains clinically consistent dynamics to determine the patients' risk of mortality.

Table 15.1 depicts the comparison of various deep-learning NLP methods developed for mortality rate prediction (MRP). The standard RNN model by Aczon et al. (2017) has shown test accuracy of about 93% in their 12 hours long observation time. An interpretable model developed by Ge et al. (2018) using RNN has included sequential and nonsequential features and has identified features that strongly contribute to predicting the mortality rate. The interpretable model was able to achieve an accuracy of 76%. Ho et al. (2019) introduced LBM to identify the features for the RNN model. The RNN-LBM model has achieved about 94% accuracy in the MRP task.

Therefore developing predictive models based on NLP plays a key role in assessing the severity of illness, which helps identify high-risk patients. This aids in making critical decisions and enables effective triage, composing ICU resources for patients in need. These models can also be used as an indicator for benchmarking ICU performance across various hospitals and help in identifying risk stratification (Power & Harrison, 2014). Thus mortality prediction models play a key factor in reducing the clinical and economic burden of ICU care resulting in quick response and systematic decisions during emergencies.

15.2.3 Social media and health care

Being social creatures, humans naturally tend toward companionship to thrive, which has a significant impact on mental health. Having a social connection with each other can reduce stress and anxiety. Moreover, this helps in boosting one's self-worth. Lacking a social connection can result in serious risks to emotional and mental health. Many people invest most of their time on social media platforms such as Twitter, Instagram, and Facebook. Studies show that in-person contact with people triggers stress-relieving hormones and helps them feel positive. On the flip side, social anxiety has been shown in certain groups of people spending excess time on these platforms, leading to loneliness and isolation. This results in mental health conditions such as anxiety and depression.

Fig. 15.11 highlights the possible links between social media use and mental health (Kelly et al., 2018). Poor sleep patterns, online bullying, negative self-esteem, and body image are examples of this. For example, poor sleep, self-esteem, and body image all play a role in the link between online bullying and harassment. In addition, poor self-esteem may play a role in the link between poor body image and depressive symptoms.

Social media can promote various negative impacts on mental health, sometimes leading to self-harm and suicidal thoughts. Images being viewed on social media can make a person feel insecure about one's looks and the thought of how other people are leading better lives. Majority of

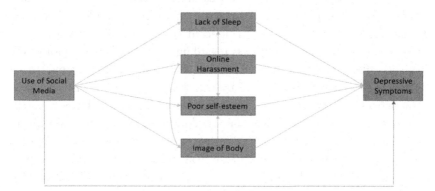

Figure 15.11 A speculation of pathways between the usage of social media and depressive symptoms among people (Kelly et al., 2018).

the young report cases of cyberbullying and are often subjected to offensive, racial, body-shaming comments. These platforms can be the hotspots for spreading vicious rumors, lies that can leave a deep scar to the person involved. The recent development of predictive techniques in deep NLP has led to identifying the presence of specific mental health issues and other symptomatology issues (Laacke et al., 2020). Sometimes, based on the information a person posted on social media platforms, the underlying mental state can be determined (Kim et al., 2020).

Sentiment analysis is the automatic determination of the user's opinion from the text (Liu, 2010a, 2010). Most methods focus on mining social media for user opinions and product reviews. Very few models have been developed to analyze a person's post's internal thoughts and emotions from a mental health perspective. The useful application of such modeling can be utilized to develop online mental health therapy services (Kim et al., 2020; Shaw et al., 2017). Shickel et al. (2017) have utilized RNN based attention model to quantify the semantic differences between public posts and private mental health journals and showed that social media data could be utilized to create robust mental health models for emotional valence prediction. Over the years, health forums have evolved and allowed more users to share their medical experiences, which helps seek guidance from people within the forum. This content contains valuable medical information and can be leveraged to produce a more structural suggestion and be utilized as a recommendation engine (Aipe et al., 2019). Along with personal medical information, people have also shared the experience of visiting medical centers stating the services and availability (Manke & Shivale, 2015). Therefore it became necessary for patients to make decisions about their related medical issues by learning from other patients' experiences. As per the statistics, around 14.1 million people in the world have been affected by cancer. To diagnose, it is important to consider various facts regarding the cognitive behavior of the patients affected with cancer. Due to increasing health forums, many patients who have been affected by this deadly disease have shared their experiences and prognosis through various portals and social media. Social media provides a lot of cancer support communities (Zhao et al., 2014). This support provided by various interactions with the people in such groups plays an important role in survival. Edara et al. (2019) have proposed an LSTM model for categorizing text and analyzing various texts related to users' opinions and analyzing each individual's mental states, which helps create a support system to uplift the affected. With more advanced text analytics growing up, the task of sentiment analysis has evolved to aspect-based called aspect-based sentiment analysis (ABSA). ABSA of health care recommends the treatments and the strong aspects/features for which the services/treatments are preferred. This takes the patient's reviews and opinions to modify their policies and directly address the said problems. Wang et al. (2016) proposed an attention-based LSTM model employed for ABSA. They achieved a state-of-the-art performance of 90.9% on ABSA classification.

The recent outbreak of novel coronavirus (in 2019) and COVID-19-related discussions have emerged in social media. Jelodar et al. (2020) have developed a model based on LSTM to analyze the COVID-19-related comments from various subreddits and used sentiment analysis for opinion analysis of COVID-19-related comments. This helped in understanding the needs and concerns of people for pandemic-related issues. Raamkumar et al. (2020) developed and evaluated deep-learning-based text classification methods for classifying social media content posted during the recent outbreak of COVID-19. A health belief model has been used to characterize user-generated social media content to analyze the behavioral health pattern of people. They developed this model toward analyzing the public perception and mental, behavioral patterns regarding the ongoing norms of COVID-19 protocols. They achieved mean accuracy rates of 0.92, 0.95, 0.91, and 0.94 for the constructs perceived susceptibility, perceived severity, perceived benefits, and perceived barriers, respectively.

Therefore, the development of NLP-based methods for various healthcare data has helped assess various mental and physical aspects that could be affected by social media usage. In addition, the positive side of social media has helped to assess various personalized experiences from users for treating health-related conditions and using deep NLP techniques to extract such information.

15.3 Conclusions and future scope

NLP is paving the way for a better future of healthcare delivery and patient engagement. It will not be long before it allows doctors to devote as much time as possible to patient care while still assisting them in making informed decisions based on real-time, reliable results. By automating workflows, NLP is also reducing the amount of time being spent on administrative tasks. With the recent advances of deep NLP, the evaluation of voluminous data has become straightforward. We have outlined the methodological aspects and how recent works for various healthcare flows can be adopted for real-world problems. This largely helps in the clinics with inexperienced physicians over an underlying condition and handling critical situations and emergencies.

From all the sections discussed in our chapter, we can say that NLP is an upcoming digitized way of analyzing the vast number of medical records generated by doctors, clinics, etc. So, the data generated from the EHRs can be analyzed with NLP and efficiently be utilized in an innovative, efficient, and cost-friendly manner. There are different techniques for preprocessing techniques, as discussed in the first sections of the chapter, including the tokenization, Stop words removal, stemming, lemmatization, and PoS tagger techniques. Further, we went through various levels of analysis that can be utilized in text representations. And then, the text can be applied to frequency-based methods, embedding-based methods, which further can be used in machine and deep-learning-based methods. Different cases and implementations are also discussed in the later parts of the chapter.

References

Aczon, M., Ledbetter, D., Ho, L., Gunny, A., Flynn, A., Williams, J., & Wetzel, R. (2017). Dynamic mortality risk predictions in pediatric critical care using recurrent neural networks. *arXiv preprint arXiv:1701.06675*.

Aipe, A., Sundararaman, M. N., & Ekbal, A. (2019). Sentiment-aware recommendation system for healthcare using social media. *arXiv preprint arXiv:1909.08686*.

Bahdanau, D., Cho, K., and Bengio, Y. 2014. Neural machine translation by jointly learning to align and translate. *arXiv preprint arXiv:1409.0473*.

Bai, T., Chanda, A. K., Egleston, B. L., & Vucetic, S. (2018). EHR phenotyping via jointly embedding medical concepts and words into a unified vector space. *BMC Medical Informatics and Decision Making, 18*, Article number: 123.

Beam, A. L., & Kohane, I. S. (2016). Translating artificial intelligence into clinical care. *JAMA: The Journal of the American Medical Association, 346*, 456.

Choi, E., Bahadori, M. T., Schuetz, A., Stewart, W. F. & Sun, J. (2016). Doctor AI: Predicting clinical events via recurrent neural networks. In: *Proceedings of the 1st Machine Learning for Healthcare Conference, in PMLR* (Vol. 56, pp. 301−318).

Charles, D., Gabriel, M., & Furukawa, M. F. (2013). Adoption of electronic health record systems among US non-federal acute care hospitals: 2008-2014. *ONC data brief, 9*, 1−9.

Chen, Q., Leaman, R., Allot, A., Luo, L., Wei, C. H., Yan, S., & Lu, Z. (2021). Artificial Intelligence in Action: Addressing the COVID-19 Pandemic with Natural Language Processing. *Annual Review of Biomedical Data Science, 4*.

Chen, Q., & Sokolova, M. (2018). Word2vec and doc2vec in unsupervised sentiment analysis of clinical discharge summaries. *arXiv preprint arXiv:1805.00352*.

Deng, J., Dong, W., Socher, R., Li, L.-J., Li, K., & Fei-Fei, L. (2009). Imagenet: A large-scale image database. In: *Computer vision and pattern recognition, 2009. CVPR 2009. IEEE conference on*.

Dudchenko, A., & Kopanitsa, G. (2019). Comparison of word embeddings for extraction from medical records. *International Journal of Environmental Research and Public Health, 16* (22), 4360.

Edara, D. C., Vanukuri, L. P., Sistla, V., & Kolli, V. K. K. (2019). Sentiment analysis and text categorization of cancer medical records with LSTM. *Journal of Ambient Intelligence and Humanized Computing.* Available from https://doi.org/10.1007/s12652-019-01399-8.

El-Rashidy, N., El-Sappagh, S., Abuhmed, T., Abdelrazek, S., & El-Bakry, H. M. (2020). Intensive care unit mortality prediction: An improved patient-specific stacking ensemble model. *IEEE Access, 8,* 133541−133564. Available from https://doi.org/10.1109/ACCESS.2020.3010556.

Esteban, C., Staeck, O., Baier, S., Yang, Y., & Tresp, V. (2016). Predicting clinical events by combining static and dynamic information using recurrent neural networks. In: *2016 IEEE international conference on healthcare informatics (ICHI)* (pp. 93−101), Chicago, IL, USA. doi: 10.1109/ICHI.2016.16.

Esteva, A., Robicquet, A., Ramsundar, B., Kuleshov, V., DePristo, M., Chou, K., Cui, C., Corrado, G., Thrun, S., & Dean, J. (2019). A guide to deep learning in healthcare. *Nature Medicine, 25*(1), 24.

Friedman, C., & Johnson, S. B. (2006). *Natural language and text processing in biomedicine* (pp. 312−343). United States of America: Springer.

Ge, W., Huh, J. W., Park, Y. R., Lee, J. H., Kim, Y. H., & Turchin, A. (2018). An interpretable ICU mortality prediction model based on logistic regression and recurrent neural networks with LSTM units. *AMIA Annual Symposium Proceedings, 2018,* 460.

Ghosh, S., Chakraborty, P., Cohn, E., Brownstein, J.S., & Ramakrishna, N. (October 24−November 28, 2016). Characterizing diseases from unstructured text: A vocabulary driven word2vec approach. In: *CIKM'16.*

Gulshan, V., Peng, L., Coram, M., Stumpe, M. C., Wu, D., Narayanaswamy, A., Venugopalan, S., Widner, K., Madams, T., Cuadros, J., Kim, R., Raman, R., Nelson, P. C., Mega, J. L., & Webster, D. R. (2016). Development and validation of a deep learning algorithm for detection of diabetic retinopathy in retinal fundus photographs. *JAMA: The Journal of the American Medical Association, 304,* 649.

He, K., Zhang, X., Ren, S., & Sun, J. (2016). Deep residual learning for image recognition. In: *Proceedings of the IEEE conference on computer vision and pattern recognition.*

Ho, L. V., Aczon, M. D., Ledbetter, D., & Wetzel, R. (2019). Interpreting a recurrent neural network model for ICU mortality using learned binary masks. *arXiv preprint arXiv:1905.09865.*

Hochreiter, S., & Schmidhuber, J. (1997). Long short-term memory. *Neural Computation, 9*(8), 1735−1780.

Jelodar, H., Wang, Y., Orji, R., & Huang, S. (2020). Deep sentiment classification and topic discovery on novel coronavirus or COVID-19 online discussions: NLP using LSTM recurrent neural network approach. *IEEE Journal of Biomedical and Health Informatics, 24*(10), 2733−2742. Available from https://doi.org/10.1109/JBHI.2020.3001216.

Karunarathna, K.M.D.M. (January 2018). Predicting ICU death with summarized patient data. In: *Proceedings of the IEEE 8th annual computing and communication workshop conference (CCWC)* (pp. 238−247).

Kelly, Y., Zilanawala, A., Booker, C., & Sacker, A. (2018). Social media use and adolescent mental health: Findings from the UK Millennium Cohort Study. *EClinicalMedicine, 6,* 59−68.

Kim, J., Lee, J., Park, E., & Han, J. (2020). A deep learning model for detecting mental illness from user content on social media. *Scientific Reports, 10.* Available from https://doi.org/10.1038/s41598-020-68764-y.

Kovacs, G., & Kardkovacs, Z. T. (2014). Finding sequential patterns with TF-IDF metrics in health-care databases. *Acta Universitatis Sapientiae Informatica, 6*(2), 287−310.

Laacke, S., Mueller, R., Schomerus, G., & Salloch, S. (2020). Artificial intelligence, social media and depression. A new concept of health-related digital autonomy. *The American Journal of Bioethics*, 1−33.

Liu, B. (2010a). Sentiment analysis and subjectivity. *Handbook of natural language processing 2* (2010), 627−666.

Liu, B. (2010b). Sentiment analysis: A multi-faceted problem. *IEEE Intelligent Systems, 25,* 76−80.

Manke, S. N., & Shivale, N. (2015). A review on: Sentiment analysis mining and sentiment analysis based on natural language processing. *International Journal of Computer Applications, 109*(4).

Power, G. S., & Harrison, D. A. (2014). Why try to predict ICU outcomes? *Current Opinion in Critical Care, 20*(5), 544−549.

Raamkumar, A. S., Tan, S. G., & Wee, H. L. (2020). Use of health belief model−based deep learning classifiers for covid-19 social media content to examine public perceptions of physical distancing: Model development and case study. *JMIR Public Health and Surveillance, 6*(3), e20493.

Resnik, P., Niv, M., Nossal, M., Schnitzer, G., Stoner, J., Kapit, A., & Toren, R. (2006). Using intrinsic and extrinsic metrics to evaluate accuracy and facilitation in computer-assisted coding. *Perspectives in Health Information Management*.

Samuel, J., Ali, G. G., Rahman, M. N., Esawi, E., & Samuel, Y. (2020). COVID-19 public sentiment insights and machine learning for tweets classification. *Information, 11*(6), 314.

Shaw, B. M., Lee, G., & Benton, S. (2017). *Work smarter, not harder: Expanding the treatment capacity of a university counseling center using therapist-assisted online treatment for anxiety* (pp. 197−204). Cham: Springer International Publishing.

Shickel, B., Heesacker, M., Benton, S., & Rashidi, P. (2017). Hashtag healthcare: from tweets to mental health journals using deep transfer learning. *arXiv preprint arXiv:1708.01372*.

Simonyan, K., & Zisserman, A. (2014). Very deep convolutional networks for large-scale image recognition. *arXiv preprint arXiv:1409.1556*.

Soutner, D., & Müller, L. (2013). Application of LSTM neural networks in language modelling. In I. Habernal, & V. Matoušek (Eds.), *Text, speech, and dialogue. TSD 2013, Lecture notes in computer science* (vol. 8082). Berlin: Springer.

Szegedy, C., Vanhoucke, V., Ioffe, S., Shlens, J., & Wojna Z. (2016). Rethinking the inception architecture for computer vision. In: *Proceedings of the IEEE conference on computer vision and pattern recognition*.

Velupillai, S., Suominen, H., Liakata, M., Roberts, A., Shah, A. D., Morley, K., Osborn, D., Hayes, J., Stewart, R., Downs, J., Chapman, W., & Dutta, R. (2016). Using clinical natural language processing for health outcomes research: Overview and actionable suggestions for future advances. *Journal of Biomedical Informatics, 88*, 11−19.

Wang, Y., Huang, M., Zhu, X., & Zhao, L. (November 2016). Attention-based LSTM for aspect-level sentiment classification. In: *Proceedings of the 2016 conference on empirical methods in natural language processing* (pp. 606−615).

Yang, Y., Fasching, P. A., Wallwiener, M., Fehm, T. N., Brucker, S. Y., & Tresp, V. (2016). Predictive clinical decision support system with RNN encoding and tensor decoding. *arXiv preprint arXiv:1612.00611*.

Zhao, K., Yen, J., Greer, G., Qiu, B., Mitra, P., & Portier, K. (2014). Finding influential users of online health communities: A new metric based on sentiment influence. *Journal of the American Medical Informatics Association: JAMIA, 21*(e2), 1. Available from https://doi.org/10.1136/amiajnl-2013-002282.

CHAPTER 16

Measurement of the effects of parks on air pollution in megacities: do parks support health betterment?

Ahmet Anıl Müngen
Department of Software Engineering, OSTIM Technical University, Ankara, Turkey

16.1 Introduction

After settled life in history, urbanization and cohabitation in cities increase almost every century. Urbanization continues faster than ever due to technology development, the increase in population, and economic and geographical reasons. As urbanization increased and people gathered in one area, cities started to emerge. On the other hand, megacities can be named when cities are now similar to countries rather than cities. In theory, megacities should not be considered as large cities only in terms of population and area. Some of the main features of megacities are that they have an economic, political, and ecological ecosystem. Thus megacities actually resemble an independent structure and country rather than a large region.

Istanbul has always been a popular settlement throughout history as it connects two continents and is on the Bosporus and trade routes. Being the capital of more than one empire, Istanbul is historically and culturally important in human history. The population of Istanbul has always increased over the centuries since the day it was founded. According to official records (TUIK, 2021), the number of citizens living in Istanbul exceeded 15 million in 2020. As such, Istanbul has a population of more than 46 states in Europe, including the Netherlands, Greece, and Belgium, which are among the leading countries of Europe. Therefore a study covering Istanbul can actually be seen as a study covering a country.

It has been proven as a result of many medical studies that air quality negatively affects human health. It has been determined that some diseases

Edge-of-Things in Personalized Healthcare Support Systems
DOI: https://doi.org/10.1016/B978-0-323-90585-5.00015-1

occur in the long term due to people being constantly exposed to polluted air. There have been periods in many parts of the world that banned people from going out due to air pollution. Air pollution is seen in countries where coal and diesel are used instead of electricity and natural gas due to toxic outputs. In some cities heated by this type of fuel with dirty output, it was difficult for people to breathe in winter, and there were times when the local authorities prohibited going outside (Sümer, 2014).

Abelshon and Stieb published their short- and long-term effects of bad weather by compiling them from dozens of different academic studies in their article investigating the diseases caused by bad air quality. Some of the prominent diseases presented in the study of Abelshon and Stieb are given in Table 16.1 (Abelsohn & Stieb, 2011).

Most of the diseases mentioned in Table 16.1 affect a certain group, such as young or the elderly people and people of all ages and professions. For example, myocardial infarction is the leading cause of death and disability worldwide (Thygesen et al., 2012). Another important point in Abelshon and Stieb's study is that children are more affected because they are exposed to polluted air more than expected. Children spend more time outdoors and are exposed to more polluted and poisonous air due to higher air intake per kilogram due to their intense activities (Thygesen et al., 2012). According to the data discussed in this study, nitrogen

Table 16.1 Short and long-term effects of air pollution (Abelsohn & Stieb, 2011).

Time type	Type	Medical negative effect
Short-term	Cardiovascular	Myocardial infarction and ischemia rates
		Aggravation of Heart Failure and Increases of Arrhythmia
	Respiratory	Increased wheezing breathing
		COPD and increasing the effect of asthma
		Increase in Respiratory Infections
Long-term	Cardiovascular	Increased myocardial infarction rates
		Accelerated development of atherosclerosis
		Increased blood clotting
		Increase in systemic inflammatory markers
	Respiratory	Increasing rates of pneumonia
		Increase in the incidence of lung cancer
		Negative effects on lung development in children
	Reproductive	Increased prevalence of preterm births
		An increase in low birth weight

dioxide (NO2), carbon monoxide (CO), and sulfur dioxide (SO2) were found to be some of the main substances of air pollution (Thygesen et al., 2012). Another study (Bergstra et al., 2018) examining the effect of air pollution on children was conducted with a questionnaire applied to more than 500 children between the ages of 7 and 13 and their families. In this survey study, the researchers revealed the negative effects of outdoor air pollution and indoor air pollution on families and children's lung development.

According to the Air Pollution Report of Canadian Government (Health Canada, 2019), it is estimated that there are 14,600 premature deaths associated with air pollution in Canada each year. According to the same report, it was determined that the incidence of acute respiratory symptoms increased to 35 million days per year. Based on the same report's data, it is estimated that the total cost of air pollution in Canada to the Canadian economy is $114 billion annually. Air pollution has negative effects on countries' economies in the long run, although not instantaneously.

Laws and regulations in countries regulate events and actions that have a social denominator. We can divide regulations and laws into two as those that apply to institutions and those that apply to individuals. Regulations and laws that apply to individuals are the most prominent in terms of visibility in the media. Again, it can be seen that media awareness arises mostly with individual campaigns. On the other hand, for many types of pollution, the pollution caused by industry and production facilities is much higher than that caused by individuals. On the other hand, it is not preferred to take measures that will create economic costs by the legislators due to the work done by the industrial and production facilities with strong lobbies and the positive effect of these workplaces on the economy and unemployment. Therefore laws and regulations are too light or flexible applied for such institutional organizations. As a result, the majority of environmental pollution is industrial, although it has a little positive effect on individuals' individual efforts.

Between 1992 and 2007, 446 commitments related to environmental issues were made at the G7/8 Conferences. Among these interviews, in 1997, 2001, and 2002, air quality commitments were made (Warren, 2018). At the G8 summit in 2008, the aim was to reduce carbon-induced air pollution. They determined it to reduce by half by 2050. They did not set a more recent goal at the summit. In addition, they could not agree on a starting point to initiate measures regarding air pollution (AreaDevelopment, 2008).

Air pollution is a common problem all over the world and is encountered in many developed countries. In 2019 according to the data by the IQAir air pollution PM 2.5 concentrations Turkey in terms of pollution has on the second floor of the World Health Organization limits and has the worst air quality in 98 countries and 43 countries (IQAir, 2020). Although 21% of all countries are measured with tiny and little air pollution according to IQAir numbers and World Health Organization calculations, more than 90% of the whole world population lives in polluted air, shown in Fig. 16.1 and Fig. 16.2.

While many municipalities allocate more budgets for parks and gardens for ecological, political, and economic reasons, some municipalities do not take any action in this regard. In a small city and a district located in a rural area, the local municipality's afforestation or green fieldwork can be predicted to have a positive effect, albeit small. On the other hand, it is unknown exactly how the high number of parks and gardens affects human health in a megacity such as Istanbul.

The general opinion of local government officials and the community is that parks help reduce air pollution. This is because the trees in the park are scientifically known to reduce air pollution. Nevertheless, the extent of parks' positive impact on metropolitan areas and whether this effect can be measured is unknown. In this study, the relationship of parks in big

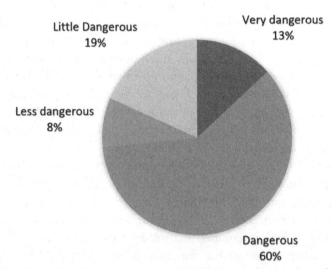

Figure 16.1 The ratio of country-based air pollution to the number of countries (IQAir, 2020).

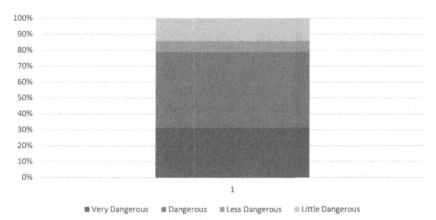

Figure 16.2 The ratio of country-based air pollution to the population of the countries (IQAir, 2020).

cities to air pollution was analyzed. In addition, this study has tried to reveal the effect of all parks in an area by looking at many other factors.

With the developing technology, Istanbul Municipality tried to measure the air quality for people and future planning by establishing dozens of air quality meter stations in different parts of Istanbul. This study has tried to measure whether the parks and gardens in Istanbul have a positive effect on air pollution by processing the parks and gardens' numbers and locations in Istanbul and the locations of the measuring devices.

16.2 Related works

This section deals with three key issues. In the first part, the previous studies on clean air and human health are compiled. In the second part, legal initiatives taken by some countries for clean air are mentioned. Finally, studies showing the relationship between trees and clean air are discussed.

There are many medical studies on clean air and human health. Kampa and Castanas (2008) also associated air pollution with respiratory and heart diseases, short0 and long-term premature deaths, and reduced life expectancy. Schraufnagel et al. (2019) showed in their study that reducing air pollution decreased the complaints of shortness of breath, cough, and sore throat within a few weeks, decreased school absenteeism and decreased the number of hospitalizations and preterm births. Gładka et al. (2018) showed that air pollution could affect respiratory diseases and the central nervous system by decreasing particles' brain barriers and leading to the early onset of diseases such as Alzheimer's or Parkinson's.

Also, there are different laws in some countries, especially the US Clean Air Act (Criteria et al., 2004) stating that clean air is a right of citizenship and should be provided by the state. Air Quality Index (AQI) is an index that publishes daily air quality and considers five main substances that deteriorate the quality of the air while calculating the quality. These parts are ground-level ozone, particle pollution, carbon monoxide, sulfur dioxide, and nitrogen dioxide and examine each of these particles' levels according to the level they should be. According to AQI, when the data obtained from the US measurements are evaluated, ground-level ozone and airborne particles are the leading pollutants that threaten human health (United States Environmental Protection Agency, 2014).

In the study of Nowak et al., the positive effect of forests and woodlands in the United States on air pollution and its effect on people getting sick was examined. It was estimated that trees and forests in the United States eliminated 17.4 million tons (t) of air pollution in 2010. When this positive effect of trees and forests is considered the decrease in the number of patients applying to the American health system, it was estimated that $6.8 billion was saved (Nowak et al., 2014) in their study. Hirabayashi and Nowak (2016) modeled the regional impact of trees and forests on human life by measuring forests and trees' performance in reducing air pollution, taking into account the changing leaf types for each province and district of the United States. Nowak et al. (2013) measured the impact of air pollution and human life for 10 US cities and investigated the relationship between changes in air pollution and mortality rates.

Jaffe et al. (2020) studied air pollution caused by forest fires and its impact on humans. It has revealed the seriousness of human exposure's health effects for long periods of fine particulate matter (PM) generated after forest fires. He conducted detailed research on the impact of these fires in the United States, where hundreds of forest fires are seen every year. Emmanuel (2000) also investigated the public health effects of many forest fires in Asia in 1997.

There are many academic studies on climate change and global warming, awareness studies of nature conservation associations, and independent organizations' measurements and forecasts. Many organizations have proposed some models for global warming. In most of the proposed models, it is predicted that the air quality decreases and decreases to the limit values for humans. Although climate change and global warming are also some of the main factors that reduce air quality, states either do not take any or very little steps in this regard. Tagaris et al. (2009) stated in their study that the decrease in air quality caused by climate change would harm two-thirds of the US continent, and premature death rates due to PM2.5 will increase. Orru et al. (2017) found that air pollution causes premature death of nearly 7 million children worldwide. At the same time, Orru has shown that climate change is a cycle that feeds itself and magnifies its negative effect in the diagram shown in Fig. 16.3 and that this effect grows both by human hands and by itself.

Although not addressed in this study, air pollution affects people because polluted air pollutes water and negatively affects people's health as polluted water. According to the study of Nagy et al. (2011), it has been explained extensively that air pollution and water pollution affect people through foods and their negative effects on human health. da Silveira et al. (2018) revealed that the effects of air pollution on foods cause obesity and negatively affect the body's healthy functioning. Sun et al. (2017) investigated the negative effects of air pollution on food during the growing process and the supply and storage process. It also

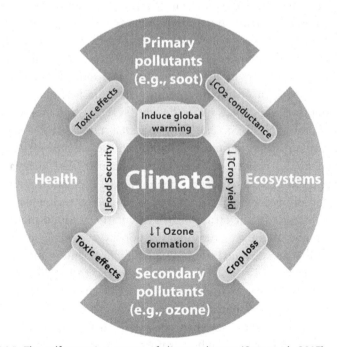

Figure 16.3 The self-nurturing nature of climate change (Orru et al., 2017).

mentions the measures and policies that can be taken to reduce the pollution in the supply chain caused by air pollution.

Yang et al. (2005) examined the number of trees in Beijing, China, which has been dealing with air pollution for a long time, and how you support reducing air pollution and calculated that it cleans more than 1000 tons of pollutants from the air annually. Nowak et al. (2006) demonstrated the effect of trees on air quality with a computer model of Cascine Park located in Florence, Italy. Since the study demonstrates the fight against air pollution based on tree species, it provides information about the type and variety of trees planted to reduce air pollution for regional and city planners. Escobedo and Nowak (2009) analyzed the relationship between air removal quality and urban forest structure measured in three socioeconomic subregions in Santiago, Chile.

There are various studies on air pollution in Istanbul. Çapraz et al. (2017) analyzed the data over a 3-year period and tried to find the relationship between air pollution in Istanbul and patients admitted to hospital with respiratory tract complaints. With these data, they grouped patients and revealed the link between contaminants and hospitalization rates. Çapraz et al. (2016) also demonstrated the link between daily air pollution and Istanbul's mortality rates. They also found that SO_2 was the most dangerous pollutant among air pollutants and showed that the excess of this substance in the air increased the mortality rates more than other substances. Onkal-Engin et al. (2004) proposed a fuzzy synthetic assessment technique for urban air quality using five different pollutant levels from five different air quality monitoring stations in Istanbul.

Air pollutant data such as sulfur dioxide (SO2), carbon monoxide (CO), nitrogen dioxide (NO2), ozone (O3), and total suspended PM collected at five different air quality monitoring stations in the western part of Istanbul were used in this assessment. The results obtained were compared with those applied to the United States Environmental Protection Agency AQI. It has been shown that fuzzy synthetic assessment techniques are very suitable techniques for air quality management. They found the highest air pollution points monthly, yearly, and seasonally according to SO2, CO, and PM in the data obtained from 10 different air pollution monitoring stations in Istanbul for 1 year. Yolsal (2016) showed that sudden changes in air pollution studies made from the data collected in Istanbul might not be realistic, and the studies should be done with the Seasonal Kendall Test.

More than 30 studies examined reveal that air pollution directly affects human health and that researchers from different parts of the world have their findings on this issue. It also reveals a one-to-one relationship between trees and forests and air pollution and that the lack of trees and forests will require the death of thousands of people and millions of people to receive hospital treatment. Again, different studies and different academicians have revealed that countries with a certain size of economy and health standards, especially the United States, put a burden on billions of dollars in budgets as a result of bad weather. Besides, it has been demonstrated that with the measures that regulators can take, it is possible to efficiently and rapidly reduce air pollution and increase air quality.

Studies have also revealed the economic effects of human health and polluted air, forests and air pollution, air and nutrient pollution, and polluted air. Despite this, studies have not examined the effects of parks and gardens in certain parts of the city on air quality in the city's relevant region.

16.3 The proposed method

Many open-source data providers can be used to measure air quality around the world. Some data providers provide data by country, some by city, and some by region. In general, it is possible to get information about a city's air pollution, as data providers are city based. On the other hand, in this study, a data set covering at least 6 months and at least a certain number of regions in the same city is required, from which the five main factors that constitute air pollution—PM10, SO2, O3, NO2, and CO—can be taken separately.

IBB Open Data Portal (Istanbul Metropolitan Municipality Air Quality Station Information Web Service, no date) has been established to provide free access to the data obtained by Istanbul Metropolitan Municipality for academic and other purposes. More than 100 datasets related to many different disciplines, including environment, health, and earth science, have been published. The data set shared by the Istanbul Municipality Environmental Protection and Control Department is available from 28 stations in the Istanbul region between June 1, 2019 and June 19, 2020. With this service, not only process but also raw data from air quality stations were obtained. With this data service, the raw data of PM10, SO2, O3, NO2, and CO measurements and the AQI of these parameters were calculated. The data acquisition and processing process are shown in Fig. 16.4.

Within the study's scope, measurements made between January 1, 2020 and June 1, 2020 were collected. In some cases, it appears that the

Figure 16.4 Data collection and data processing steps.

Table 16.2 Some statistics about data.

Data type	Number
Number of Air Quality Stations	28
Number of Parks and Gardens	3.626
Number of Collected Air Quality Data	96.634
Number of Invalid Data	2.692

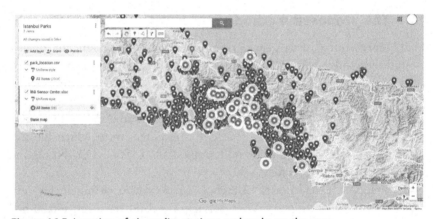

Figure 16.5 Location of air quality stations and parks on the map.

data are not recorded due to technical reasons in the measuring device. In such cases, the relevant records have not been taken into account. Information on the number of records is included in Table 16.2.

Another service on the IBB Open portal is the service where the names and locations of the parks and gardens in Istanbul are available. Parks are marked with red markers and measurement stations with blue markers on the map are shown in Fig. 16.5.

As shown in Fig. 16.5, the locations of the parks and the locations of the measurement stations are not distributed homogeneously. Therefore it is not possible to match a homogeneous number of parks to each measuring station. Also, it was thought that matching a park to a measurement station could also mislead the data. Although it can be said that parks increase, especially in places where there are more settlements, it can be predicted that there will be pollution due to the carbon gases created by people from there. In Fig. 16.6, park distributions in a 10 km area are shown.

Fig. 16.6 is a small section of Fig. 16.5. In Fig. 16.3, park distributions in a coastal area of only 10 km are shown. While calculating, the number of parks in the 2, 5, and 10 km radius of the measurement stations and the measurement station's data were calculated. The data were divided monthly and focused on higher than normal values taken monthly. In Fig. 16.7, it is written how many parks the stations cover at 2−5 and 10 km distances.

The US AQI was used to measure air's impact on human health. In the AQI, five main pollutants are included in the calculation. These are ground-level ozone, particulate pollution, carbon monoxide, sulfur

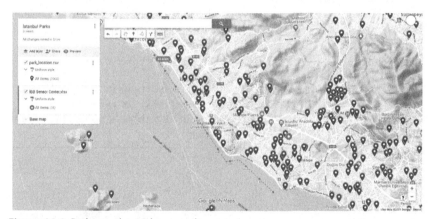

Figure 16.6 Parks on the 10-km coastline.

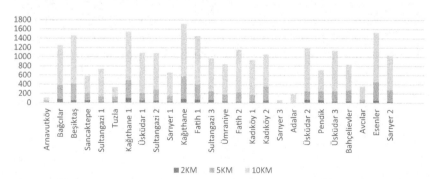

Figure 16.7 The proximity of the measuring stations to the parks according to the selected radio.

dioxide, and nitrogen dioxide. The AQI shows that the higher the AQI value, the higher the air pollution, a measure ranging from 0 to 500. Health criteria according to AQI are shown in Table 16.3. Each independent value is marked as a yellow risk if higher than 50 and an orange risk if higher than 100.

Fig. 16.8 shows that air pollution occurs on all days but mostly on weekends. Although it is not possible to determine the exact reason for this, it is thought that people have mobility with their private vehicles and intensive activities in the open area.

In Fig. 16.9, air pollution, which increases toward 12 at night, decreases with the minimum mobility. With the morning traffic and

Table 16.3 AQI quality criteria.

Air quality type	Score	Description
Green	0−50	Air quality is good, no risk
Yellow	51−100	Air quality is good but maybe a risk for some people
Orange	101−150	Air quality can be dangerous for vulnerable groups
Red	151−200	Air quality is very serious for vulnerable groups
Purple	201−300	Poor air quality, risky for everyone
Maroon	301	The air quality is terrible, it poses a big risk for everyone

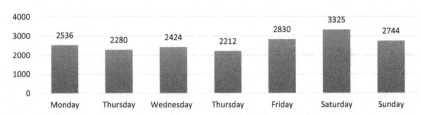

Figure 16.8 The number of Poor Air Quality Record for the day of the week.

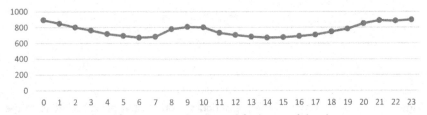

Figure 16.9 Number of Poor Air Quality Record for hours of the day.

activities, it rises with the sudden work of the factories in the mornings, it partially falls with the effect of photosynthesis of the trees, and continues to increase with the increase in mobility

16.3.1 Results

In Fig. 16.10, while NO2 and PM10 values are high even when there is no risk, an average increase of almost 100% can be seen when they enter the risk status. O3 level is normally very low, but it increases by 500% on average, creating risky air in some cases. SO2 is the value with the least proportional change for measurements classified with or without risk.

In Fig. 16.11, the number of parks and trees with a diameter of 2 km and the rates of the points that warn of air pollution are presented. As can be seen from the graph, as the number of parks and trees decreases, it

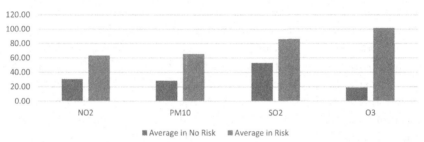

Figure 16.10 Comparison of the number of parks in a 2-km radius and 5-km radius via air pollution records.

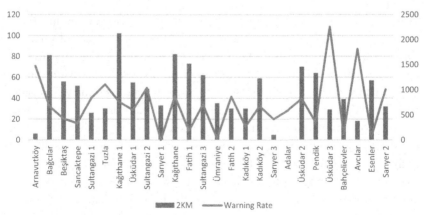

Figure 16.11 The relationship between the number of parks in a 2-km radius and air pollution records.

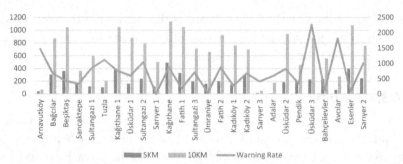

Figure 16.12 Comparison of the number of parks in a 5-km radius and 10-km radius over air pollution records.

gives more pollution warnings. This is the most important result showing that increasing the number of graphic parks and trees will reduce air pollution. It is thought that there are exceptions in Kağıthane and Bağcılar, which can be seen as exceptions in the graph since there are more production and industrial facilities compared to other districts.

The reverse ratio seen in 2 km between parks and risk numbers is not seen in 5 and 10 km. With this calculation, there is no evidence that parks and gardens effectively reduce air pollution up to 2 km but effective for 5 and 10 km, as shown in Fig. 16.12. Also, the PM10 violation accounts for more than half of all violations. The second most common violation is the NO2 violation.

16.4 Discussion

However, some issues are not addressed in this study, considering that it will increase the success rate if added. It is essential to take into account the size of each park first. On the other hand, there is no information about the park's size in the data obtained from the Istanbul Metropolitan Municipality service. In future studies, approximate size analysis of parks can be made by detecting green areas from satellite images. There is no information about the trees in these parks and the age and species of the trees. On the other hand, it is known that different types of trees affect air pollution at different rates. This problem can be overcome by classifying trees from wide-angle photographs of the parks. The number of sensor collection stations was less than it should be, both for Istanbul and according to the number of parks. Increasing the number of air quality measurement stations can increase the accuracy of future studies. In the study, variables that affect air pollution measurement metrics such as air temperature and humidity were not addressed. These variables should also be considered in finding the relationship between trees and air pollution to make them healthier. Since the data collection period coincides with when people use fewer vehicles, less movement, and factories work less due to the COVID-19 epidemic, air pollution seems less polluted than expected and values found in previous academic studies. However, it is predicted that air pollution will worsen after COVID-19.

By combining this study with the number of patients admitted to the hospital and treated with air pollution, the study's accuracy can be checked from another perspective. With these data obtained from health institutions, a new control mechanism may arise for working.

16.5 Conclusion

As a result of the study, the location and number of urban parks positively affect air quality. However, seasonal or periodical changes in carbon emissions, which are thought to be released from vehicles and solid fuels, are greater than the effects of parks and gardens. Therefore, parks and gardens are not effective solutions for cleaning the air on their own. It has been observed that permanent results can be obtained if working together by creating awareness for individuals with air pollution and determining policies for industrial production facilities. The park's size, the type of trees, the severity of the diseases, and the number of patients admitted to the hospital, which can change the study results, could not be used due to the lack of data. However, in future studies, disease and hospital data and regional estimates can increase the success rate. In addition, data such as the frequency of use of parks can be used to measure how parks affect the healthy living.

References

Abelsohn, A. R. & Stieb, D. M. (2011). Health effects of outdoor air pollution, *Canadian Family Physician*. College of Family Physicians of Canada. Available at: https://pmc/articles/PMC3155438/, Accessed 25.02.21.

AreaDevelopment (2008). *G-8 leaders agree to pollution cuts by 2050 - area development*. Available at: https://www.areadevelopment.com/newsItems/7−8-2008/g8leadersagreetopollutioncuts.shtml, Accessed 25.02.21.

Bergstra, A. D., Brunekreef, B., & Burdorf, A. (2018). The effect of industry-related air pollution on lung function and respiratory symptoms in school children. *Environmental Health: A Global Access Science Source*, 17(1), 30. Available from https://doi.org/10.1186/s12940-018-0373-2, BioMed Central Ltd.

Çapraz, Ö., Deniz, A., & Doğan, N. (2017). Effects of air pollution on respiratory hospital admissions in İstanbul, Turkey, 2013 to 2015. *Chemosphere, 181*, 544—550. Available from https://doi.org/10.1016/j.chemosphere.2017.04.105, Elsevier Ltd.

Çapraz, Ö., Efe, B., & Deniz, A. (2016). Study on the association between air pollution and mortality in İstanbul, 2007—2012. *Atmospheric Pollution Research, 7*(1), 147—154. Available from https://doi.org/10.1016/j.apr.2015.08.006, Dokuz Eylul Universitesi.

Criteria, A. Q., Matter, P., & Ii, V. (2004). Air quality criteria for particulate matter volume II of II. *Environmental Protection.*

da Silveira, C. G., et al. (2018). Subchronic air pollution exposure increases highly palatable food intake, modulates caloric efficiency and induces lipoperoxidation. *Inhalation Toxicology, 30*(9—10), 370—380. Available from https://doi.org/10.1080/08958378.2018.1530317, Taylor and Francis Ltd.

Emmanuel, S. C. (2000). Impact to lung health of haze from forest fires: The Singapore experience. *Respirology (Carlton, Vic.).* Available from https://doi.org/10.1046/j.1440-1843.2000.00247.x.

Escobedo, F. J., & Nowak, D. J. (2009). Spatial heterogeneity and air pollution removal by an urban forest. *Landscape and Urban Planning.* Available from https://doi.org/10.1016/j.landurbplan.2008.10.021.

Gładka, A., Rymaszewska, J., & Zatoński, T. (2018). Impact of air pollution on depression and suicide. *International Journal of Occupational Medicine and Environmental Health,* 711—721. Available from https://doi.org/10.13075/ijomeh.1896.01277, Nofer Institute of Occupational Medicine.

Health Canada (2019). *Health impacts of air pollution in canada: Estimates of morbidity and premature mortality outcomes.* Available at: http://publications.gc.ca/site/eng/9.874080/publication.html, Accessed 25.02.21.

Hirabayashi, S., & Nowak, D. J. (2016). Comprehensive national database of tree effects on air quality and human health in the United States. *Environmental Pollution.* Available from https://doi.org/10.1016/j.envpol.2016.04.068.

IQAir (2020). *World's most polluted countries in 2019 - PM2.5 ranking | AirVisual.* Available at: https://www.iqair.com/world-most-polluted-countries, Accessed 25.02.21.

Istanbul Metropolitan Municipality Air Quality Station Information Web Service (no date). Available at: https://data.ibb.gov.tr/tr/dataset/hava-kalitesi-istasyon-bilgileri-web-servisi, Accessed 26.02.21.

Jaffe, D. A., et al. (2020). Wildfire and prescribed burning impacts on air quality in the United States. *Journal of the Air and Waste Management Association.* Available from https://doi.org/10.1080/10962247.2020.1749731.

Kampa, M., & Castanas, E. (2008). Human health effects of air pollution. *Environmental Pollution,* 362—367. Available from https://doi.org/10.1016/j.envpol.2007.06.012, Environ Pollut.

Nagy, R. C., et al. (2011). Water resources and land use and cover in a humid region: The Southeastern United States. *Journal of Environmental Quality.* Available from https://doi.org/10.2134/jeq2010.0365.

Nowak, D. J., et al. (2013). Modeled PM2.5 removal by trees in ten U.S. cities and associated health effects. *Environmental Pollution.* Available from https://doi.org/10.1016/j.envpol.2013.03.050.

Nowak, D. J., et al. (2014). Tree and forest effects on air quality and human health in the United States. *Environmental Pollution.* Available from https://doi.org/10.1016/j.envpol.2014.05.028.

Nowak, D. J., Crane, D. E., & Stevens, J. C. (2006). Air pollution removal by urban trees and shrubs in the United States. *Urban Forestry and Urban Greening.* Available from https://doi.org/10.1016/j.ufug.2006.01.007.

Onkal-Engin, G., Demir, I., & Hiz, H. (2004). Assessment of urban air quality in Istanbul using fuzzy synthetic evaluation. *Atmospheric Environment*, *38*(23), 3809−3815. Available from https://doi.org/10.1016/j.atmosenv.2004.03.058, Pergamon.

Orru, H., Ebi, K. L., & Forsberg, B. (2017). the interplay of climate change and air pollution on health. *Current Environmental Health Reports*, 504−513. Available from https://doi.org/10.1007/s40572-017-0168-6, Springer.

Schraufnagel, D. E., et al. (2019). Health benefits of air pollution reduction. *Annals of the American Thoracic Society*, 1478−1487. Available from https://doi.org/10.1513/AnnalsATS.201907-538CME, American Thoracic Society.

Sümer, G. Ç. (2014). Hava Kirliği Kontrolü: Türkiye'de Hava Kirliliğini Önlemeye Yönelik Yasal Düzenlemele-rin ve Örgütlenmelerin İncelenmesi. *Uluslararası İktisadi ve İdari İncelemeler Dergisi. International Journal of Economics and Administrative Studies, 13* (13). Available from https://doi.org/10.18092/ulikidince.232135, pp. 37−37.

Sun, F., Dai, Y., & Yu, X. (2017). Air pollution, food production and food security: A review from the perspective of food system. *Journal of Integrative Agriculture. Chinese Academy of Agricultural Sciences*, 2945−2962. Available from https://doi.org/10.1016/S2095-3119(17)61814-8.

Tagaris, E., et al. (2009). Potential impact of climate change on air pollution-related human health effects. *Environmental Science and Technology*. Available from https://doi.org/10.1021/es803650w.

Thygesen, K., et al. (2012). Third universal definition of myocardial infarction. *European Heart Journal*, *33*(20), 2551−2567. Available from https://doi.org/10.1093/eurheartj/ehs184.

TUIK (2021) *address based population registration statistics*. Available at: https://data.tuik.gov.tr/Bulten/Index?p = Adrese-Dayalı-Nüfus-Kayıt-Sistemi-Sonuçları-2020-37210&dil = 1, Accessed 25.02.21.

United States Environmental Protection Agency (2014). Air Quality Index: A guide to air quality and your health', *EPA*.

Warren, B. (2018). *G7/8 commitments on environment, 1992-2017*. Available at: http://www.g7.utoronto.ca/evaluations/g7-commitments-environment.html, Accessed 25.02.21.

Yang, J., et al. (2005). The urban forest in Beijing and its role in air pollution reduction. *Urban Forestry and Urban Greening*. Available from https://doi.org/10.1016/j.ufug.2004.09.001.

Yolsal, H. (2016). Estimation of the air quality trends in Istanbul. *İktisadi ve İdari Bilimler Dergisi. M.U. İktisadi ve İdari Bilimler Dergisi*, *38*(1), 375. Available from https://doi.org/10.14780/iibd.98771.

CHAPTER 17

Internet of Things use case applications for COVID-19

Mohammad Nasajpour[1], Seyedamin Pouriyeh[1], Reza M. Parizi[2], Liang Zhao[1] and Lei Li[1]

[1]Department of Information Technology, Kennesaw State University, Marietta, GA, United States
[2]Department of Software Engineering and Game Development, Kennesaw State University, Marietta, GA, United States

17.1 Introduction

The Internet of Things (IoT) is an arising technology enabling communications between various types of sensors or devices that would enhance the quality of our daily life in different domains (Jara et al., 2009). Recent statistics predict that the number of IoT devices will grow exponentially in the coming years and reach 125 billion IoT devices by 2030 (Statista, 2019). These devices transmit data by using network technologies, which enables easy communication between machines and humans. IoT technology could achieve substantial improvements by integrating various technologies such as Machine Learning (ML) (Adi et al., 2020; Aledhari et al., 2020), blockchain (Ekramifard et al., 2020; Połap et al., 2020; Reyna et al., 2018; Yazdinejad et al., 2021), etc., which opens great areas for applying IoT technology in different domains such as education (Gul et al., 2017), industrial settings (Da et al., 2014), healthcare (Islam et al., 2015), smart homes (Kamaludeen et al.), security (Saharkhizan et al., 2020), etc.

Healthcare is one of the domains that has received significant benefits from using IoT technology by reducing costs, improving health services, and enhancing the user's experience (Islam et al., 2015; Qi et al., 2017). Recent studies have also confirmed the role of IoT technology in this domain where the market value of IoT technology is expected to reach more than USD 188 billion in 2025 (Marketsandmarkets, 2020).

In December 2019, a severe contagious disease caused by coronavirus 2, called COVID-19, was appeared in Wuhan, China for the first time (Sohrabi et al., 2020). COVID-19 is a deadly airborne disease transmitted in respiratory droplets. Due to the high infection and death rates of COVID-19, the World Health Organization (WHO) declared this virus a

Edge-of-Things in Personalized Healthcare Support Systems
DOI: https://doi.org/10.1016/B978-0-323-90585-5.00016-3

pandemic after only 3 months (Cucinotta & Vanelli, 2020). This novel severe respiratory syndrome coronavirus 2 has killed a large number of people (more than 2.6 million), which makes it one of the deadliest pandemics in history (WHO, 2021). COVID-19 symptoms could be close to the flu's, such as, fever, muscle pain, and sore throat. In addition to flu symptoms, COVID-19 could also demonstrate loss of taste or smell, symptoms that are crucial to diagnose the virus early (CDC, 2021a). The incubation period of a patient infected with COVID-19 could last up to 14 days, which increases the chance of virus transmission (Lauer et al., 2020). Surprisingly, some people could be infected without showing any specific symptoms. As a result, quarantining infected patients and asymptomatic patients would be essential to stop the chain of transmission (Güner et al., 2020). Moreover, for the COVID-19 dilemma, dealing with confirmed/suspected cases is critically important, which has caused health authorities to apply different guidelines such as wearing masks and washing hands more frequently. Although these guidelines have efficiently reduced the negative impacts of coronavirus, applying different technologies, including Artificial Intelligence (AI) (Vaishya et al., 2020), ML (Shahid et al., 2021), IoT (Nasajpour et al., 2020), etc., could greatly assist with mitigating this virus. In particular, IoT has empowered authorities with different remote capabilities such as monitoring and diagnosing to lessen the number of infections and deaths.

In this chapter, we focus on the state-of-art IoT applications for the current pandemic and how those devices can support both patients and healthcare professionals in combating COVID-19 in different phases. In our study, we investigate the role of IoT devices and applications in five main categories: monitoring, diagnosing, tracing, disinfecting, and vaccinating. We focus on how IoT technology could improve the performance of healthcare services and mitigate the impacts of COVID-19 on patients, healthcare providers, government authorities, and communities. Table 17.1 demonstrates the perspective of the proposed approaches in these task areas.

The remainder of this chapter is organized as follows: we first demonstrate the key role of IoT within the current and future pandemics in Section 17.2. Then, Section 17.3 covers the IoT applications to diagnose and detect patients infected with COVID-19. Section 17.4 demonstrates the monitoring aspect of COVID-19. Other tasks of tracing, disinfecting, and vaccinating will be discussed in Sections 17.5, 17.6, and 17.7 respectively.

Application	Concentration	Ref.	Purpose	Sensor(s) / Method(s)	Communication	Task
Indoor monitoring	Safety rules	Petrović and Kocić (2020)	Cost-effective guidelines monitoring	Temperature and thermal camera	MQTT	–
		Fazio et al. (2020)	Cost-effective navigation system	Mobile sensors	BLE	–
		Bashir et al. (2020)	Cost-effective SOP monitoring	Temperature and ToF	WebSocket	–
	Air quality	Mumtaz et al. (2021)	Regularly air quality capturing	Gas and particle/NN and LSTM	ATmega328P and NodeMCU WiFi Chip	–
	Surface contamination	Stolojescu-Crisan et al. (2020)	Contamination reducing	Temperature and motion	WiFi	–
	Occupancy monitoring	Fernández-Caramés et al. (2020)	Securely occupancy level monitoring	Bluetooth and NFC	MQTT and Node-RED	–
Symptoms monitoring	Blood saturation	Miron-Alexe (2020)	Vital signs remote monitoring	Oximetry	WiFi	–
	Respiratory signs	Valero et al. (2020)	Cost-effective respiratory signs monitoring	Pressure	WiFi	–
		Al-Shalabi (2020)	Respiratory signs monitoring	Temperature	WiFi	–
		Chloros and Ringas (2020)	Mapping potential contamination areas	Temperature	Bluetooth	–
Airport maintenance	Controlling factors	Sales Mendes et al. (2020)	Airports spread preventing	PIR and ToF and temperature and humidity	LoRa	–
Health condition systems	Cloud-based IoT	Akhbarifar et al. (2020)	High risk patients monitoring	Bio-sensors/K-Star and RF and SVM and MLP	MQTT	–
	Robotic system	Kanade et al. (2021)	Assisting hospitals with better patient's management	Temperature	Bluetooth	–
	COVID-SAFE	Vedaei et al. (2020)	Distance and vital signs monitoring	Temperature/decision tree and SVM and PPG	WiFi and cellular and LoRa	–

(Continued)

Table 17.1 (Continued)

Application	Concentration	Ref.	Purpose	Sensor(s) / Method(s)	Communication	Task
Symptomatic patient	Finger-touch	Hasan (2021)	Fever detecting for stopping the spread	Ultrasonic and RFID and temperature	MQTT	II
	Bracelet	Cacovean et al. (2020)	Infected patients detecting	Temperature and heart rate and GPS	WiFi	II
	Smart helmet	Mohammed, Syamsudin, Al-Zubaidi et al. (2020)	High-temperature detecting of passers-by	Thermal camera	WiFi and GPS	II
	Smart glasses	Mohammed, Hazairin, Syamsudin et al. (2020)	High-temperature detecting of passers-by	Temperature	WiFi	II
	Drone	Mohammed, Hazairin, Al-Zubaidi et al. (2020)	High-temperature detecting of passers-by	Temperature	WiFi	II
	Robotic	Karmore et al. (2020)	Cost-effective system diagnosing patients	Temperature and ABS and GSR and ECS and EMS	WiFi	II
Asymptomatic patient	Smart hospital	Abdulkareem et al. (2021)	Hospital's workload reducing and infected patients detecting	NB and SVM and RF	Cloud	II
	Medical imaging	Ahmed et al. (2020)	Reducing workload and detecting infected patients	Faster-RCNN and ResNet-101	–	II
	Health condition systems	Arun et al. (2020)	Detecting suspected cases	Oximetry and blood pressure	GSM	II
		Thangamani et al. (2020)	Detecting and preventing infections	Temperature and sound and blood pressure	GSM and WiFi	II

	Wearable device			CNN	Cellular	
Social distancing violators		Kumbhar et al. (2020)	Identifying social distancing violators			II
Smartphone-based	Automated tracing	Rajasekar (2021)	Tracing suspected cases	RFID	NFC	III
	IoTrace	Tedeschi et al. (2020)	Cost-effective secure tracing system	Totem	BLE	III
Wearable-based	Digital PPE	Woodward et al. (2020)	Performing proper social distancing	Received signal strength indicator	BLE and WiFi	III
	IoT–Q–Band	Singh, Chandna et al. (2020)	Preventing quarantine absconding	ESP32 kit	WiFi and GPS	III
	EasyBand	Tripathy et al. (2020)	Performing proper social distancing	Temperature and biosensor	BLE	III
Robotic technology	Sanitizer spraying	Mohammed, Arif et al. (2020)	Disinfecting areas without human interactions	Ultrasonic	GSM and GPS	IV
	Hand sanitizer dispenser	Eddy et al. (2020)	Providing disinfectant for people nearby Infrared Distance	Ultrasonic	GSM and GPS	IV
Drone	Sanitizer spraying	Mohammed, Syamsudin, Hazairin et al. (2020)	Disinfecting areas without human interactions	Optical camera	GPS	IV

ABS, Airflow Breathing Sensor; *BLE*, Bluetooth Low Energy; *ECS*, Electrocardiography Sensor; *EMS*, Electromyogram Sensor; *GSM*, Global System for Mobile communications; *GPS*, Global Positioning System; *GSR*, Galvanic Skin Response sensor; *MLP*, MultiLayer Perception; *MQTT*, Message Queuing Telemetry Transport; *NFC*, Near-Field Communication; *PPG*, Photoplethysmogram sensor; *ToF*, Time of Flight.

17.2 IoT key role in COVID-19

The adverse effects of coronavirus on different parts of society have created an essential need for different techniques to mitigate this virus. This could be achieved by first stopping the spread and then vaccinating people (Zhang et al., 2020). However, manually monitoring and diagnosing infected patients could result in latency, high costs, or even more infections.

IoT, as an emerging technology in healthcare, has delivered superior results by reducing costs and errors, improving user experiences and treatment procedures (Bhatt & Bhatt, 2017; Zhang et al., 2020). It has also been recognized as a reliable technology that can be utilized in different stages of COVID-19.

In this study as illustrated in Fig. 17.1, we investigate the role of IoT technology in combating COVID-19 in five separate stages: monitoring, diagnosing, tracing, disinfecting, and vaccination.

The world is dealing with a deadly pandemic, which has caused more than 2.7 million deaths as of March 20, 2021 (WHO, 2021). This opens an essential area for avoiding further contamination, where IoT provides various available remote monitoring applications. Basically, these applications are aimed to monitor people based on different measurements such as respiratory signs, social distancing practices, mask-wearing, and so on. In Section 17.3, we demonstrate the various IoT applications performing the monitoring task.

One of the first major tasks for mitigating COVID-19 is to diagnose people infected with this virus. This is being done by performing a combination of testing measures such as nucleic acid testing, protein testing, and Computed Tomography (CT) (Udugama et al., 2020). The earlier the virus is diagnosed, the faster and the better the treatment is. Moreover, isolating the patient would be more effective and as a result, the contamination rates would decrease dramatically. IoT technology could assist both healthcare professionals and patients in diagnosing COVID-19 and enhance the efficiency of the COVID-19 diagnosis procedure. In Section 17.4, the applicable IoT systems for diagnosing or detecting the coronavirus will be discussed.

In general, tracking infected or suspected patients plays an important role in combating COVID-19. This strategy is more highlighted during quarantine time and even after lockdown. Using different digital contact tracing techniques has shown promising results in the current pandemic

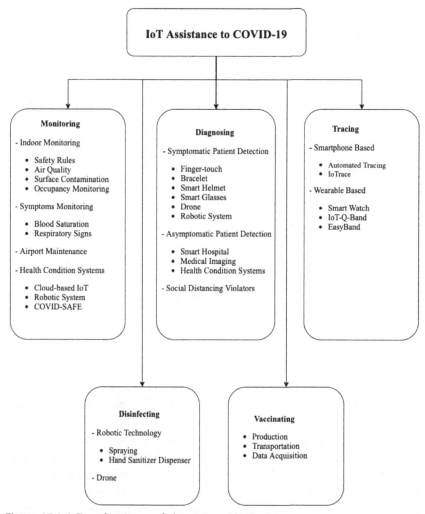

Figure 17.1 IoT applications to fight against COVID-19.

during which these technologies assist health authorities by detecting people who were in proximity of contaminated cases (Ferretti et al., 2020). Various interfaces such as Bluetooth Low Energy (BLE), Global Navigation Satellite System, etc. could be adopted for broadcasting captured data within the systems. In Section 17.5, we focus on the contact tracing applications that are embedded within the IoT devices to assist healthcare authorities with tracing tasks for different phases.

Another important task of IoT technology during the COVID-19 pandemic is the cleaning and disinfecting public spaces, surfaces, devices, etc. to prevent further transmission of the virus, which could happen either directly or indirectly (WHO, 2020a). As a result, different IoT-based devices such as robots and drones have been designed to accomplish those tasks. Section 17.6 will focus on the applicable utilization of IoT in this area.

The last key phase of the COVID-19 pandemic is vaccinating, which was first allowed for emergency use on December 11, 2020 (Oliver et al., 2020). Regarding the rapid production of COVID-19 vaccines and their effectiveness, the world might get back to normal life at the end of 2021 (Powell, 2020). However, the opposite might be true because of two factors. First, the vaccination timeline for the second dose might not be met by patients, and also the temperature of vaccines might not be monitored properly as it is crucial for some vaccines such as the Pfizer COVID-19 vaccine (Kim et al., 2021). Consequently, applying IoT technology to prevent such challenges could benefit and speed up the vaccination phase.

17.3 Monitoring

In the past several years, IoT technology has been widely deployed in different domains, particularly healthcare. A variety of IoT-based monitoring systems have been utilized for executing different tasks such as abnormal blood pressure detection, oxygen saturation, and glucose level. However, monitoring applications are not limited to vital signs, and the recent COVID-19 pandemic has increased the need for screening applications with respect to safety guidelines. The initial monitoring components during COVID-19 can be divided into indoors, symptoms, air maintenance, and health system monitoring applications as follows in the sections below. Taken together, these applications could reduce the workload, errors, and costs of manually monitoring (Haleem et al., 2020), which could potentially slow the spread of COVID-19.

17.3.1 Indoor monitoring

One of the major topics to be investigated during COVID-19 is the methods for stopping the chain of transmission among people. This virus could be transmitted from public surfaces located outdoors or even indoors. Although this remains an open problem in dealing with COVID-19, governments required measures such as social distancing, wearing masks, etc. to limit the spread of coronavirus. Moreover, the quality of consumable air by people has gained attention as well. We further focus on other applicable approaches to indoor monitoring by considering safety rules, air quality, surface contamination, and occupancy of a building.

17.3.1.1 Safety rules

Due to the required safety measurements from governments, various technologies have been developed to follow them accordingly for indoor areas. That being so, a safety monitoring system based on IoT technology was designed by Petrović and Kocić (2020). This system enables monitoring of three important aspects: temperature, mask-wearing, and physical distance. Additionally, the proposed system demonstrated an average accuracy of 87.5% and 69% for correctly detected mask-wearing and physical distance. This cost-effective system applies sensors and computer vision technology to monitor if the rules are being followed. In this case, the safety rules violators inside a building could be captured and alerted (Petrović & Kocić, 2020).

Another study was conducted on monitoring the proximity between people to ensure the practice of social distancing by people inside a building (Fazio et al., 2020). The authors of this research study adopted a low-cost IoT-based indoor navigation system for the prevention of gatherings. BLE was utilized for communicating among the users' smartphones using their beacons. As a result, an appropriate path will be provided based on the recorded beacons, so that the user could safely navigate inside the building. While the proposed configuration achieved a descent coverage and decreased interference, this approach can cover a broader zone using lower BLE Beacons. This not only reduces the costs but also enhances the energy efficiency.

Although the recent lockdowns effectively reduced the spread of COVID-19, they also caused the economic recession and unemployment soar (Bartik et al., 2020). However, reopening businesses was allowed for these reasons, it might increase the contamination level among society (Gregory et al., 2020). Therefore social distancing scenarios have been proposed to prevent the contamination and control the pandemic (Baqaee et al., 2020). In another attempt by Bashir et al. (2020), an IoT-based system equipped with a variety of sensors was proposed to ensure that standard operating procedures (SOP) against COVID-19 are being practiced. This low-cost system performed several tasks, including counting people, measuring temperature, and calculating the proximity. As a result, a real-time monitoring system with a centralized server was adopted for businesses and offices. Additionally, collected data from people did not include personal data, which preserved the privacy of users. Overall, this system is applicable to various sectors of SOP using different functionalities of contact tracing or distance violators detecting (Bashir et al., 2020).

17.3.1.2 Air quality

As most people of the world are following social distancing and stay-at-home orders, businesses have demonstrated a potential opportunity for remote working from home. Although this avoids expanding the contamination, it has impacted indoor air quality due to the widely applied lockdowns. This has caused a high increase in the pollutant levels of indoor areas compared to outdoor air quality (Kumari & Toshniwal, 2020). Since the coronavirus could be transferred through air, it is crucial to ensure buildings, including houses, offices, medical centers, etc., maintain a decent Air Quality Index. According to the (CDC, 2020a), the main pollutants could be carbon monoxide, nitrogen oxide, particle matter, etc. These pollutants can possibly enhance the risks of infection. Consequently, the concentration of these pollutants could deteriorate the death ratio of COVID-19 infections. Additionally, air quality monitoring is essential for vulnerable people with respiratory conditions (EPA, 2020).

With respect to the possible risks of air pollutants, indoor air quality monitoring has been carefully considered so that the risks for air consumers inside buildings could be lessened. During the COVID-19 pandemic, indoor contamination poses a crucial challenge for monitoring and enhancing air quality. Along with an IoT monitoring system for this purpose, Mumtaz et al. (2021) also proposed an ML analytic system to predict the rate of pollutants in the near future. In this study, continuous reports of the air quality conditions will be sent to a server, which could be monitored on a web interface and mobile application. This is where ML algorithms are applied to classify the captured air quality conditions, which Neural Network (NN) outperformed the rest of the algorithms. Besides, Long Term Short Memory (LSTM) was applied for the prediction of two other purposes including the air pollutants concentration and air quality. Altogether, the proposed sensing system achieved the highest accuracy of 99.1% and 99.37% for NN and LSTM, respectively. However, the sensors' life period and their calibration could be major challenges for the long-term use of this system (Mumtaz et al., 2021).

17.3.1.3 Surface contamination

After governments decided to reopen businesses, different requested rules were not enough to stop the spread of this virus. Disinfecting surfaces to prevent infections that could be caused by touching contaminated surfaces has gained considerable attention. However, it could be tough to minimize the contamination of surfaces only by disinfecting. To address that, Stolojescu-Crisan et al. (2020) proposed an approach for preventing the people from touching surfaces. This IoT-based system called qToggle adopts various technologies (sensors, Raspberry Pi, and smartphones) to better manage the surface contamination inside offices. qToggle enables various technologies, including objects and devices connected together on a robust Application Programming Interface. This system could perform tasks related to the objects inside a building, for example, opening and closing a door, turning on/off the lights, etc. Since users perform these tasks using a smartphone, surface contamination could be avoided to lessen the impacts of pandemic. In addition, qToggle could monitor the power usage of different appliances, which mainly helps with lowering the energy consumption during the quarantine (Stolojescu-Crisan et al., 2020).

17.3.1.4 Occupancy monitoring

With respect to the various preventive procedures (social distancing, wearing face masks, etc.) after reopening businesses, monitoring the occupancy of people within specific places could be crucial. This could matter more for overcrowded places such as public transportation. However, counting technologies (e.g., automated passenger counters) might not be efficient within overcrowded places due to counting errors. Consequently, an efficient alternative could be the adoption of various wireless identification technologies such as Radio Frequency Identification (RFID), Quick Response, etc. Fernández-Caramés et al. (2020) proposed an IoT system embedded with blockchain, which aims to monitor occupancy during the pandemic. The autonomous wireless devices were adopted to enable the system to be independent of active actions from users. The collected data do not maintain any personal information from users, which could enhance user privacy. Also, the use of blockchain within a decentralized manner could avoid security attacks. To evaluate the proposed system, estimating the occupancy level was considered within two facilities. The results demonstrated a sufficient accuracy for the estimated occupancy while having a low delay time. Overall, such an occupancy level monitoring system could ensure maintaining social distancing measures.

17.3.2 Symptoms monitoring

As mentioned, COVID-19 could be transmitted more quickly among people if the social distancing measures are not practiced. Since there are various symptoms that could appear in infected patients, there is also a need for monitoring these symptoms. However, it is critical to ensure normal people remain with good hygiene. This leads to a path of monitoring people who have been in close contact with confirmed cases because of the higher chance of infections they have. In the following sections, we will focus on proposed applications for monitoring blood saturation and respiratory signs of suspected/confirmed COVID-19 patients.

17.3.2.1 Blood saturation

Blood oxygen saturation has been considered as one of the major factors for evaluating the health condition of the lungs (Nocturnal Oxygen Therapy Trial Group, 1980). To assess the oxygen level of the blood, various tests have been adopted. However, pulse oximetry has been widely adopted as an indirect method for evaluating the health condition of breathing (Ortega et al., 2011). With respect to the current pandemic, monitoring infected patients, who are asked to be quarantined until recovery, is critical. For this reason, Miron–Alexe (2020) proposed an IoT application utilizing a pulse oximetry sensor to check on the heart rate and oxygen level of the infected quarantined patient. This system keeps health staff from being infected by preventing them from having physical contact with patients. The system provides virtual monitoring for physicians to ensure the patients' health. In comparison to a commercial pulse oximeter, the proposed approach demonstrated good accuracy while it could have lower energy consumption.

17.3.2.2 Respiratory signs

In addition to the aforementioned types of symptoms that infected patients could show, respiratory signs can be considered when diagnosing COVID-19. This could be potentially due to the damage that has been done to the patients' lungs. An effective method of evaluating the lungs' health is to perform pulmonary function tests inside hospitals. However, as a result of massive numbers of infected patients with severe symptoms, it is important to keep the suspected cases or patients with mild symptoms mostly isolated inside their homes. Consequently, monitoring these signs is essential for isolated patients during COVID-19. The R-Mon tool was proposed to monitor respiratory signs of isolated patients (Valero et al., 2020). Pressure sensors within this tool enable recording of the respiration rates of patients. This low-cost tool provides the collected data (including patients' conditions) using a cloud platform to be monitored and analyzed by physicians. In general, the authors demonstrated an mhealth tool that can be applied for real-time monitoring of patients that are suffering from pulmonary deterioration during COVID-19.

Similarly, a study based on IoT and Wireless Sensor Network was adopted by Al-Shalabi (2020) to monitor quarantined suspected cases, especially elderly people. This approach enables health providers to effectively monitor the captured health data, including the changes in the body temperature. Consequently, if the system detects any symptoms, it will alert the health provider regarding the patient's health situation. In this case, further actions could be applied to save the patient's life. The authors deployed a visualization technology called ThingSpeak to better clarify the data for physicians. The live monitoring of such patients enables physicians to assist more cases on a larger scale (Al-Shalabi, 2020).

As COVID-19 side effects are mainly similar to flu's, Chloros and Ringas (2020) proposed an IoT application, called Fluspot, for symptom monitoring. The main objective of Fluspot is to avoid the widespread of flu and also COVID-19. This application was developed by adopting three major technologies, including a wearable device, mobile application, and cloud. In addition, Fluspot provides a map of potential infections using collected temperature data from users. This makes users aware of possible contamination areas so that they can move on in a precautionary manner. Unlike most of the recently adopted applications, the proposed system could be applied worldwide even where there are internet limitations.

17.3.3 Airport maintenance

One of the major places that could easily worsen the spread of the virus are domestic and international airports where in some cases more than 300,000 travelers are transferred in one single day from a gateway. One of the big concerns in the airports with respect to COVID-19 contamination is the restrooms where it can easily be a part of transmitting the virus. As a result, maintenance and disinfection of those places are highly required within the airports. The crowd of people could increase the chance of contamination for each traveler, which means restrooms must be cleaned regularly. Sales Mendes et al. (2020) proposed an approach for toilets maintenance, which basically focuses on the status of various factors. Controlling factors such as soap level, distances, and temperature have been adopted in this system. Several sensors and technologies communicating with the LoRa protocol are enabled with a multiagent system. Additionally, the proposed approach demonstrated a successful monitoring system for those aforementioned factors inside an airport by minimizing the workload of the airport services and enhancing the cleanliness of the airport resulting in the users' satisfaction. Altogether, the proposed could prevent the spread of the virus among travelers inside the airport with minimized cost and energy consumption (Sales Mendes et al., 2020).

17.3.4 Health condition monitoring systems

With regard to the severity of COVID-19 treatment, it is important to keep patients with specific diseases safe from this virus. In general, COVID-19 could have severe impacts on patients dealing with serious diseases such as cancer, diabetes, and hypertension (Zaki et al., 2020). The need for computer-aided technologies has enabled researchers to apply various technologies such as IoT and AI to help those patients with underlying conditions. The major advancements of IoT and sensors within a cloud have been successfully implemented for different aspects of the healthcare domain. Consequently, different health systems have been developed for COVID-19. We will focus on the monitoring systems in the sections below.

17.3.4.1 Cloud-based IoT

To achieve a prediction system building on IoT and the cloud, data mining techniques, and ML algorithms have been applied for superior monitoring or diagnosing of patients with various health conditions (Kumar et al., 2018). Accordingly, Akhbarifar et al. (2020) proposed a monitoring system based on IoT and the cloud while the data collected by the cloud will be adopted to build an ML model. The model is focused on diagnosing patients with heart disorders and hypertension disease for whom COVID-19 could possibly be life-threatening. The analytical results are basically achieved using various algorithms including J48, Support Vector Machine (SVM), Random Forest (RF), etc., which will be sent to health staff for further actions. The authors also applied a lightweight encryption method within the system to maintain security. Ultimately, the outcomes of the evaluation demonstrated K-star as the best classifier with an accuracy of 95%, which demonstrates an effective and secure health monitoring system using the IoT and cloud (Akhbarifar et al., 2020).

17.3.4.2 Robotic system

Kanade et al. (2021) have proposed a robot-based approach, which can maintain detecting and remote monitoring of patients with a high risk of infection. The solution applied several technologies, including computer vision, Natural Language Processing (NLP), thermal camera, and autonomous navigation, to provide monitoring benefits to healthcare workers. The authors proposed two different scenarios during COVID-19 for managing patients in hospitals. The developed robot is empowered to perform different tasks within these two scenarios. Overall, this remote system could perform monitoring a patient's condition and providing virtual meetings with loved ones.

17.3.4.3 COVID-SAFE

AI-assisted technologies have been adopted within IoT systems, which could enhance the efficiency of real-time monitoring of pandemics. According to Vedaei et al. (2020), an IoT framework using AI was developed for monitoring the distance and vital signs of users. This framework (called COVID-SAFE) is built on three main elements, including a wearable device, smartphone application, and ML algorithms. The wearable device, empowered with various sensors, measures different health parameters, including body temperature, heart rate, etc., and also social distance. The smartphone demonstrates the results from those collected data regarding the health condition and contamination risk of users. Additionally, the data analyzed by the ML model considers the real-time spread of this contagious virus. In comparison with two other algorithms, including decision tree and SVM, the proposed method achieved a slightly better accuracy of 74.7%. Communications among these elements are placed by cellular data or LoRa enabling local communication within restricted areas (Vedaei et al., 2020).

17.3.5 Challenges

In the near future, IoT technology could widely assist the healthcare domain with its monitoring capabilities. This could be applied for various purposes as we discussed above. However, several challenges should be considered when developing a monitoring system based on IoT. Security and privacy of the collected data are critically important to assess for all sectors, especially healthcare. Regarding the safety guidelines applications, several important concepts could potentially lower the efficiency of the approaches. Briefly, the cost of deployment and installation in large scales, sensors maintenance, high communication bandwidth, and high energy consumption should be considered to achieve superior and reliable IoT systems. Taken together, the healthcare domain could be efficiently assisted using such devices delivering accurate results in the fight against COVID-19.

17.4 Diagnosing

A key point for combating all diseases is to first diagnose or detect them, which opens the area for physicians to manage treatment procedures of patients. As this increases the number of patients who are receiving care, there will be fewer infections in the society (Sabeti, 2020). Regarding the COVID-19, various types of measurements, including respiratory signs, lung nodules, have been identified for diagnosing infected cases. However, some of these symptoms, such as fever, cough, difficulty breathing, loss of taste, appear at the first stage of the disease (CDC, 2020b), These types of symptoms could be adopted to identify symptomatic patients. On the other hand, the asymptomatic COVID-19 patients could be detected using measurements such as changes in the oxygen saturation and respiratory rates. In the following sections, we further discuss recent research on the diagnosing task of IoT based on two types of symptomatic and asymptomatic patients. Then, we briefly demonstrate the importance of social distancing violators' detection.

17.4.1 Symptomatic patient detection

It is important to diagnose suspected cases of COVID-19 to stop the spread. One of the main symptoms of COVID-19 infection is fever. Therefore the rise in the infection and mortality rate of COVID-19 caused authorities to use thermometer guns to identify people with high temperatures, especially in public places such as airports and malls (WHO, 2020b). However, this could potentially increase the infection risk of health officers due to one-by-one screening in close contact. Additionally, if the number of people increases unexpectedly, health officers might not be able to screen all people in a crowd (Mohammed, Syamsudin, Al-Zubaidi et al., 2020). A better solution for detecting people with fever could be the adoption of smart thermometers. The advantage of these smart devices over manual temperature detecting devices is to expose staff to less risk of contamination. Industrial examples of such wearable devices on the market are Kinsa or Tempdrop. While these devices are worn by users, the temperature is recorded regularly to identify any new changes. In the following sections, we will focus on recent diagnostic IoT systems for symptomatic COVID-19 patients based on different applicable technologies.

17.4.1.1 Finger-touch

An alternative method for fever detection was proposed by Hasan (2021), which is basically an IoT system for remote detection of fever. The proposed system uses various types of tools within a cloud to efficiently avoid spreading the virus. Additionally, a microcontroller connected with a finger-touch temperature sensor captures any signs of fever. Also, this system is equipped with a motion detection tool to monitor the passersby close to the user. The monitoring manager is alerted via SMS regarding any new fever detection. Using the ThingSpeak platform, the captured data were presented for better clarifications for the users Hasan (2021). Overall, the application of Finger-touch devices could enhance the efficiency of symptomatic patients' detection.

17.4.1.2 Bracelet

An IoT system for detecting patients infected with the coronavirus was implemented by Cacovean et al. (2020). Various technologies have been adopted within this system to collaborate with health authorities to better fight the COVID-19 pandemic. The first is a wearable bracelet device equipped with three types of sensors to capture temperature, heart rate, and location. Using the assigned cloud database within this IoT system, collected information from patients is sent to physicians and health authorities. Moreover, different ML models are applied to analyze the cloud built-in model based on the patients' information. The highest accuracy was achieved by the Logistic Regression model, where 81% of the samples were accurately classified whether they are suspected of COVID-19. As the model demonstrated efficient results in COVID-19 diagnosis, it could be used for combating COVID-19 within national health systems.

17.4.1.3 Smart helmet

A smart helmet, a wearable device, enables detecting the high temperature of passers-by within the crowd. Mohammed, Syamsudin, Al-Zubaidi et al. (2020) developed a smart helmet for monitoring the temperature of people by using thermal and optical cameras. This IoT-based system aims to lessen human involvement for better control of this pandemic. Additionally, real-time data collected from suspected cases, including the person's face image, body temperature, and Global Positioning System (GPS), will be shown on an assigned smartphone so that authorities can take proper actions regarding the infected case. The deployed software in this system along with various proposed methods had reduced the human error while capturing temperature quickly and efficiently (Mohammed, Syamsudin, Al-Zubaidi et al., 2020).

17.4.1.4 Smart glasses

Another application could be the use of smart glasses, which reduces close contacts and interactions among healthcare professionals and people. An IoT-based system equipped with face detection technology was developed by Mohammed, Hazairin, Syamsudin et al. (2020) to detect suspected cases that demonstrate signs of fever. Then, the captured data by health officers using these glasses (temperature and face image) will be sent to health authorities for further action. In addition to the reliability of the collected data, smart glasses could significantly reduce the spread of COVID-19 due to the lower interactions. One industrial example is Rokid, which enables monitoring and detecting people with fever in a crowd.

17.4.1.5 Drone

Slowing the spread of coronavirus by detecting the users with high temperature could be practical. In addition to the two previous wearable devices, an Unmanned Aerial Vehicle (UAV) could be developed for early diagnosis of COVID-19. Mohammed, Hazairin, Al-Zubaidi et al. (2020) proposed an IoT-based system empowered with sensors, thermal and optical cameras, etc. Authorities could be empowered by this system having the benefit of accessing hard-to-access areas. Also, they will be notified by the alerts from the UAV device. Since this system is not dependent on a health officer for temperature screening in the crowd, it potentially prevents COVID-19 from spreading. Moreover, virtual reality could be a satisfactory additional tool for visualizing collected data. In

Canada, the Pandemic Drone has shown decent results for not only fever detection in a crowd but also monitoring various respiratory measures (Cozzens, 2020).

17.4.1.6 Robotic system

In a study, authors aimed to focus on an IoT-based robotic system to assist the health community for better diagnosis of COVID-19 (Karmore et al., 2020). They claimed this humanoid system is a cost-effective approach for medical practitioners while it could stop the spread of the virus by performing a complete test considering the COVID-19 infection of the user. The main principles of this approach could be divided into various technologies, including sensors, medical imaging, cameras, blood samples, and autonomous navigation. Utilizing real data from adopted technologies for feeding the ML models, this system could be used as a diagnostic tool for assisting healthcare authorities to fight against this pandemic. Although ResNet50 achieved a decent accuracy for diagnosing COVID-19, it is predicted that developing a decentralized NN algorithm would enhance the accuracy while reducing security issues.

17.4.2 Asymptomatic patient detection

COVID-19 can be diagnosed using the aforementioned patient symptoms. However, infected patients are not always symptomatic. This disease could also infect people without showing any symptoms, e.g., fever, cough, etc. Considering the possibilities of asymptomatic infections, COVID-19 could be more dangerous for patients due to the latency of diagnosis. Also, an asymptomatic patient could easily spread the virus among people whom he/she has been in close contact with (Nishiura et al., 2020). In the following sections, we focus on different IoT-based approaches for diagnosing asymptomatic patients infected with COVID-19.

17.4.2.1 Smart hospital

The next generation of hospitals could be smart hospitals, which are impacted by the emergence of IoT technology. This kind of hospital enables various functions including diagnosing, treating, managing, and decision-making that are linked together (Yu et al., 2012). Additionally, implementing an actual smart hospital for a community requires the hospital to conform to different concepts of informative, intelligent, and digital hospital (Jinjun, 2010; Yu et al., 2012).

With respect to the advancements within ML and IoT, a study demonstrated the effectiveness of these two technologies within a smart hospital environment during this pandemic (Abdulkareem et al., 2021). The authors proposed an ML model using three different algorithms of RF, Naive Bayes, and SVM to diagnose patients infected with COVID-19. The impact of IoT within this model is to collect the data from RT-PCR, CT-scan, and X-Ray images, then it transfers them to the appropriate storage. This allows ML models to diagnose COVID-19 patients. After applying the algorithms on a normalized dataset along with a feature selection technique, SVM performed more efficiently with higher accuracy of 95% achieving better diagnosis results. In addition, this model can help medical staff to avoid overcrowding within a hospital and reduce the workload (Abdulkareem et al., 2021).

17.4.2.2 Medical imaging
One of the main methods of diagnosing and screening COVID-19 patients is using medical images including X-ray and CT scan (Kassani et al., 2020). This enables great resources of imaging data for further analysis with ML technologies. Ahmed et al. (2020) proposed an IoT-based DL framework for diagnosing patients infected with COVID-19. The framework has different capabilities such as reducing the workload pressure, managing the pandemic, and, more importantly, diagnosing infections. The DL architecture is built upon two developed benchmarks: Faster-RCNN and ResNet-101 (He et al., 2016; Ren et al., 2016). The authors adopted the required data from various medical imaging datasets, which were stored accordingly, enabling the IoT-based DL framework to detect COVID-19 contaminated cases. As the study claims, the developed model outperformed the recently adopted DL frameworks with an accuracy of 98% for early diagnosis of COVID-19 (Ahmed et al., 2020).

17.4.2.3 Health condition systems
Several systems based on IoT have been developed to focus on detecting infected patients. These health condition approaches are basically considering the respiratory signs using various sensors. In a study, authors (Arun et al., 2020) proposed an IoT system, so that these suspected cases could be detected and quarantined faster. Various technologies, including Raspberry Pi, sensors, GPS, etc., have been adopted for asymptomatic patient detection and health condition monitoring. Based on the results captured by the pulse oximeter, the system determines the need for

isolation if a patient is suspected to COVID-19. If a patient is isolated, the COVID-19 infection will be considered based on the collected data, including blood pressure, and heart rate, from adopted sensors without having any close contact. In addition, the proposed approach enables physicians to monitor those patients during their quarantine period. The essential need for such approaches could also prevent potentially serious damage to the lungs (Arun et al., 2020).

With respect to the COVID-19 chain of transmission, it is important to detect and prevent infections. To do so, Thangamani et al. (2020) developed an IoT system for predicting the symptomatic or asymptomatic infections to this virus. The adopted system is built on various sensors for capturing different vital characteristics of patients, including body temperature, cough detection prototype, and blood pressure. From all these metrics, the chance of infection from COVID-19 can be indicated. For example, if the patient's cough rate is higher than normal, an alert will be sent to health providers for further follow-ups. The authors used Arduino to connect and manage the adopted sensors. Moreover, they provided a last examination using pulse oximeter sensors to discover asymptomatic patients. This could possibly detect if sufficient oxygen is being transmitted to different parts of the body. By predicting infected cases, the spread of this virus could be better handled (Thangamani et al., 2020).

17.4.3 Social distancing violators

In addition to COVID-19 negative effects on societies, it also has caused a recession in almost every country's economy. Therefore businesses have to reopen as soon as possible to avoid further downturns. This caused authorities to assign social distancing guidelines to reduce infections. However, the chain of transmission could not be stopped due to some people that violate these guidelines.

Kumbhar et al. (2020) developed a paradigm that is based on IoT wearable technology. They applied several technologies, including surveillance cameras, wearable devices, and cellular devices. Taken together, they could identify violators of social distancing requirements in a timely manner. Moreover, areas with a high risk of infection can be detected. The authors evaluated the efficiency of their model using object detection with Deep Learning (DL) and a simulation by Python for spread detection. The adopted model using the Convolutional NN method achieved 90% accuracy for distance violators detection. Consequently, users' activities could be monitored and traced for better reducing the chain of transmission.

17.4.4 Challenges

Applying IoT technology for early detecting and diagnosing COVID-19 patients could bring numerous exponential benefits to the healthcare system. However, various challenges have to be considered to enhance the efficacy of the applications. Since COVID-19 was started in late 2019, the number of datasets containing real-world data is limited. Also, the existing ones could be wrongly labeled. Such datasets could impact the results of the diagnosis. Additionally, choosing the right algorithms and methods for making proper predictions could be critical. Due to the massive use of IoT devices for symptom detection (such as pulse oximeter, thermometer), another challenge in this area could be considering the battery life of the devices.

17.5 Tracing

The widespread of the virus among people within public areas or business offices has led researchers to focus on implementing different IoT-based approaches using contact tracing technology. This provides major methods to the healthcare system for tracing the infected or suspected cases.

With respect to the exceptional benefits of adopting IoT along with contact tracing technology, a study introduced an IoT-based contact tracing system (IoT-CTS) for facing pandemics (Hu, 2020). The authors determined the various applicable sensors based on their type, including location and vision. The main perspectives for designing an architecture of IoT-CTS have been addressed within this study. In the following two sections, we discuss different proposed IoT and contact tracing applications based on smartphones and wearables.

17.5.1 Smartphone-based

During the recent COVID-19 outbreak, numerous smartphone applications have been developed for different phases of the pandemic (Nasajpour et al., 2020). These contact tracing applications such as Stop Corona (Stopcorona, 2020), TraceTogether (Tracetogether, 2020), Hamagen (Stub, 2020), etc. were mainly designed to slow and stop the spread of the coronavirus. Embedding IoT with smartphone devices could greatly benefit the tracing phase of COVID-19. We further discuss two recently proposed tracing systems based on IoT and smartphones.

17.5.1.1 Automated tracing
The vast number of confirmed and suspected cases could lead to needing an automated model with low cost of adoption where it can enable authorities to better stop the spread by avoiding new infections. However, contact tracing itself might not be able to achieve the aforementioned advantages. This persuaded the authors to propose an IoT-based approach using two different technologies (Rajasekar, 2021). They implemented the RFID Tag for labeling suspected cases, so they could be traced by the authorities. Moreover, a smartphone application was implemented, so that the user could be alerted whenever a case crosses a smartphone (within a specific radius). This system outperforms the manual tracing by demonstrating higher efficacy, requiring less workers, decreasing the chance of absconding. Also, it could enable real-time monitoring of the cases. Taken together, the pandemic could be managed to reduce the negative effects of COVID-19.

17.5.1.2 IoTrace
The authors proposed a privacy-preserving IoT-based architecture for performing contact tracing tasks using a smartphone (Tedeschi et al., 2020). IoTrace was basically developed to address different challenges associated with the IoT contact tracing systems such as users' privacy and communication's latency. Briefly, privacy and security, communication, and computation costs were considered in this distributed model, which enhanced the privacy of the user's collected data by adopting k-anonymity (Wang et al., 2019). They have adopted a threat model for demonstrating the effectiveness of their architecture against privacy attacks in two areas: location and health status. Moreover, it reduces the overhead communications between devices, which makes it more cost-effective than the regular contact tracing approaches.

17.5.2 Wearable-based
Wearable devices have been widely adopted within healthcare sectors for mainly self-monitoring patients (Appelboom et al., 2014). These devices could be extremely helpful for isolating patients during COVID-19. According to the CDC (CDC, 2021b), isolating people who show symptoms of COVID-19 such as fever, coughing, shortness of breath, etc. is

essential. People who have been in close contact with confirmed cases should also be quarantined. On the other hand, after authorities decided to reopen businesses, people were asked to follow social distancing rules. Moreover, contact tracing of people could potentially reduce the spread of the virus by alerting the people who are violating the rules. These two main objectives have created a great opportunity for applying wearable devices. Here we discuss three recently developed IoT applications utilized within this sector.

17.5.2.1 Digital personal protective equipment—smart watch

Due to the need to reopen businesses, governments attempted to adopt different technologies to provide safer environments. One of the main practices during this pandemic is to maintain social distance to mitigate contamination. To ensure the correct practice of this method, Woodward et al. (2020) proposed a low-cost approach using wearable IoT technology and tracing sensors. They implemented personal protective equipment (PPE) using the BLE interface, which can be adopted within different sectors of daily life including public areas and work offices. The model, called Digital PPE, is equipped with a smart watch. Overall, the proposed system alerts users about any violation of social distancing with an accuracy of 90%. Digital PPE could possibly omit any privacy concerns, high costs for adoption, etc. compared to the contact tracing approaches using smartphone applications.

17.5.2.2 IoT-Q-Band

IoT-Q-Band was designed by Singh, Chandna et al. (2020) to track people who are not following isolation principles. This wearable band prevents isolated patients from leaving their assigned areas. The web interface connected to these devices enables authorities to continuously track the patients. Moreover, losing or taking off the band will also notify the authorities so that they can follow the conditions of patients by alerting or calling them. This device is connected to a smartphone application through Bluetooth. The authors validated their approach with the performance of four different samples based on whether the cloud system detects any radius changes from a specific distance. Consequently, the system performed accurately by tracing absconding cases. Similarly, real-time use of electronic wristbands/bracelets has been adopted in Hong Kong and the United States (Hui, 2020; Izaguirre, 2020).

17.5.2.3 EasyBand

Regarding the huge number of people coming back to work, it is important to ensure required physical distancing measures are being followed. EasyBand was developed for this purpose based on the Internet of Medical Things (Tripathy et al., 2020). The workflow of this wearable device is to demonstrate any possible risks of infection by tracking the user's close contacts with other people. The users are given no warnings as long as they maintain the appropriate distance. However, if users meet at a distance less than 4 m, the device will alert the users with a careful or critical warning based on the distance. This is performed by the BLE, LED, etc. equipped within the system. EasyBand is independent of using smartphones, and it could be adopted as a standalone device for tracking cases. Similarly, various industrial wearable tools such as Safe Spacer (Safespacer, 2021) and Proximity Trace (New Equipment, 2021) have been adopted among different workplaces to effectively alert violators of social distancing measures.

17.5.3 Challenges

During the COVID-19 pandemic, the healthcare domain demonstrated an essential need for tracing suspected and confirmed cases. This encouraged researchers to develop IoT applications to lower the contamination and spread of this deadly virus. However, such approaches could not be as cost-effective as the manual tracing systems. Moreover, most of the approaches require people to have a smartphone. Also, many approaches make users to carry or wear devices. As a result, this could be understood as a privacy concern to users, which requires the developers to consider privacy and security issues within their implementations. Finally, many devices might not be reused for another case due to the chance of contamination.

17.6 Disinfecting

COVID-19 has brought many challenges to healthcare and government authorities. It can be transmitted between all ages of people through droplets produced by sneezing or coughing from infected cases. These droplets containing the virus do not vary based on the symptomatic or asymptomatic nature of patients; they can be present in both types (Velavan & Meyer, 2020). Most importantly, they could easily survive on various object surfaces and in the air for various amounts of time. As an example, droplets could survive on stainless steel materials for 3 days (Van Doremalen et al., 2020). With respect to the importance of COVID-19 mitigation, disinfecting areas could prevent the spread of coronavirus. Manual disinfection could be practical, but it still exposes people to the risks of contamination. Various approaches based on IoT technology have been proposed, which we focus on in the next sections.

17.6.1 Robotic technology

As we discussed, breaking the chain of COVID-19 transmission is crucial for authorities. This opens an area for disinfecting methods using various technologies such as IoT, AI, and so on. One important technology that has been adding value to healthcare is robotics. It has been applied for various purposes including different surgical operations, radiosurgery, and so on (Bogue, 2011). Here, we focus on adopting the system for disinfecting areas contaminated by coronavirus.

17.6.1.1 Sanitizer spraying

Mohammed, Arif et al. (2020) designed a system for disinfecting areas utilizing IoT and robotic technology. They performed the disinfection by spraying sanitizer on surfaces. Additionally, the system is equipped with the ability of spraying long distances, which is moderated by an optical camera and an autonavigation vehicle. This robot was developed to reduce human involvement in disinfecting surfaces.

17.6.1.2 Hand sanitizer dispenser

As the authorities adopted several approaches to reduce human interactions, the need for cleaning hands regularly is crucial. However, as we mentioned, touching surfaces could pose a high risk of infection. Here, this is due to the use of sanitizer bottles. Similar to other approaches, human interactions should have been reduced to avoid further spread of this virus. Accordingly, robotic technology was designed based on the IoT system, so that hand sanitizers could be dispensed among people without touching surfaces (Eddy et al., 2020). Additionally, the proposed robot enables spraying the sanitizer whenever anyone waves his/her hands close to the robot's infrared sensor. In this case, the robot will detect the hands and perform the sanitization.

17.6.2 Drone

The chance of infections from touching surfaces has led to increased sanitizing and washing of hands. However, another possible way to get infected is by airborne transmission. According to the CDC, droplets produced by coughing or sneezing from infected people could survive hours suspended in the air (CDC, 2021c). Utilizing drones within large buildings for disinfecting could be a potential solution.

An IoT-based system was developed for disinfecting areas using UAVs technology (Mohammed, Syamsudin, Hazairin et al., 2020). The authors implemented their system aiming to lessen the workload for maintenance staff by remotely sanitizing the areas. To efficiently moderate these IoT-based drones, they were equipped with optical cameras. Also, communicating with the people close to the drone was enabled using a speaker, so that they could be notified about the disinfected areas. Finally, the locations of these areas will be sent to the system for better management. An industrial example of this type of drone is Spanish authorities have adopted Da-Jiang Innovations (DJI), a UAV-based company, for disinfecting contaminated areas.

17.6.3 Challenges

As the COVID-19 virus could last a significant amount of time on various surfaces, it is important to develop different IoT-based approaches to facilitate manually disinfecting and sanitizing surfaces. Although various tools and methods have been proposed using IoT for this purpose, it is important to consider the existing challenges. As we discussed previously, battery life could be one major issues of working with IoT devices. It could be highlighted when considering disinfection of large areas. Finally, the high price of deployment could limit applications only in essential areas such as hospitals.

17.7 Vaccinating

After a year of confronting COVID-19, the first vaccine was issued by the US Food and Drug Administration in December 2020 (FDA, 2020). As the world is trying to combat this virus, various types of vaccines (including Pfizer-BioNTech, Moderna, Johnson & Johnson, etc.) have been approved to prevent further infections of COVID-19 (CDC, 2021d). Although vaccinating people has reduced the contamination and death toll, there are some factors that need to be considered to get the best results from vaccines. Briefly, the challenges are production, transportation, and management of injections. Regarding the importance of vaccination, various technologies, especially IoT and Blockchain, could greatly assist the healthcare system in dealing with these challenges. It is important to note that since vaccination is in its first stage, the research conducted on this area is limited. We will demonstrate three potential parts of vaccination.

17.7.1 Production

Regarding the widespread of coronavirus, medical manufacturers are trying to produce vaccines to quicken the process of vaccinating people. However, social distancing measures have reduced the number of workers, which opens an area for automation. Moreover, a huge amount of time is required to produce a vaccine, which could be lessened by adopting IoT, AI, etc. As discussed, IoT devices have been adopted in various fields of healthcare with respect to COVID-19. The IoT sensors could possibly reduce the amount of work by measuring various parameters during the vaccine setup. Screening physical distances and air quality within the workplace would also be other advantages of IoT during vaccine production (Sisto, 2021).

17.7.2 Transportation

The next stage of vaccination is the distribution of vaccines to the health sectors. Here, the major challenge for the vaccine rollouts is to store them safely during transportation. Therefore it is critical to maintain the required temperature for the cold chain of vaccines (Hanson et al., 2017; Matthias et al., 2007). Temperature changes of the vaccines could lead to weakness or could even be dangerous for the patient (Hibbs et al., 2018). Since the quality of COVID-19 vaccines is extremely dependent on temperature, different technologies could be applied for vaccine transportation and distribution. Using IoT sensors, the cloud, blockchain, etc., authorities can efficiently monitor the temperature changes of the stored vaccines. This enables healthcare facilities to capture the locations of inappropriate storage temperature (Kaplan, 2020). For this purpose, different types of systems could be implemented to ensure the safe transport of vaccines. Cloudleaf and Varcode, examples of IoT technology, are respectively using Bluetooth and barcode to safely distribute the vaccines by monitoring different measures and alerting about essential actions (Smith, 2020; Whipple, 2021). Moreover, Fulzele et al. (2020) also designed an IoT-based cold box for maintaining the appropriate temperature. This box is also equipped with location monitoring as well.

17.7.3 Data acquisition

The last stage of dealing with public vaccination is to make sure people get their vaccines. Some vaccines, including Pfizer and Moderna, require two doses to be effective. Regarding the date for the second shot, users should be notified about getting it on time. Consequently, IoT wearable devices could be applied to remind them of their shot (Marsh, 2021). Additionally, the hospital's record-keeping could be improved by these devices for providing treatments for the future. Moreover, the adoption of vaccine passports for different places such as airports and business offices could possibly be a solution for getting back to normal life.

17.7.4 Challenges

As IoT devices could be proposed to reduce the rate of degradation due to the temperature drop during the production and transportation of the vaccines, there are still security challenges that need to be considered. One main challenge is to ensure the IoT sensors will work securely and properly. An example of security concerns could be adversary attacks, which attempt to perform malicious activities such as deleting or changing the collected data from sensors and making false transactions (Singh, Dwivedi et al., 2020). Battery life of the deployed sensors is critical to consider while applying inside the cold boxes for temperature monitoring. Finally, countries started to define vaccine passports for travelers who are vaccinated. As these passports will require private information, it is essential to consider privacy concerns (Albano, 2021).

17.8 Conclusion

As governments are planning to vaccinate people, COVID-19 will possibly become endemic. This means the virus will circulate seasonally similar to the flu (Denworth, 2020). As a result, it is likely that the world has to overcome COVID-19 each year, which could be achieved by adopting different technologies. As we noted, IoT technology could be extremely helpful during pandemics by assisting healthcare professionals, authorities, and patients in different stages such as monitoring, diagnosing, tracing, disinfecting, and vaccinating. In each task, we reviewed the IoT proposed technologies or systems based on different concepts applicable to each task. Although the current IoT applications have positively enhanced the efficiency of the fight against COVID-19, it is essential to improve these proposed systems regarding different possible challenges. Privacy and security of the data collected from patients are extremely important to be considered in future IoT systems.

References

Abdulkareem, K. H., Mohammed, M. A., Salim, A., Arif, M., Geman, O., Gupta, D., & Khanna, A. (2021). Realizing an effective COVID-19 diagnosis system based on machine learning and IOT in smart hospital environment. *IEEE Internet of Things Journal.*

Adi, E., Anwar, A., Baig, Z., & Zeadally, S. (2020). Machine learning and data analytics for the IoT. *Neural Computing & Applications, 32,* 16205–16233.

Ahmed, I., Ahmad, A., & Jeon, G. (2020). An IoT based deep learning framework for early assessment of Covid-19. *IEEE Internet of Things Journal.*

Akhbarifar, S., Javadi, H. H. S., Rahmani, A. M., & Hosseinzadeh, M. (2020). A secure remote health monitoring model for early disease diagnosis in cloud-based IoT environment. *Personal and Ubiquitous Computing,* 1–17.

Albano, C. (2021). Digital vaccine passports and COVID-19: What privacy concerns should you be thinking about?.

Aledhari, M., Razzak, R., Parizi, R. M., & Dehghantanha, A. (2020). A deep recurrent neural network to support guidelines and decision making of social distancing. In 2020 IEEE international conference on big data (Big Data).p. 4233–40.

Al-Shalabi, M. (2020). COVID-19 symptoms monitoring mechanism using internet of things and wireless sensor networks. *IJCSNS, 20*(8), 16.

Appelboom, G., Camacho, E., Abraham, M. E., Bruce, S. S., Dumont, E. L. P., Zacharia, B. E., et al. (2014). Smart wearable body sensors for patient self-assessment and monitoring. *Archives of Public Health, 72*(1), 1–9.

Arun, M., Baraneetharan, E., Kanchana, A., Prabu, S., et al. (2020). Detection and monitoring of the asymptotic COVID-19 patients using IoT devices and sensors. *International Journal of Pervasive Computing and Communications.*

Baqaee, D., Farhi, E., Mina, M. J., & Stock, J. H. (2020). Reopening scenarios.

Bartik, A. W., Bertrand, M., Cullen, Z., Glaeser, E. L., Luca, M., & Stanton, C. (2020). The impact of COVID-19 on small business outcomes and expectations. *Proceedings of the National Academy of Sciences of the United States of America, 117*(30), 17656–17666.

Bashir, A., Izhar, U., & Jones, C. (2020). IoT-based COVID-19 SOP compliance and monitoring system for businesses and public offices. In *Engineering proceedings.* p. 14.

Bhatt, Y., & Bhatt, C. (2017). *Internet of things in healthcare. Internet of things and big data technologies for next generation HealthCare* (pp. 13–33). Springer.

Binti, A., Kamaludeen, N., Lee, S. P., & Parizi, M. (2019). R. Guideline-based approach for IoT home application development. In 2019 *international conference on internet of things (ithings) and IEEE green computing and communications (GreenCom) and IEEE cyber, physical and social computing (CPSCom) and IEEE smart data (SmartData).* p. 929–36.

Bogue, R. (2011). Robots in healthcare. *Industrial Robot: An International Journal.*

Cacovean, D., Ioana, I., & Nitulescu, G. (2020). IoT system in diagnosis of Covid-19 patients. *Information Economics, 24*(2), 75−89.

CDC, (2020a). Air Pollutants. [Internet]. Available from: https://www.cdc.gov/air/pollutants.html, Accessed 25.03.21.

CDC, (2020b). Symptoms [Internet]. Available from: https://www.cdc.gov/coronavirus/2019-ncov/symptoms-testing/symptoms.html, Accessed 10.03.21.

CDC (2021a). What is the difference between Influenza (Flu) and COVID-19? [Internet]. 2021. Available from: https://bit.ly/3vGojqS, Accessed 18.03.21.

CDC, (2021b). When to Quarantine. [Internet]. Available from: https://www.cdc.gov/coronavirus/2019-ncov/if-you-are-sick/quarantine.html, Accessed 30.03.21.

CDC, (2021c). SARS-CoV-2 is transmitted by exposure to infectious respiratory fluids. [Internet]. Available from: https://bit.ly/3pd9kDn, Accessed 13.03.21.

CDC, (2021d). Different COVID-19 Vaccines. [Internet]. Available from: https://www.cdc.gov/coronavirus/2019-ncov/vaccines/different-vaccines.html, Accessed 13.03.21.

Chloros, D., & Ringas, D. (2020). Fluspot: Seasonal flu tracking app exploiting wearable IoT device for symptoms monitoring. In 2020 5th South-East Europe design automation, computer engineering, computer networks and social media conference (SEEDA-CECNSM).p. 1−7.

Cozzens, T. (2020). Pandemic drones to monitor, detect those with COVID-19.

Cucinotta, D., & Vanelli, M. (2020). WHO declares COVID-19 a pandemic. *Acta Biomedica Atenei Parmensis, 91*(1), 157.

Da Xu, L., He, W., & Li, S. (2014). Internet of things in industries: A survey. *IEEE Transactions on Industrial Informatics, 10*(4), 2233−2243.

Denworth, L. (2020). How the COVID-19 pandemic could end.

Eddy, Y., Mohammed, M. N., Daoodd, I. I., Bahrain, S. H. K., Al-Zubaidi, S., Al-Sanjary, O. I., et al. (2020). 2019 Novel coronavirus disease (Covid-19): Smart contactless hand sanitizer-dispensing system using IoT based robotics technology. *Revista Argentina de Clinica Psicologica, 29*(5), 215.

Ekramifard, A., Amintoosi, H., Seno, A. H., Dehghantanha, A., & Parizi, R. M. (2020). A systematic literature review of integration of blockchain and artificial intelligence. In K.-K. R. Choo, A. Dehghantanha, & R. M. Parizi (Eds.), *Blockchain cybersecurity, trust and privacy* (pp. 147−160). Cham: Springer International Publishing.

EPA, (2020). Indoor air and coronavirus (Covid-19). [Internet]. Available from: https://www.epa.gov/coronavirus/indoor-air-and-coronavirus-covid-19, Accessed 25.03.21.

Fazio, M., Buzachis, A., Galletta, A., Celesti, A., & Villari, M. (2020). A proximity-based indoor navigation system tackling the COVID-19 social distancing measures. In 2020 *IEEE symposium on computers and communications* (ISCC).p. 1−6.

FDA Takes key action in fight against COVID-19 by issuing emergency use authorization for first COVID-19 vaccine. [Internet] (2020). Available from: https://www.fda.gov/news-events/press-announcements/fda-takes-key-action-fight-against-covid-19-issuing-emergency-use-authorization-first-covid-19, Accessed 30.03.21.

Fernández-Caramés, T. M., Froiz-Miguez, I., & Fraga-Lamas, P. (2020). An IoT and blockchain based system for monitoring and tracking real-time occupancy for COVID-19 public safety. *Engineering Proceedings*, 67.

Ferretti, L., Wymant, C., Kendall, M., Zhao, L., Nurtay, A., Abeler-Dörner, L., et al. (2020). Quantifying SARS-CoV-2 transmission suggests epidemic control with digital contact tracing. *Science, 368*(6491).

Fulzele, D. P., Kumbhare, A., Mangde, A., Gaidhane, D., Palsodkar, D., Narkhede, P., et al. (2020). An IoT enabled convenient vaccine cold box for biomedical use. *European Journal of Molecular & Clinical Medicine, 7*(7), 1576−1585.

Gregory, V., Menzio, G., & Wiczer, D. G. (2020). Pandemic recession: L or V-shaped?

Gul, S., Asif, M., Ahmad, S., Yasir, M., Majid, M., Malik, M. S. A., et al. (2017). A su. vey on role of internet of things in education. *International Journal of Computer Science and Network Security*, *17*(5), 159−165.

Güner, H. I. R., Hasanoglu, I., & Aktas, F. (2020). COVID-19: Prevention and control measures in community. *Turkish Journal of Medical Sciences*, *50*(SI-1), 571−577.

Haleem, A., Javaid, M., Vaishya, R., Vaish, A., et al. (2020). Role of internet of things for health-care monitoring during COVID-19 pandemic. *ApolloMed*, *17*(5), 55.

Hanson, C. M., George, A. M., Sawadogo, A., & Schreiber, B. (2017). Is freezing in the vaccine cold chain an ongoing issue? A literature review. *Vaccine*, *35*(17), 2127−2133.

Hasan, M. W. (2021). Covid-19 fever symptom detection based on IoT cloud. *International Journal of Electrical and Computer Engineering*, *11*(2).

He, K., Zhang, X., Ren, S., & Sun, J. (2016). Deep residual learning for image recognition. In *Proceedings of the IEEE conference on computer vision and pattern recognition*. p. 770−778.

Hibbs, B. F., Miller, E., Shi, J., Smith, K., Lewis, P., & Shimabukuro, T. T. (2018). Safety of vaccines that have been kept outside of recommended temperatures: Reports to the vaccine adverse event reporting system (VAERS), 2008−2012. *Vaccine*, *36*(4), 553−558.

Hu, P. (2020). *IoT-based contact tracing systems for infectious diseases: Architecture and analysis.* arXiv Prepr arXiv200901902.

Hui, M. (2020). Hong Kong is using tracker wristbands to geofence people under coronavirus quarantine.

Islam, S. M. R., Kwak, D., Kabir, M. D. H., Hossain, M., & Kwak, K.-S. (2015). The internet of things for health care: a comprehensive survey. *IEEE Access*, *3*, 678−708.

Izaguirre, A. (2020). Judge OKs ankle monitors for virus scofflaws.

Jara, A. J., Zamora, M. A., & Skarmeta, A. F. G. (2009). {HWSN6}: Hospital wireless sensor networks based on 6LoWPAN technology: Mobility and fault tolerance management. In 2009 *International conference on computational science and engineering*. 2009. p. 879−84.

Jinjun, M. (2010). A brief talk on the problem in integration of hospital intelligence and information and developing directionof intelligence. *Intell Build & City Inf*, *158*, 94−96.

Kanade, P., Akhtar, M., & David, F. (2021). Remote monitoring technology for COVID-19 patients. *European Journal of Electrical & Computer Engineering*, *5*(1), 44−47.

Kaplan, D. A. (2020). Why cold chain tracking and IoT sensors are vital to the success of a COVID-19 vaccine.

Karmore, S., Bodhe, R., Al-Turjman, F., Kumar, R. L., & Pillai, S. (2020). IoT based humanoid software for identification and diagnosis of Covid-19 suspects. *IEEE Sensors Journal*.

Kassani, S. H., Kassasni, P. H., Wesolowski, M. J., Schneider, K. A., & Deters, R. (2020). *Automatic detection of coronavirus disease (covid-19) in x-ray and ct images: A machine learning-based approach.* arXiv Prepr arXiv200410641.

Kim, J. H., Marks, F., & Clemens, J. D. (2021). Looking beyond COVID-19 vaccine phase 3 trials. *Nature Medicine*, *27*(2), 205−211.

Kumar, P. M., Lokesh, S., Varatharajan, R., Babu, G. C., & Parthasarathy, P. (2018). Cloud and IoT based disease prediction and diagnosis system for healthcare using Fuzzy neural classifier. *Future Generation Computer Systems*, *86*, 527−534.

Kumari, P., & Toshniwal, D. (2020). Impact of lockdown on air quality over major cities across the globe during COVID-19 pandemic. *Urban Climate*, *34*, 100719.

Kumbhar, F. H., Hassan, S. A., & Shin, S. Y. (2020). *New normal: Cooperative paradigm for Covid-19 timely detection and containment using internet of things and deep learning.* arXiv Prepr arXiv200812103.

Lauer, S. A., Grantz, K. H., Bi, Q., Jones, F. K., Zheng, Q., Meredith, H. R., et al. (2020). The incubation period of coronavirus disease 2019 (COVID-19) from publicly reported confirmed cases: Estimation and application. *Annals of Internal Medicine, 172* (9), 577–582.

Marketsandmarkets, (2020). *IIoT in healthcare market by component (Medical Device, Systems & Software, Services, and Connectivity Technology), application (Telemedicine, Connected Imaging, and Inpatient Monitoring), end user, and region - global forecast to 2025.*

Marsh, J. (2021). COVID-19 vaccine: The role of IoT.

Matthias, D. M., Robertson, J., Garrison, M. M., Newland, S., & Nelson, C. (2007). Freezing temperatures in the vaccine cold chain: A systematic literature review. *Vaccine, 25*(20), 3980–3986.

Miron-Alexe, V. (2020). *IoT pulse oximetry status monitoring for home quarantined COVID-19 patients.*

Mohammed, M. N., Hazairin, N. A., Al-Zubaidi, S., Sairah, A., Mustapha, S., & Yusuf, E. (2020). Toward a novel design for coronavirus detection and diagnosis system using IoT based drone technology. *International Journal of Psychosocial Rehabilitation, 24*(7), 2287–2295.

Mohammed, M. N., Arif, I. S., Al-Zubaidi, S., Bahrain, S. H. K., Sairah, A. K., Eddy, Y., et al. (2020). Design and development of spray disinfection system to combat coronavirus (Covid-19) using IoT based robotics technology. *Revista Argentina de Clinica Psicologica, 29*(5), 228.

Mohammed, M. N., Hazairin, N. A., Syamsudin, H., Al-Zubaidi, S., Sairah, A. K., Mustapha, S., et al. (2020). 2019 Novel coronavirus disease (Covid-19): Detection and diagnosis system using IoT based smart glasses. *International Journal of Advanced Science and Technology, 29*(7 Special Issue).

Mohammed, M. N., Syamsudin, H., Al-Zubaidi, S., Aks, R. R., & Yusuf, E. (2020). Novel COVID-19 detection and diagnosis system using IOT based smart helmet. *International Journal of Psychosocial Rehabilitation, 24*(7), 2296–2303.

Mohammed, M. N., Syamsudin, H., Hazairin, N. A., Haki, M., Al-Zubaidi, S., Sairah, A. K., et al. (2020). Toward a novel design for spray disinfection system to combat coronavirus (Covid-19) using IoT based drone technology. *Revista Argentina de Clinica Psicologica, 29*(5), 240.

Mumtaz, R., Zaidi, S. M. H., Shakir, M. Z., Shafi, U., Malik, M. M., Haque, A., et al. (2021). Internet of things (IoT) based indoor air quality sensing and predictive analytic—A COVID-19 perspective. *Electronics, 10*(2), 184.

Nasajpour, M., Pouriyeh, S., Parizi, R. M., Dorodchi, M., Valero, M., & Arabnia, H. R. (2020). Internet of Things for current COVID-19 and future pandemics: An exploratory study. *Journal of Healthcare Informatics Research*, 1–40.

New Equipment, (2021). Contact Tracing IoT Solution. [Internet]. Available from: https://www.directory.newequipment.com/classified/contact-tracing-iot-solution-253439.html, Accessed 10.03.21.

Nishiura, H., Kobayashi, T., Miyama, T., Suzuki, A., Jung, S., Hayashi, K., et al. (2020). Estimation of the asymptomatic ratio of novel coronavirus infections (COVID-19). *International Journal of Infectious Diseases, 94*, 154.

Nocturnal Oxygen Therapy Trial Group. (1980). Continuous or nocturnal oxygen therapy in hypoxemic chronic obstructive lung disease: A clinical trial. *Annals of Internal Medicine, 93*(3), 391–398.

Oliver, S. E., Gargano, J. W., Marin, M., Wallace, M., Curran, K. G., Chamberland, M., et al. (2020). The advisory committee on immunization practices' interim recommendation for use of Pfizer-BioNTech COVID-19 vaccine—United States, December 2020. *Morbidity and Mortality Weekly Report, 69*(50), 1922.

Ortega, R., Hansen, C. J., Elterman, K., & Woo, A. (2011). Pulse oximetry. *The New England Journal of Medicine, 364*(16), e33–e36.

Petrović, N., & Kocić, D. (2020). IoT-based system for COVID-19 indoor safety monitoring. *Prepr IcETRAN, 2020*, 1–6.

Połap, D., Srivastava, G., Jolfaei, A., & Parizi, R. M. (2020). Blockchain technology and neural networks for the internet of medical things. In *IEEE INFOCOM 2020 - IEEE conference on computer communications workshops (INFOCOM WKSHPS)*. p. 508–13.

Powell, A. (2020). Fauci says herd immunity possible by Fall, 'Normality' by end of 2021. [Internet]. Available from: https://bit.ly/3GDw9pv, Acessed 23.03.21.

Qi, J., Yang, P., Min, G., Amft, O., Dong, F., & Xu, L. (2017). Advanced internet of things for personalised healthcare systems: A survey. *Pervasive and Mobile Computing, 41*, 132–149.

Rajasekar, S. J. S. (2021). An enhanced IoT based tracing and tracking model for COVID-19 cases. *SN Computer Science, 2*(1), 1–4.

Ren, S., He, K., Girshick, R., & Sun, J. (2016). Faster R-CNN: Towards real-time object detection with region proposal networks. *IEEE Transactions on Pattern Analysis and Machine Intelligence, 39*(6), 1137–1149.

Reyna, A., Martín, C., Chen, J., Soler, E., & Díaz, M. (2018). On blockchain and its integration with IoT. Challenges and opportunities. *Future Generation Computer Systems, 88*, 173–190.

Sabeti, P. (2020). Early detection is key to combating the spread of coronavirus. Time.

Safespacer, (2021). Keep people safe and workplaces open. [Internet]. Available from: https://www.safespacer.net, Accessed 10.03.21.

Saharkhizan, M., Azmoodeh, A., Dehghantanha, A., Choo, K.-K. R., & Parizi, R. M. (2020). An ensemble of deep recurrent neural networks for detecting IoT cyber attacks using network traffic. *IEEE Internet of Things Journal, 7*(9), 8852–8859.

Sales Mendes, A., Jiménez-Braedao, D. M., Navarro-Cáceres, M., Reis Quietinho Leithardt, V., & Villarrubia González, G. (2020). Multi-agent approach using LoRaWAN devices: An airport case study. *Electronics, 9*(9), 1430.

Shahid, O., Nasajpour, M., Pouriyeh, S., Parizi, R. M., Han, M., Valero, M., et al. (2021). Machine learning research towards combating COVID-19: Virus detection, spread prevention, and medical assistance. *Journal of Biomedical Informatics [Internet]*, 103751. Available from https://www.sciencedirect.com/science/article/pii/S1532046421000800.

Singh, R., Dwivedi, A. D., & Srivastava, G. (2020). Internet of things based blockchain for temperature monitoring and counterfeit pharmaceutical prevention. *Sensors, 20*(14), 3951.

Singh, V., Chandna, H., Kumar, A., Kumar, S., Upadhyay, N., & Utkarsh, K. (2020). IoT-Q-Band: A low cost internet of things based wearable band to detect and track absconding COVID-19 quarantine subjects. *EAI Endorsed Trans Internet Things, 6*(21).

Sisto, A. (2021). The role of IoT in scaling up vaccine manufacturing.

Smith, J. (2020). Temperature-tracking tools take center stage in Covid-19 vaccine rollout.

Sohrabi, C., Alsafi, Z., O'neill, N., Khan, M., Kerwan, A., Al-Jabir, A., et al. (2020). World Health Organization declares global emergency: A review of the 2019 novel coronavirus (COVID-19). *International Journal of Surgery (London, England), 76*, 71–76.

Statista, (2019). Internet of Things - number of connected devices worldwide 2015–2025 [Internet]. Available from: https://www.statista.com/statistics/471264/iot-number-of-connected-devices-worldwide, Accessed 18.03.21.

Stolojescu-Crisan, C., Butunoi, B.-P., & Crisan, C. (2020). IoT based intelligent building applications in the context of COVID-19 pandemic. In *2020 international symposium on electronics and telecommunications (ISETC)*. p. 1–4.

.topcorona, (2020). United against coronavirus! Stopcorona App. [Internet]. Available from: https://stopcorona.app, Accessed 30.03.21.

Stub, S. T. Israeli phone apps aim to track coronavirus, guard privacy. [Internet] (2020). Available from: https://www.usnews.com/news/best-countries/articles/2020-04-20/new-tech-apps-in-israel-aim-to-track-coronavirus-guard-privacy, Accessed 28.03.21.

Tracetogether, (2020). TraceTogether, safer together. [Internet]. Available from: https://www.tracetogether.gov.sg, Accessed 30.03.21.

Tedeschi, P., Bakiras, S., & Di Pietro, R. (2020). *IoTrace: A flexible, efficient, and privacy-preserving IoT-enabled architecture for contact tracing.* arXiv Prepr arXiv200711928.

Thangamani, M., Ganthimathi, M., Sridhar, S. R., Akila, M., Keerthana, R., & Ramesh, P. S. (2020). Detecting coronavirus contact using internet of things. *International Journal of Pervasive Computing and Communications.*

Tripathy, A. K., Mohapatra, A. G., Mohanty, S. P., Kougianos, E., Joshi, A. M., & Das, G. (2020). EasyBand: A wearable for safety-aware mobility during pandemic outbreak. *IEEE Consumer Electronics Magazine, 9*(5), 57−61.

Udugama, B., Kadhiresan, P., Kozlowski, H. N., Malekjahani, A., Osborne, M., Li, V. Y. C., et al. (2020). Diagnosing COVID-19: The disease and tools for detection. *ACS Nano, 14*(4), 3822−3835.

Vaishya, R., Javaid, M., Khan, I. H., & Haleem, A. (2020). Artificial Intelligence (AI) applications for COVID-19 pandemic. *Diabetes & Metabolic Syndrome: Clinical Research & Reviews, 14*(4), 337−339.

Valero, M., Shahriar, H., & Ahamed, S. I. (2020). R-Mon: An mhealth tool for real-time respiratory monitoring during pandemics and self-isolation. *2020 IEEE World Congress on Services (SERVICES),* 17−21.

Van Doremalen, N., Bushmaker, T., Morris, D. H., Holbrook, M. G., Gamble, A., Williamson, B. N., et al. (2020). Aerosol and surface stability of SARS-CoV-2 as compared with SARS-CoV-1. *The New England Journal of Medicine, 382*(16), 1564−1567.

Vedaei, S. S., Fotovvat, A., Mohebbian, M. R., Rahman, G. M. E., Wahid, K. A., Babyn, P., et al. (2020). COVID-SAFE: An IoT-Based System for Automated Health Monitoring and Surveillance in Post-Pandemic Life. *IEEE Access, 8,* 188538−188551.

Velavan, T. P., & Meyer, C. G. (2020). La epidemia de COVID-19. *Tropical Medicine & International Health.*

Wang, J., Cai, Z., & Yu, J. (2019). Achieving personalized k-Anonymity-Based content privacy for autonomous vehicles in CPS. *IEEE Transactions on Industrial Informatics, 16* (6), 4242−4251.

Whipple, K. (2021). How a digital visibility platform can increase COVID-19 vaccine distribution success.

WHO, (2020a). Modes of transmission of virus causing COVID-19: Implications for IPC precaution recommendations: scientific brief [Internet]. Available from: https://bit.ly/32rLUS8, Accessed 24.03.21

WHO, (2020b). Coronavirus disease 2019 (COVID-19): Situation report, 82.

WHO, (2021). WHO Coronavirus (COVID-19) Dashboard [Internet]. Available from: https://covid19.who.int, Accessed 18.03.21.

Woodward, K., Kanjo, E., Anderez, D. O., Anwar, A., Johnson, T., & Hunt J. (2020). DigitalPPE: low cost wearable that acts as a social distancingreminder and contact tracer. In *Proceedings of the 18th conference on embedded networked sensor systems.* p. 758−759.

Yazdinejad, A., Parizi, R. M., Dehghantanha, A., Karimipour, H., Srivastava, G., & Aledhari, M. (2021). Enabling drones in the internet of things with decentralized blockchain-based security. *IEEE Internet of Things Journal, 8*(8), 6406−6415.

Yu, L., Lu, Y., & Zhu, X. (2012). Smart hospital based on internet of things. *Journal of Networks, 7*(10), 1654.

Zaki, N., Alashwal, H., & Ibrahim, S. (2020). Association of hypertension, diabetes, stroke cancer, kidney disease, and high-cholesterol with COVID-19 disease severity and fatality: A systematic review. *Diabetes & Metabolic Syndrome: Clinical Research & Reviews*, *14*(5), 1133–1142.

Zhang, S. X., Wang, Y., Rauch, A., & Wei, F. (2020). Unprecedented disruption of lives and work: Health, distress and life satisfaction of working adults in China one month into the COVID-19 outbreak. *Psychiatry Research*, *288*, 112958.

Index

Note: Page numbers followed by "*f*" and "*t*" refer to figures and tables, respectively.

Printed in the United States
by Baker & Taylor Publisher Services